ENCLOSURE FIRE DYNAMICS

Environmental and Energy Engineering Series

Series Editors:

Ashwani K. Gupta
Department of Mechanical Engineering
University of Maryland
College Park, Maryland

David G. Lilley
School of Mechanical and Aerospace Engineering
Oklahoma State University
Stillwater, Oklahoma

Published Titles

Integrated Product and Process Design and Development,
 Edward B. Magrab

ENCLOSURE FIRE DYNAMICS

BJÖRN KARLSSON

JAMES G. QUINTIERE

CRC Press
Boca Raton London New York Washington, D.C.

Library of Congress Cataloging-in-Publication Data

Karlsson, Björn.
    Enclosure fire dynamics / Björn, Karlsson, James G. Quintiere.
       p.   cm.
    Includes bibliographical references and index.
    ISBN 0-8493-1300-7 (alk. paper)
    1. Enclosure fires.   I. Quintiere, James G.   II. Title.
TH9195.K37 1999
693.8′2—dc21

99-32642
CIP

Visit the CRC Press Web site at www.crcpress.com

© 2000 by CRC Press LLC

No claim to original U.S. Government works
International Standard Book Number 0-8493-1300-7
Library of Congress Card Number 99-32642
Printed in the United States of America   0
Printed on acid-free paper

Preface

The body of engineering knowledge that defines the discipline of fire safety engineering (or fire protection engineering) is relatively broad. For educational purposes this discipline can be divided into a number of fundamental courses. One of these is *enclosure fire dynamics*: the study of how to estimate the environmental consequences of a fire in an enclosure.

For over a decade the Department of Fire Safety Engineering at Lund University, Sweden, has been teaching a course on this topic. Professor Sven Erik Magnusson initiated and defined the contents of the course and collected material to be taught from a variety of sources. The course literature until now has consisted of a collection of journal articles, scientific papers, and selected chapters from various handbooks and textbooks. The lack of completeness and homogeneity in the course literature has been a source of frustration for both teachers and students.

The authors of this textbook have to a considerable extent followed the framework provided by Professor Magnusson but they have added some new topics and expanded others.

Fire safety science is a rapidly developing field and fire safety engineering a relatively young engineering discipline. The scientific and engineering communities have therefore not fully standardized the assignments of symbols and units used in the field. Further, the contents of the fundamental courses—including enclosure fire dynamics—have not been standardized and may vary greatly from one educational institution to another.

The purpose of this textbook is not only to act as course literature for fire safety engineering students, but also to offer educators in the field the opportunity to comment on the contents of the course, its organization, and the definitions of dimensions and symbols used. It is the hope of the authors that this book will contribute toward some standardization of both educational material and terminology used in the fire safety engineering discipline.

The Authors

Björn Karlsson, Ph.D., received his B.Sc. (Honours) in Civil Engineering from Heriot-Watt University, Edinburgh, Scotland, in 1985; his Degree of Licentiate in Fire Safety Engineering from Lund University, Sweden, in 1989; and his Ph.D. in Fire Safety Engineering, also from Lund University, in 1992. His research has focused on modeling ignition, flame spread, and fire growth on solids; the development of an expert system for risk analysis in industry; the under-ventilated fire; two-zone fire models; performance-based codes for fire safety engineering; and performance-based test methods for the reaction-to-fire of products. He has initiated and managed several projects in cooperation with the Swedish Rescue Services, most recently on the phenomena of flashover, backdraft, and smoke gas explosion.

Dr. Karlsson was Swedish representative to ISO/TTC92/SC4/WG2 and the Nordic Committee for Building Regulations. He was actively involved in ISO/TC92/SC1/WG8 and contributed to work on CIB W14. He also contributed to work of the ASTM Committee E5. He was an organizer of the January 1995 CIB W14 workshop "Flame Spread and Fire Growth on Products," held in Espoo, Finland. He has been an invited speaker at a national meeting of the American Chemical Society and has been involved in research projects in Japan, Europe, and the United States.

In 1996, Dr. Karlsson was a Visiting Associate Professor in the Department of Fire Protections Engineering at the University of Maryland, College Park. He is currently Senior Lecturer in the Department of Fire Safety Engineering, Lund University, and project manager for a 2-year research project with the aim of establishing a risk index method for assessing the fire safety of timber-frame multistory apartment buildings.

James G. Quintiere, Ph.D.M.E., received his B.S.M.E. from the New Jersey Institute of Technology in 1962; his M.S.M.E. from New York University in 1966; and his Ph.D.M.E., also from New York University, in 1970. He has worked as an aerospace engineer with NASA Lewis Research Center; a research scientist with the American Standard Research-Development Center in Piscataway, New Jersey; a mechanical engineer with the Center for Fire Research; and a program analyst with the National Engineering Laboratory. He was chief of the Fire Science and Engineering Division, Center for Fire Research Technology, National Institute of Standards and Technology (formerly the National Bureau of Standards), Gaithersburg, Maryland.

Dr. Quintiere has served as Chairman, K-11 Committee on Fire and Combustion, Heat Transfer Division, 1986–1989; Papers/Program Co-Chairman, Eastern Section of the Combustion Institute, 1978–1981; and Chairman of the International Association for Fire Safety Science. He is an editorial board member of the *Journal of Fire Protection Engineering, Fire Technology,* and *Fire Safety Journal.*

The author of more than 78 publications, 42 reports, and one book, Dr. Quintiere is the recipient of the Howard W. Emmons Lecture Award (1986); the Jack Bono Communications Award, SFPE Best Paper, *Journal of Fire Protection Engineering*, Vol. VII (1996); and the Harry C. Bigglestone Award, *Fire Technology*, Vol. 33, 1997 (1998). Since 1990, he has been Professor of Fire Protection Engineering, University of Maryland, College Park.

Acknowledgments

The pioneering work of Professor Sven Erik Magnusson, who initiated and developed the course on Enclosure Fire Dynamics taught at Lund University, is gratefully acknowledged. Prof. Magnusson provided the basic framework, identified the subject matter, and gave guidance on where to strike the balance between basic and applied knowledge.

We extend our thanks to the staff at the Department of Fire Safety Engineering at Lund University and the Department of Fire Protection Engineering at the University of Maryland for their moral support. The department heads at both institutions—Robert Jönsson of Lund University and Steven Spivak of the University of Maryland—made it possible for Björn Karlsson to stay at the University of Maryland as a Visiting Associate Professor for 3 months in 1996, where the first part of the book was written.

Bödvar Tomasson drew most of the figures and helped set up the distance learning course that accompanies this textbook. Without his considerable efforts and patience with respect to the artwork, the preparation of the manuscript would have been a far more onerous task. His invaluable assistance and encouragement is deeply appreciated.

Johan Lundin is gratefully acknowledged for collecting and summarizing information on Internet resources for Fire Safety Engineering, presented in Appendix A.

Jonas Nylén is gratefully acknowledged for writing a short user guide (in Swedish) to the computer program CEdit and allowing its translation and inclusion in this textbook as Appendix C.

We also extend our thanks to Stefan Särdqvist, who assisted with layout, Maria Andersson and Ann Bruhn, who typed part of the text, and BRANDFORSK for providing a travel grant.

Finally, a number of scholars in the field of fire safety science have read and commented on selected sections of the text. While the responsibility for the material presented is entirely in the hands of the authors, we gratefully acknowledge the assistance of Ove Pettersson, Gunnar Heskestad, Jim Shields, Andy Buchanan, Craig Beyler, and Dougal Drysdale.

Table of Contents

Chapter 1
Introduction...1

Chapter 2
A Qualitative Description of Enclosure Fires......................................11

Chapter 3
Energy Release Rates 25

Chapter 4
Fire Plumes and Flame Heights...47

Chapter 5
Pressure Profiles and Vent Flows for Well-Ventilated Enclosures 81

Chapter 6
Gas Temperatures in Ventilated Enclosure Fires 115

Chapter 7
Heat Transfer in Compartment Fires...141

Chapter 8
Conservation Equations and Smoke Filling.......................................181

Chapter 9
Combustion Products...227

Chapter 10
Computer Modeling of Enclosure Fires...255

Appendix A
Fire Safety Engineering Resources on the Internet............................279

Appendix B
Suggestions for Experiments and Computer Labs 289

Appendix C
A Simple User's Guide to CEdit..299

Index...307

List of Symbols

A	Area (m^2)
A_c	Area of ceiling vent (m^2)
A_D	Area of opening for in-flowing gases (Chapter 8) (m^2)
A_E	Area of opening for out-flowing gases (Chapter 8) (m^2)
A_f	Horizontal burning area of fuel (m^2)
A_l	Area of lower opening (Chapter 5) (m^2)
A_o	Area of opening (Chapter 6) (m^2)
A_t	Total surface area bounding an enclosure (m^2)
A_T	Total internal enclosure surface area (minus opening area) (m^2)
A_u	Area of upper opening (Chapter 5) (m^2)
A_w	Enclosure surface area in contact with hot gases (Chapter 8) (m^2)
b	Plume radius (m)
c	Specific heat (Chapter 6) (kJ/(kg K))
c_p	Specific heat at constant pressure (kJ/(kg K))
c_v	Specific heat at constant volume (kJ/(kg K))
C	Correction factor for mean beam length (Chapter 7) (-)
C_d	Flow coefficient (-)
C_1	Constant in ideal plume equations (-)
C_2	Constant in ideal plume equations (m$^{4/3}$/s)
D	Diameter (m)
e	Specific total energy (Chapter 8) (J/kg)
E	Total energy (Chapter 8) (J)
E	Emissive power (Chapter 7) (W/m^2)
E_b	Black-body emissive power (Chapter 7) (W/m^2)
f	Fuel mixture fraction (Chapter 9) (-)
F	Force (N)
F	Configuration factor (Chapter 7) (-)
Fr	Froude number (-)
g	Acceleration due to gravity (m/s^2)
G	Irradiance (Chapter 7) (W/m^2)
h	Specific enthalpy (Chapter 8) (J/kg)
h	Height (m)
h_c	Convective heat transfer coefficient (W/(m^{2}K))
h_k	Effective heat conduction coefficient (W/(m^{2}K))
h_l	Height from neutral layer to center of lower opening (Chapter 3) (m)
h_u	Height from neutral layer to center of upper opening (Chapter 3) (m)
H	Enthalpy (Chapter 8) (J)
H	Height between centerlines of two openings (Chapter 5) (m)
H_D	Height of smoke layer from some reference point (m)

H_N	Height of neutral layer from some reference point (m)
H_o	Height of opening (m)
I	Radiative intensity (Chapter 7) (W/m^2)
I_b	Black-body radiative intensity (Chapter 7) (W/m^2)
J	Radiosity (Chapter 7) (W/m^2)
k	Thermal conductivity (W/(mK))
$k\beta$	Material constant for liquid fuels (Chapter 3) (m^{-1})
K_f	Multiplying factor (Chapter 6) (-)
L	Length (m)
L	Flame height (Chapter 4) (m)
L_f	Flame height (Chapter 7) (m)
L_0	Geometric mean beam length (Chapter 7) (m)
L_m	Mean beam length (Chapter 7) (m)
m	Mass (kg)
$\dot{m}$	Mass flow rate or mass burning rate (kg/s)
$\dot{m}_a$	Mass flow rate of ambient air (kg/s)
$\dot{m}_b$	Mass burning rate (kg/s)
$\dot{m}_c$	Mass flow rate of hot gases through a ceiling vent (kg/s)
$\dot{m}_d$	Mass flow rate of air in through an opening (Chapter 8) (kg/s)
$\dot{m}_e$	Mass flow rate of gases exiting from an opening (kg/s)
$\dot{m}_g$	Mass flow rate of hot gases (kg/s)
$\dot{m}_o$	Fan mass flow rate (Chapter 8) (kg/s)
$\dot{m}_p$	Plume mass flow rate (kg/s)
$\dot{m}''$	Mass flow rate or mass burning rate per unit area (kg/(s m^2))
$\dot{m}''_f$	Mass flow rate of fuel (kg/(s m^2))
$\dot{m}''_\infty$	Fuel burning rate for a large diameter (kg/(s m^2))
M	Molecular mass (kg/mol)
n	Fire growth rate exponent (Chapter 8) (-)
Nu	Nusselt Number (Chapter 7) (-)
P	Pressure (Pa, N/m^2)
P_a	Ambient pressure (Chapter 8) (Pa, N/m^2)
P_0	Ambient pressure (Chapter 5) (Pa, N/m^2)
Pr	Prantl Number (Chapter 7) (-)
q	Heat (kJ)
$\dot{q}$	Heat flow rate (kW)
$\dot{q}_B$	Rate of heat storage in the gas volume (Chapter 6) (kW)
$\dot{q}_{loss}$	Rate of heat lost to compartment boundaries (kW)
$\dot{q}_L$	Rate of heat lost due to replacement of hot gases by cold (Chapter 6) (kW)
$\dot{q}''_r$	Radiative heat flux (kW/m^2)
$\dot{q}_R$	Rate of heat lost by radiation through openings (Chapter 6) (kW)
$\dot{q}_W$	Rate of heat lost to compartment boundaries (Chapter 6) (kW)
$\dot{q}''$	Heat flux, or heat flow rate per unit area (kW/m^2)
$\dot{q}''_c$	Convective heat flux (kW/m^2)
Q	Energy (kJ)
Q	Fire load (Chapter 6) (kJ)
Q''	Fire load density (Chapter 6) (kJ/m^2)

Q_f''	Fire load density per unit floor area of enclosure (Chapter 6) (kJ/m^2)
Q_t''	Fire load density per total enclosure surface area (Chapter 6) (kJ/m^2)
Q	Energy release rate (kW)
Q_c	Convective part of energy release rate (kW)
Q_{ch}	Chemical energy release rate (Chapter 8) (kW)
Q_{FO}	Energy release rate sufficient to cause flashover (kW)
Q_{max}	Maximum energy release rate (kW)
Q^*	Dimensionless energy release rate (-)
r	Stoichiometric fuel to oxygen ratio (-)
r_f	Radial flame extension under a ceiling (m)
R	Gas Constant for air (J/(kg K))
R	Distance (Chapter 7) (m)
Ra	Rayleigh Number (Chapter 7) (-)
R_0	Universal Gas Constant (J/(mol K))
R_0	Distance (Chapter 7) (m)
S	Floor area (Chapter 8) (m^2)
t	Time (s)
t_0	Starting time (Chapter 3) (s)
t_p	Thermal penetration time (s)
t_s	Duration of the steady phase of the fire (Chapter 3) (s)
T	Temperature (°C or K)
T_a	Ambient air temperature (°C or K)
T_e	Temperature of gases exiting an enclosure (Chapter 8) (°C or K)
T_f	Film temperature (Chapter 7) (°C or K)
T_g	Temperature of hot gases (°C or K)
T_{max}	Maximum ceiling jet temperature (°C or K)
T_0	Centerline plume temperature (°C or K)
T_∞	Ambient air temperature (°C or K)
u	Specific internal energy (Chapter 8) (J/kg)
u	Velocity (m/s)
u_{max}	Maximum ceiling jet velocity (m/s)
u_0	Centerline plume velocity (m/s)
U	Internal energy (Chapter 8) (J)
v	Velocity (m/s)
v_a	Velocity of ambient gases (m/s)
v_c	Velocity of hot gases through a ceiling vent (m/s)
v_g	Velocity of hot gases (m/s)
V	Volume (m^3)
$\dot{V}$	Volumetric flow rate (m^3/s)
W	Width of opening (m)
W	Work done (Chapter 8) (J)
$\dot{W}$	Rate of work done (W)
X_1	Dimensionless variable (Chapter 6) (-)
X_2	Dimensionless variable (Chapter 6) (-)
y	Dimensionless height of smoke layer interface above floor (Chapter 8) (-)
y_i	Yield of species i (Chapter 9) (-)

$y_{i,max}$	Maximum yield of species i (Chapter 9) (-)
$y_{i,Wv}$	Unlimited air yield of species i (Chapter 9) (-)
$y_{i,\infty}$	Unlimited air yield of species i (Chapter 9) (-)
Y_i	Mass fraction of species i (Chapter 9) (-)
Y_p	Combustion product mass fraction (Chapter 7) (-)
z	Height in a fire plume (Chapter 4) (m)
z	Height when integrating over an opening (Chapter 5) (m)
z	Height of smoke layer interface above floor (Chapter 8) (m)
z_0	Height of virtual origin of fire plume (m)

α	Fire growth factor (Chapter 3) (kW/s^2)
α	Thermal diffusivity (Chapter 6) (m^2/s)
α	Absorptivity (Chapter 7) (-)
β	Expansion coefficient (Chapter 7) (K^{-1})
δ	Thickness (m)
ΔH_g	Heat of gasification (kJ/kg)
ΔH_c	Complete heat of combustion (kJ/kg)
ΔH_{eff}	Effective heat of combustion (kJ/kg)
ΔP	Pressure difference (Pa, N/m^2)
ΔP_l	Pressure difference across lower opening (Pa, N/m^2)
ΔP_u	Pressure difference across upper opening (Pa, N/m^2)
ε	Emissivity (-)
θ_0	Angle (Chapter 7) (radians)
κ	Constant in plume equation (Chapter 4) (units vary)
κ	Extinction coefficient (Chapter 7) (m^{-1})
λ	Wavelength (Chapter 7) (μm)
η	Constant in plume equation (Chapter 4) (-)
ν	Kinematic viscosity (Chapter 7) (m^2/s)
ρ	Density (kg/m^3)
ρ_a	Ambient air density (kg/m^3)
ρ_g	Density of hot gases (kg/m^3)
ρ_∞	Ambient air density (kg/m^3)
σ	Stefan–Boltzmann Constant (W/(m^2 K^4))
τ	Dimensionless time (Chapter 8) (-)
ϕ	Equivalence ratio (-)
χ	Combustion efficiency (-)
χ_r	Fraction of total energy radiated (Chapter 7) (-)

A general discussion of symbols is given in Chapter 1. The units shown here are those most commonly used; some variations with respect to the SI decimal relationships kilo and mega may occur.

1 Introduction

This chapter begins with a brief description of the state of fire safety engineering design and explains the context in which this textbook was written. It discusses the core curriculum of fire safety engineering and places the material presented in the book into context with other topics within the fire safety engineering discipline. Enclosure fire models currently used in fire safety engineering design are briefly discussed. Finally, an overview of the contents of the book is presented and general notations are explained.

CONTENTS

1.1   Background 1
1.2   Core Curriculum in Fire Safety Engineering 2
1.3   Engineering Models for Enclosure Fires...3
    1.3.1   Energy Evolved and Species Generated...5
    1.3.2   Fire-Induced Environment 5
    1.3.3   Heat Transfer...6
1.4   Contents of This Textbook..6
1.5   A Note on Dimensions, Units, and Symbols 7
References 9

1.1   BACKGROUND

Fire safety regulations can have a major impact on many aspects of the overall design of a building, including layout, aesthetics, function, and cost. Rapid developments in modern building technology in the last decades often have resulted in unconventional structures and design solutions. The physical size of buildings increases continually; there is a tendency to build large underground car parks, warehouses, and shopping complexes. The interior design of many buildings—with large light shafts, patios, and covered atriums within buildings connected to horizontal corridors or malls—introduces new risk factors concerning spread of smoke and fire. Past experiences or historical precedents (which form the basis of current prescriptive building codes and regulations) rarely provide the guidance necessary to deal with fire hazards in new or unusual buildings.

At the same time there have been great strides in the understanding of fire processes and their interrelationship with humans and buildings. Advancement has been particularly rapid in the area of analytical fire modeling. Several different types of such models, with varying degrees of sophistication, have been developed in recent years and are used by engineers in the design process.

As a result, we have a worldwide movement to replace prescriptive building codes with ones based on performance. Instead of prescribing exactly which protective measures are required (such as prescribing a number of exits for evacuation purposes), the performance of the overall system is presented against a specified set of design objectives (such as stating that satisfactory escape should be effected in the event of fire). Fire modeling and evacuation modeling can often be used to assess the effectiveness of the protective measures proposed.

The need to take advantage of the new emerging technology, both with regard to design and regulatory purposes, is obvious. The increased complexity of the technological solutions, however, requires higher levels of academic training for fire protection engineers and a higher level of continuing education during their careers.

Some excellent textbooks, handbooks, and design guides have been produced for this purpose, including *An Introduction to Fire Dynamics* by Drysdale,[1] *The SFPE Handbook of Fire Protection Engineering*,[2] and *Design of Smoke Control Systems* by Klote and Milke,[3] to name only a few.

Apart from the book by Drysdale and the one by Shields and Silcock,[4] textbooks on fire safety engineering specifically written for engineering students have been scarce. Design guides and handbooks generally list engineering problems and provide methodologies by which these problems can be solved using specific calculational procedures. The equations used are seldom derived from first principles, and little information is given on the assumptions made or the validity of the approach. To fully understand the effect these assumptions may have in a specific design situation and to be confident of the validity of the chosen calculational procedure, the engineer at some point must have derived the equations from first principles.

The purpose of this textbook is not to act as a design guide or a list of equations that can be applied to specific scenarios, but rather to show how engineering equations for certain applications can be arrived at from first principles, to state the assumptions clearly, and to show how the resulting analytical equations compare to experimental data. In this way the reader will get a strong feeling for validity and applicability of a wide range of commonly used engineering equations and models.

This textbook specifically examines enclosure fire dynamics, the study of how the outbreak of a fire in a compartment causes changes in the environment of the enclosure. Before introducing the contents of the book we shall discuss the fire safety engineering core curriculum. In Section 1.4 we briefly discuss engineering models currently used for calculating the environmental consequences of a fire in an enclosure. Finally, we discuss some symbols and units.

1.2   CORE CURRICULUM IN FIRE SAFETY ENGINEERING

The field of fire safety engineering encompasses topics from a wide range of engineering disciplines as well as material of unique interest to fire safety engineering. It is not immediately obvious which of these topics of interest should be addressed in a textbook for students.

When identifying the subject area of the current textbook the authors were greatly assisted by the publication "A Proposal for a Model Curriculum in Fire Safety Engineering," by Magnusson et al.[5], which identifies the contents of the background, fundamental, and applied courses that may be taught within the discipline of fire safety engineering.

The fundamental courses are divided into five modules:

- Fire fundamentals
- Enclosure fire dynamics
- Active fire protection
- Passive fire protection
- Interaction between fire and people

This textbook deals mainly with the second module, enclosure fire dynamics. The modules, however, are interlinked to a considerable extent, and it is often a question of preference where to include borderline topics and where to present a summarized background. The book by Drysdale[1] is an excellent text for a course on fire fundamentals that emphasizes the basic chemistry and physics of fire, but the book also touches upon several topics within the other modules listed above.

Also, it is not obvious where to strike the balance between material presented in the fundamental modules and material assumed to be prerequisite knowledge from basic courses in physics, chemistry, fluid mechanics, etc. We assume that the student has a basic knowledge of mathematics, physics, and chemistry.

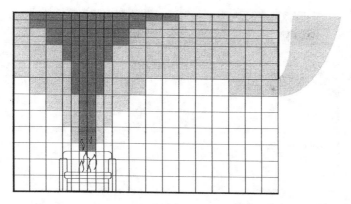

FIGURE 1.1   Computational fluid dynamics models divide the enclosure into a large number of sub-volumes.

This textbook does not attempt to provide an in-depth study of the phenomena, but rather to present the most dominating mechanisms controlling an enclosure fire and to derive some simple analytical relationships that can be used in practice. In view of the increased use of calculational procedures and computer models in building fire safety engineering design, the main purpose of this textbook is to:

- provide an introductory, basic understanding of the phenomena of interest and present some examples where these can be used in practice;
- derive the equations from first principles in order to give the reader a true sense of the validity of the procedures in each design situation; and
- compare the derived equations with experimental data to provide a sense of confidence in the analytical results.

1.3   ENGINEERING MODELS FOR ENCLOSURE FIRES

The rapid progress in the understanding of fire processes and their interaction with buildings has resulted in the development of a wide variety of models that are used to simulate fires in compartments. The models can be classified as either probabilistic or deterministic. Probabilistic models do not make direct use of the physical and chemical principles involved in fires; rather, they make statistical predictions about the transition from one stage of fire growth to another. Such models will not be discussed further here. The deterministic models can roughly be divided into three categories: CFD models, zone models, and hand-calculation models.

CFD models: The most sophisticated of these are termed "field models" or computational fluid dynamics (CFD) models. The CFD modeling technique is used in a wide range of engineering disciplines. Generally, the volume under consideration is divided into a very large number of sub-volumes, and the basic laws of mass, momentum, and energy conservation are applied to each sub-volume. Figure 1.1 shows a schematic of how this may be done for a fire in an enclosure. The governing equations contain as further unknowns the viscous stress components in the fluid flow. Substitution of these into the momentum equation yields the so-called Navier–Stokes equations, the solution of which is central to any CFD code.

The myriad engineering problems that can be addressed by CFD models are such that no single CFD code can incorporate all of the physical and chemical processes that are of importance. There is only a handful of CFD codes that can be used for problems involving fire. These, in turn, use a number of different approaches to the sub-processes that need to be modeled. Some of the most important of these sub-processes can be considered to be turbulence modeling, radiation and soot modeling, pyrolysis and flame spread modeling, and combustion modeling. The sub-processes are

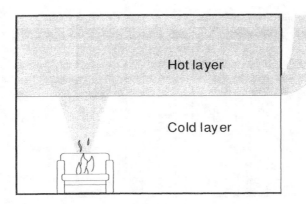

FIGURE 1.2   Two-zone modeling of a fire in an enclosure.

usually modeled at a relatively fundamental level, the understanding of which requires expert knowledge in a number of specialized fields of physics and chemistry. Cox[6] provides an excellent summary of the main issues. A description of the fundamental laws of physics and chemistry contained in CFD models is outside the scope of this textbook.

Use of CFD models requires considerable computational capacity as well as expert knowledge, not only in physics and chemistry, but also in numerical methods and computer science. In addition, it is a very time consuming and costly process to set up the problem, run it on the computer, extract the relevant output, and present the results, so practical use of this methodology for fire safety engineering design is relatively rare. However, such a modeling methodology can be very useful when dealing with complex geometries, and it may be the only way to proceed with certain design problems.

Two-zone models: A second type of deterministic fire model divides the room into a limited number of control volumes or zones. The most common type is the "two-zone model," where the room is divided into an upper, hot zone and a lower, cold zone (Figure 1.2). The equations for mass and energy conservation are solved numerically for both zones for every time step. The momentum equation is not explicitly applied; instead, information needed to calculate velocities and pressures across openings comes from analytically derived expressions where a number of limiting assumptions have been made. Several other sub-processes, such as plume flows and heat transfer, are modeled in a similar way. The section on hand calculations below lists a number of these processes, and later chapters of this book derive the equations and introduce the assumptions made.

Many two-zone models have been described in the literature. Some of these only simulate a fire in a single compartment; others simulate fires in several compartments, linked by doors, shafts, or mechanical ventilation. Additionally, the degree of verification, documentation, and user-friendliness varies greatly between these models.

In recent years there has been an upsurge in the use of two-zone models in fire safety engineering design. This is partly due to the increasing availability and user-friendliness of computer programs. However, any serious use of such models demands that the user be well acquainted with the assumptions made and the limitations of the models, i.e., that the user has had some training in the subject of enclosure fire dynamics. This textbook aims to provide the necessary background.

Hand-calculation models: A third way to analytically describe some basic fire processes is to use simple hand-calculation methods. These are basically a collection of simplified solutions and empirical methods to calculate flame heights, mass flow rates, temperature and velocities in fire plumes, time to sprinkler activation, room overpressure, and many other variables.

The remainder of this section describes the hand-calculation models. The methods discussed below can, for convenience, be divided into three categories: those that deal with combustion; those that estimate the resulting environmental conditions; and those that involve heat transfer.

Applications of these methods have greatly varying limitations, and the user must have some knowledge of classical physics in order to apply them correctly.

1.3.1   ENERGY EVOLVED AND SPECIES GENERATED

Calculating fire growth and the amount of energy evolved from the primary fire source requires knowledge of the type and amount of fuel involved. Typical burning rates and the heat of combustion for a range of liquid fuels burning in the open have been experimentally determined and are provided in the literature. This allows the energy evolved to be calculated if the area of the liquid spill is known. If the amount of spilled liquid is known then the time to burn-out can also be calculated. Fire growth information for solids and other burning objects is available from several sources. Energy release rates for many items of furniture, curtains, and different types of materials are available. Such values are also available for species production rates, which allows calculation of species concentrations.

The rate of energy evolved in a compartment is also dependent on the rate of supply of oxygen. Knowledge of the ventilation conditions can therefore be used to evaluate the maximum rate of energy release inside a compartment. Any excess, unburned fuel will then be burned outside the fire compartment, where oxygen is available.

Computer programs with material databases are also available to assist the user in choosing an appropriate energy release rate curve.

1.3.2   FIRE-INDUCED ENVIRONMENT

The basic principles used to calculate the environmental conditions due to a fire in a compartment are the conservation of mass, energy, and momentum. The application of the conservation laws will lead to a series of differential equations. By making certain assumptions about the energy and mass transfer in and out of the compartment boundaries, the laws of mass and energy conservation can result in a relatively complete set of equations. The complexity and the large number of equations involved make a complete analytical solution impossible, so one must resort to numerical analysis through computer programs.

However, analytical solutions can be derived by using results from experiments and a number of limiting approximations and assumptions. Such solutions have generated numerous expressions, which may be used to predict a variety of environmental factors in a fire room. Several examples are given below.

The buoyant gas stream rising above a burning fuel bed is often referred to as the fire plume. The properties of fire plumes are important in dealing with problems related to fire detection, fire venting, heating of building structures, smoke filling rates, etc. By using dimensional analysis, the conservation equations and data from experiments, expressions for various plume properties have been developed. These include expressions for plume temperature, mass flow, and gas velocities at a certain height above the fire as well as flame height. Similar expressions have been derived for the jet that results when the plume gases impinge on a ceiling.

Mass flow in and out of compartment openings can be calculated since the pressure differences across the opening can be estimated. The use of classical hydraulics and experimentally determined flow coefficients has resulted in hand calculation expressions for such mass flows.

The gas temperature in a naturally or mechanically ventilated compartment can be calculated by hand, using regression formulae based on experimentally measured gas temperatures in a range of fire scenarios and a simplified energy and mass balance. Such expressions are available for both pre- and post-flashover fires. By using similar expressions, the onset of flashover can be estimated.

By combining the expressions for gas temperature, plume flows, and vent flows, the descent of the smoke layer as a function of time can be calculated. The resulting expressions are usually in the form of differential equations, but certain limiting cases can be solved by hand. Such solutions usually require an iteration process or the use of precalculated curves or tables.

Several other types of hand calculation expressions have been developed, including expressions for mass flow through roof openings, buoyant pressure of hot gases, species concentration, fire-induced room pressures, flame sizes from openings, etc. Some such expressions have been collected in relatively user-friendly computer programs.

1.3.3 HEAT TRANSFER

There are three mechanisms by which heat is transferred from one object to another: radiation, convection, and conduction. Classical textbooks on heat transfer provide innumerable hand-calculation expressions for calculating heat fluxes to and from solids, liquids, and gases, as well as expressions for estimating the resulting temperature profiles in a target. These analytical expressions are usually arrived at by setting up the energy balance, by assuming constant properties and homogeneity in the media involved, and by ignoring the heat transfer mechanisms that seem to be of least importance in each case.

The radiative heat flux from flames, hot gases, and heated surfaces impinging on a solid surface can be estimated using classical heat transfer and view factors. The same applies for convective heat transfer to solids and conductive heat transfer through solids.

The surface temperature of a solid subjected to a radiative, convective, or conductive heat flux can be calculated by hand assuming the solid is either semi-infinite or behaves as a thermally thin material. Numerous types of heat transfer problems can be solved in this way. A few examples are given below.

Assuming that a secondary fuel package is subjected to a known heat flux and that it has a certain ignition temperature and constant thermal properties, then the time to ignition can be calculated. Similarly, if the activation temperature of a sprinkler bulb is known, the activation time can be estimated. Several other problems can be addressed in this way, including temperature profiles in building elements, flame spread over flat solids, heat detector activation, spread of fire from one building to another, etc. Analytical solutions to such problems can be found in standard textbooks on heat transfer.

1.4 CONTENTS OF THIS TEXTBOOK

The previous section summarized deterministic models for enclosure fire calculations available to the fire safety engineer. This textbook addresses these issues in the following order:

Chapter 2: A qualitative description of enclosure fires. This chapter contains a general, qualitative description of the chemical and physical phenomena associated with fires in enclosures and the environmental conditions that result. The different stages in enclosure fire development are discussed. Terms essential for the subsequent treatment of the subject are identified and defined.

Chapter 3: Energy release rates. In order to calculate the environmental consequences of a fire in an enclosure, the rate at which the fuel releases energy must be known. This chapter outlines the methods commonly used to estimate the energy release rate produced by a burning fuel package.

Chapter 4: Fire plumes and flame heights. The buoyant gas flow above a fire source is called a plume. As the hot gases rise, cold air will be entrained into the plume. This mixture of combustion products and air then rises to the ceiling of an enclosure and causes formation of a hot upper layer. This chapter discusses the most fundamental properties of fire plumes; gives expressions for calculating variables associated with them; and examines flame heights and analytical expressions for estimating these heights in certain given scenarios.

Chapter 5: Pressure profiles and vent flows for well-ventilated enclosures. Knowledge of the flow of gases exiting and entering an enclosure provides information needed for the mass and energy balance in the enclosure and thus allows us to calculate several important environmental consequences of compartment fires. In this chapter we derive engineering equations used to calculate pressure differences across openings, as well as equations for calculating the mass flow of gases in and out through vents for several common enclosure fire scenarios.

Chapter 6: Gas temperatures in ventilated enclosure fires. Knowledge of the temperature of the hot smoke in an enclosure can be used to assess when hazardous conditions for humans will arise, when flashover may occur, when structural elements are in danger of collapsing, and the thermal feedback to fuel sources or other objects. This chapter derives and reviews a few analytical methods that have been developed to predict temperatures in both the pre- and post-flashover phases of well-ventilated enclosure fires.

Chapter 7: Heat transfer in compartment fires. The enclosure energy balance is greatly affected by transfer of heat from the flames and the hot gases to the enclosure surfaces and out through the enclosure openings. The heat transferred from these sources toward a fuel package will control, to a considerable extent, the rate at which fuel evaporates and heat is released. This chapter focuses on radiative heat transfer in enclosures and briefly discusses convection heat transfer as applied to enclosure fires.

Chapter 8: Conservation equations and smoke filling. This chapter states the conservation laws for mass and energy and introduces some commonly applied assumptions that allow the derivation of analytical solutions and iterative methods, which can be applied to problems related to the smoke filling process. Two types of ventilation conditions are considered: compartments that are closed or have small leakage vents and compartments with openings large enough to prevent the build-up of pressures due to gas expansion. The conservation equations are applied to calculate smoke filling time and derive smoke control methodologies for several cases.

Chapter 9: Combustion products. The ability to estimate the toxic hazards of combustion gases in a fire compartment enables us to estimate the toxic hazard to humans. This chapter discusses methods for estimating the amount of each toxic species produced per unit fuel burnt, i.e., the species yield. Once the production term is known, the concentration in the fire gases can be estimated. The generation of combustion products is a very complex issue, and the engineer must rely on measurement and approximate methods for estimating the yield of a product. This chapter introduces some methods available to the engineer for estimating the yield of a species and discusses methods for calculating species concentrations.

Chapter 10: Computer modeling of enclosure fires. This chapter summarizes how the methods discussed in the previous chapters are used in compartment fire modeling set up on a computer. The main part of this chapter is devoted to the zone modeling technique (and is partly based upon Quintiere's chapter in the *SFPE Handbook of Fire Protection Engineering*[7]). CFD models are also discussed. The final section of the chapter lists some Internet addresses from where computer models can be downloaded.

Appendix A: Fire safety engineering resources on the Internet. Appendix A provides a list of Internet addresses of special interest fire safety engineering professionals. Johan Lundin, Department of Fire Safety Engineering, Lund University, collected and summarized this material.

Appendix B: Suggestions for experiments and computer labs. This appendix provides a quick introduction to a well-known two-zone model, describes full scale and 1/3 scale experiments that can be carried out, and suggests how these can be simulated using hand calculation and two-zone models.

Appendix C: A simple user's guide to CEdit. This appendix gives a quick introduction to the input part of the well-known two-zone model CFAST.

1.5 A NOTE ON DIMENSIONS, UNITS, AND SYMBOLS

General: A physical quantity may be characterized by dimensions. The arbitrary magnitudes assigned to the dimensions are called units. The dimensions and units may be assigned certain

TABLE 1.1
Some Standard Prefixes in SI Units

Prefix	Multiple	Symbol
10^{12}	tera	T
10^9	giga	G
10^6	mega	M
10^3	kilo	k
10^{-2}	centi	c
10^{-3}	milli	m

TABLE 1.2
The Seven Fundamental Dimensions and Their Units in SI

Dimension	Symbol	Unit	Unit Symbol
Length	L	meter	m
Mass	m	kilogram	kg
Time	t	second	s
Temperature	T	Kelvin	K
Electric current	*	ampere	A
Amount of light	*	candela	c
Quantity of matter	*	mole	mol

* Not used in this textbook.

symbols. For example, mass is a dimension often assigned the symbol m, and often expressed in kilograms, which is assigned the symbol kg.

Two sets of unit systems are in common use today: the British system, also known as the U.S. Customary System (USCS), and the metric SI (from Le Systéme International d'Unités). The SI, a simple and logical system based on a decimal relationship between the various units, is used for engineering work in most industrialized nations, including Britain. The U.S. is the only industrialized country that has not yet fully converted to the metric system; it uses both systems.

Since this textbook uses the SI system, we shall discuss the basis of the SI and provide some conversion factors to the British system.

Table 1.1 shows the SI decimal relationship between some standard units, along with the assigned symbols.

There are seven fundamental dimensions in the SI, and all other dimensions in the SI can be expressed in terms of these fundamental dimensions. Table 1.2 shows the dimensions and their assigned symbols, the corresponding units, and the unit symbols.

Units and symbols used in fire safety engineering: Fire safety science is a rapidly developing field, and fire safety engineering is a relatively young engineering discipline. The scientific and engineering communities have therefore not yet fully standardized the assignments of symbols and units used in this field. The variation in symbols and units used is often a source of confusion and irritation to the engineering student.

In fire safety engineering, energy release rates are commonly given in kilowatts (kW), whereas according to SI they should be given in watts (W). We shall follow the fire safety engineering tradition and express energy release rates in kW.

Some of the equations that have been derived semi-empirically also require that nonstandard SI units be used as input, since the constants in the expressions are derived using nonstandard SI

TABLE 1.3
Some Dimensions, Units, and Symbols Used in This Textbook

Dimension	Symbol Often Used	Conversion
Length	L	1 m = 3.2808 ft
Area	A	1 m² = 10.7639 ft²
Density	r	1 kg/m³ = 0.06243 lb/ft³
Mass	m	1 kg = 2.2046 lb
Mass flow rate	$\dot{m}$	1 kg/s = 2.2064 lb/s
Mass flow rate per unit area	$\dot{m}''$	1 kg/(s m²) = 0.205 lb/(s ft²)
Energy	Q	1 kJ = 0.94738 Btu
Heat	q	1 kJ = 0.94738 Btu
Energy release rate	$\dot{Q}$	1 kW = 3412.1 Btu/hr
Heat flow rate	$\dot{q}$	1 kW = 3412.1 Btu/hr
Heat flux, or heat flow rate per unit area	$\dot{q}''$	1 kW/m² = 3.17 Btu/(hr ft²)

units. For example, the equations given in this book for calculating plume mass flow rates require that the energy release rate used in the expressions be given in kW.

Units and symbols used in this book: Below we list some of the most commonly used units and symbols in this book.

A symbol with a dot above it denotes a quantity per unit time. Thus, the symbol $\dot{m}$ represents mass per unit time, given in kg/s, and the symbol $\dot{Q}$ denotes energy release rate, given in kJ/s, i.e., kW.

A symbol followed by a double prime sign denotes a quantity per unit area. For example, the symbol $\dot{m}''$ represents mass per unit time per unit area and is given in kg/(s m²).

Table 1.3 shows some dimensions, symbols, and units used in this book, specifically those related to fire safety engineering, and provides some conversion factors to the British system.

Most other symbols and units used in this book are commonly known and widely accepted. The symbols are also defined in the text.

Temperature conversions:
°F: degree Fahrenheit: T(°F) = T(°C) · 1.8 + 32
°R: degree Rankine: T(°R) = T(°F) + 459.69
K: Kelvin T(K) = T(°C) + 273.15

Pressure conversions:
Pressure, P: 1 atm = 1.01325 · 10⁵ N/m² = 14.69595 lbf/in²

Some physical constants:
Universal Gas Constant, R_0 = 8.314 J/(mol K)
Gas Constant for air, R = 287 J/(kg K)
Stefan–Boltzmann Constant, σ = 5.67·10⁻⁸ W/(m² K⁴)

The *SFPE Handbook of Fire Protection Engineering*[2] provides a comprehensive list of common units and conversion factors.

REFERENCES

1. Drysdale, D., *An Introduction to Fire Dynamics*, Wiley-Interscience, New York, 1992.
2. *The SFPE Handbook of Fire Protection Engineering*, 2nd ed., National Fire Protection Association, Quincy, MA, 1995.

3. Klote, J.H. and Milke, J.A., *Design of Smoke Management Systems*, American Society of Heating, Refrigeration and Air-Conditioning Engineers, Atlanta, GA, 1992.
4. Shields, T.J. and Silcock, G.W.H., *Buildings and Fire*, Longman Scientific and Technical, London, 1987.
5. Magnusson, S.E., Drysdale, D.D., Fitzgerald, R.W., Mowrer, F., Quintiere, J., Williamson, R.B., and Zalosh, R.G., A Proposal for a Model Curriculum in Fire Safety Engineering, *Fire Safety Journal*, Vol. 25, 1995.
6. Cox, G., "Compartment Fire Modelling," Chapter 6 in *Combustion Fundamentals of Fire*, Cox, G., Ed., Academic Press, London, 1995.
7. Quintiere, J.G., Compartment Fire Modeling, in *The SFPE Handbook of Fire Protection Engineering*, 2nd ed., National Fire Protection Association, Quincy, MA, 1995.

2 A Qualitative Description of Enclosure Fires

This chapter provides an overview and a qualitative description of the development of a fire in an enclosure and discusses the resulting environmental conditions that arise. It lays groundwork for the more calculation-intensive chapters that follow. We introduce and define a number of terms necessary for our subsequent treatment of the subject of enclosure fires. The chapter begins with a general description of the process of combustion and a description of a typical development of a fire in an enclosure. The fire development is commonly divided into different stages; these are identified and discussed. Finally, we discuss a number of factors that influence enclosure fire development.

CONTENTS

2.1   Terminology 11
2.2   Introduction 12
     2.2.1   General Description of the Process of Combustion......................................12
     2.2.2   General Description of Fire Growth in an Enclosure 14
2.3   Stages in Enclosure Fire Development..17
     2.3.1   Fire Development in Terms of Enclosure Temperatures..............................17
     2.3.2   Fire Development in Terms of Flow through Openings 19
     2.3.3   Other Common Terms Describing Enclosure Fire Stages 19
2.4   Factors Influencing Fire Development in an Enclosure..21
2.5   Summary 24
References 24

2.1 TERMINOLOGY

Backdraft — Limited ventilation during an enclosure fire can lead to the production of large amounts of unburnt gases. When an opening is suddenly introduced, the inflowing air may mix with these, creating a combustible mixture of gases in some part of the enclosure. Any ignition sources, such as a glowing ember, can ignite this flammable mixture, resulting in an extremely rapid burning of the gases. Expansion due to the heat created by the combustion will expel the burning gases out through the opening and cause a fireball outside the enclosure. The phenomenon can be extremely hazardous.

Flashover — The transition from the fire growth period to the fully developed stage in the enclosure fire development. The formal definition from the International Standards Organization is "the rapid transition to a state of total surface involvement in a fire of combustible material within an enclosure." In fire safety engineering the word is used to indicate the demarcation point between two stages of a compartment fire, i.e., pre-flashover and post-flashover.

Fuel-controlled fire — After ignition and during the initial fire growth stage, the fire is said to be fuel-controlled, since in the initial stages there is sufficient oxygen available for combustion and the growth of the fire entirely depends on the characteristics of the fuel and its geometry. The fire can also be fuel-controlled in later stages.

Fully developed fire — Synonymous with the post-flashover fire, which lasts from the occurrence of flashover through the decay stage to extinction. During most of this period the fire is ventilation-controlled; at some point during the decay period it becomes fuel-controlled again.

Post-flashover fire — When the objective of fire safety engineering design is to ensure structural stability and safety of fire fighters, the post-flashover fire is of greatest concern. The design load in this case is characterized by the temperature–time curve assumed for the fully developed fire stage.

Pre-flashover fire — The growth stage of a fire, where the emphasis in fire safety engineering design is on the safety of humans. The design load is in this case characterized by an energy release rate curve, where the growth phase of the fire is of most importance.

Ventilation-controlled fire — As the fire grows it may become ventilation-controlled, when there is not enough oxygen available to combust most of the pyrolyzing fuel. The energy release rate of the fire is then determined by the amount of oxygen that enters the enclosure openings.

2.2   INTRODUCTION

Fire is a physical and chemical phenomenon that is strongly interactive by nature. The interactions between the flame, its fuel, and the surroundings can be strongly nonlinear, and quantitative estimation of the processes involved is often complex. The processes of interest in an enclosure fire mainly involve mass fluxes and heat fluxes to and from the fuel and the surroundings. Figure 2.1 shows a schematic of these interactions, indicating the complexity of the mass and heat transfer processes occurring in an enclosure fire.

In order to introduce the most dominant of these processes, this section includes a general qualitative description of the chemical and physical phenomena associated with fire. The discussion is divided into two parts: a general and summary discussion on the process of combustion, and a qualitative description of the development of a fire in an enclosure and the effects it has on the enclosure environment.

2.2.1   General Description of the Process of Combustion

The study of combustion is a complex subject; it includes a number of disciplines such as fluid mechanics, heat and mass transport, and chemical kinetics. In fire safety engineering, a course on fire fundamentals deals extensively with the subject of combustion.

In our introductory discussion on combustion we will use a candle as our fuel; the discussion is therefore limited to a laminar, steady flame on a solid substrate. The study of a burning candle, however, is very illustrative of the natural processes in which we are interested. The 19th century scientist Michael Faraday gave lectures on "The Chemical History of a Candle" at the so-called Christmas Lectures at the Royal Institution in London.[2] He claimed that there was no better way one could introduce the study of natural philosophy than by considering the physical phenomenon of a candle.

Consider Figure 2.2, which shows a burning candle and the temperature distribution through the flame. An ignition source, a match for example, heats up the wick and starts melting the solid wax. The wax in the wick vaporizes, and the gases move, by the process of diffusion, out into a region where oxygen is found. The gases are oxidized in a complex series of chemical reactions, in regions where the oxygen–fuel mixture is flammable. The candle flame is then stable; it radiates energy to the solid wax, which melts. Since the wax vaporizes and is removed from the wick, the melted wax moves up the wick, vaporizes, burns, and the result is a steady combustion process.

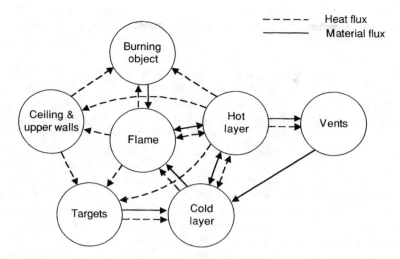

FIGURE 2.1 Schematic of the heat fluxes and mass fluxes occurring in an enclosure fire. (Adapted from Friedman[1]. With permission.)

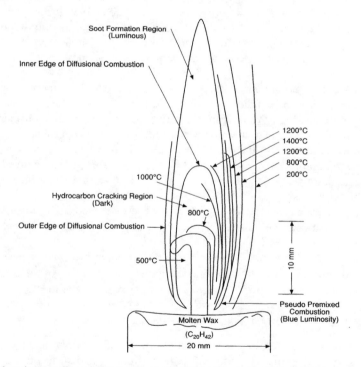

FIGURE 2.2 A burning candle and the temperature distribution in the flame. (From Lyons[3].)

The processes occurring in the flame involve the flow of energy and the flow of mass. **The flow of energy** occurs by the processes of radiation, convection, and conduction. The dominant process is that of radiation; it is mainly the soot particles produced by combustion that glow and radiate heat in all directions. The radiation down toward the solid is the main heat transfer mode, which melts the solid, but convection also plays a role. The convective heat flux is mainly upward, transferring heat up and away from the combustion zone. The larger and more luminous the flame, the quicker the melting process.

The radiative energy reaching the solid is, however, not sufficient to vaporize the wax, only to melt it. The wick is therefore introduced as a way to transport the melted wax up into the hot gases, where the combined processes of radiation, convection, and conduction supply sufficient energy to vaporize the melted wax.

The mass transfer and the phase transformations are also exemplified by the burning candle. The fuel transforms from solid to liquid state. The mass balance requires that the mass that disappears from the wick by vaporization be replaced, and thus the liquid is drawn up into the wick by capillary action. Once there, the heat transfer from the flame causes it to vaporize, and the gases move away from the wick by the process of diffusion. The inner portion of the flame contains insufficient oxygen for full combustion, but some incomplete chemical reactions occur, producing soot and other products of incomplete combustion. These products move upward in the flame due to the convective flow and react there with oxygen. At the top of the flame nearly all the fuel has combusted to produce water and carbon dioxide; the efficiency of the combustion can be seen by observing the absence of smoke emanating from the top of the candle flame.

This self-sustained combustion process can most easily be changed by altering the dimensions and properties of the wick, and thereby the shape and size of the flame. A longer and thicker wick will allow more molten wax to vaporize, resulting in a larger flame and increased heat transfer to the solid. The mass and heat flows will quickly enter a balanced state, with steady burning as a result.

Other solid fuels: Without the wick, the candle will not sustain a flame, as is true for many other solid fuels. Factors such as the ignition source, the type of fuel, the amount and surface area of the fuel package determine whether the fuel can sustain a flame. A pile of wooden sticks may sustain a flame, while a thick log of wood may not do so. Once these factors are given, the processes of mass and energy transport will determine whether the combustion process will decelerate, remain steady, or accelerate.

Also, the phase transformations of other solid fuels may be much more complicated than the melting and vaporizing of the candle wax. The solid fuel may have to go through the process of decomposition before melting or vaporizing, and this process may require considerable energy. The chemical structure of the fuel may therefore determine whether the burning is sustained.

For fuels more complex than the candle, it is difficult to predict fire growth. The difficulty is not only due to the complexity of the physical and chemical processes involved, but also due to the dependence of these processes on the geometric and other fuel factors mentioned above and the great variability in these.

When the fuel package is burnt in an enclosure, the fire-generated environment and the enclosure boundaries will interact with the fuel, as seen in Figure 2.1, making predictions of fire growth even more complex.

It is currently beyond the state of the art of fire technology to predict the fire growth in an enclosure fire with any generality, but reasonable engineering estimates of fire growths in buildings are frequently obtained using experimental data and approximate methods.

2.2.2   GENERAL DESCRIPTION OF FIRE GROWTH IN AN ENCLOSURE

A fire in an enclosure can develop in a multitude of different ways, mostly depending on the enclosure geometry and ventilation and the fuel type, amount, and surface area. The following is a general description of the various phenomena that may arise during the development of a typical fire in an enclosure.

Ignition: After ignition, the fire grows and produces increasing amounts of energy, mostly due to flame spread. In the early stages the enclosure has no effect on the fire, which then is fuel-controlled. Besides releasing energy, a variety of toxic and nontoxic gases and solids are produced. The generation of energy and combustion products is a very complex issue, as mentioned above, and the engineer must rely on measurements and approximate methods in order to estimate energy

FIGURE 2.3   Schematic of the development and descent of a hot smoke layer. (From Cooper.[4] With permission.)

release rates and the yield of combustion products. The issue of energy release rate is examined in Chapter 3 and the issue of combustion product yields is discussed in Chapter 9.

Plume: The hot gases in the flame are surrounded by cold gases, and the hotter, less dense mass will rise upward due to the density difference, or rather, due to buoyancy. The buoyant flow, including any flames, is referred to as a fire plume.

As the hot gases rise, cold air will be entrained into the plume. This mixture of combustion products and air will impinge on the ceiling of the fire compartment and cause a layer of hot gases to be formed. Only a small portion of the mass impinging on the ceiling originates from the fuel; the greatest portion of this mass originates from the cool air entrained laterally into the plume as it continues to move the gases toward the ceiling. As a result of this entrainment, the total mass flow in the plume increases, and the average temperature and concentration of combustion products decreases with height. Methods for estimating the mass flow rate in the plume as well as other plume properties are discussed in Chapter 4.

Ceiling jet: When the plume flow impinges on the ceiling, the gases spread across it as a momentum-driven circular jet. The velocity and temperature of this jet is of importance, since quantitative knowledge of these variables will allow estimates to be made on the response of any smoke and heat detectors and sprinkler links in the vicinity of the ceiling. Methods for estimating the velocity and temperature of the ceiling jet are discussed in Chapter 4.

The ceiling jet eventually reaches the walls of the enclosure and is forced to move downward along the wall, as shown in Figure 2.3. However, the gases in the jet are still warmer than the surrounding ambient air, and the flow will turn upward due to buoyancy. A layer of hot gases will thus be formed under the ceiling.

Gas temperatures: Experiments have shown, for a wide range of compartment fires, that it is reasonable to assume that the room becomes divided into two distinct layers: a hot upper layer consisting of a mixture of combustion products and entrained air, and a cold lower layer consisting of air. Further, the properties of the gases in each layer change with time but are assumed to be uniform throughout each layer. For example, it is commonly assumed when using engineering methods that the temperature is the same throughout the hot layer at any given time. Methods for estimating the upper layer temperature as a function of time are discussed in Chapter 6.

The hot layer: The plume continues to entrain air from the lower layer and transport it toward the ceiling. The hot upper layer therefore grows in volume, and the layer interface descends toward the floor. The smoke-filling process and methods for calculating smoke-filling times are discussed in Chapter 8.

Heat transfer: As the hot layer descends and increases in temperature, the heat transfer processes are augmented. Heat is transferred by radiation and convection from the hot gas layer to the ceiling and walls that are in contact with the hot gases. Heat from the hot layer is also radiated toward the floor and the lower walls, and some of the heat will be absorbed by the air in the lower layer. Additionally, heat is transferred to the fuel bed, not only by the flame, but to an increasing extent by radiation from the hot layer and the hot enclosure boundaries. This leads to an increase in the burning rate of the fuel and the heating up of other fuel packages in the enclosure. These heat transfer processes are discussed in Chapter 7.

Vent flows: If there is an opening to the adjacent room or out to the atmosphere, the smoke will flow out through it as soon as the hot layer reaches the top of the opening. Often, the increasing heat in the enclosure will cause the breakage of windows and thereby create an opening. Methods for calculating the mass flow rates through vents are discussed in Chapter 5.

Flashover: The fire may continue to grow, either by increased burning rate, by flame spread over the first ignited item, or by ignition of secondary fuel packages. The upper layer increases in temperature and may become very hot. As a result of radiation from the hot layer toward other combustible material in the enclosure, there may be a stage where all the combustible material in the enclosure is ignited, with a very rapid increase in energy release rates. This very rapid and sudden transition from a growing fire to a fully developed fire is called *flashover*. The fire can thus suddenly jump from a relatively benign state to a state of awesome power and destruction.

The solid line in Figure 2.4 shows the initiation of the transition period at Point A, resulting in a fully developed fire at Point B. Once Point B has been reached (the fully developed fire), flashover is said to have taken place. Flashover is further described in Section 2.3.

The fully developed fire: At the fully developed stage, flames extend out through the opening and all the combustible material in the enclosure is involved in the fire. The fully developed fire can burn for a number of hours, as long as there is sufficient fuel and oxygen available for combustion. Section 2.3 further discusses the fully developed fire and the phenomenon of flashover.

Oxygen starvation: For the case where there are no openings in the enclosure or only small leakage areas, the hot layer will soon descend toward the flame region and eventually cover the flame. The air entrained into the combustion zone now contains little oxygen and the fire may die out due to oxygen starvation. The dotted line in Figure 2.4 shows that a fire may reach Point A and start the transition period toward flashover, but due to oxygen depletion the energy release rate decreases, as well as the gas temperature.

Even though the energy release rate decreases, the pyrolysis may continue at a relatively high rate, causing the accumulation of unburnt gases in the enclosure. If a window breaks at this point, or if the fire service create an opening, the hot gases will flow out through the top of the opening and cold and fresh air will flow in through its lower part. This may diminish the thermal load in the enclosure, but the fresh air may cause an increase in the energy release rate. The fire may then grow toward flashover, as shown by the dotted line in Figure 2.4.

Backdraft: As a worst case, the inflowing air may mix up with the unburnt pyrolysis products from the oxygen-starved fire. Any ignition sources, such as a glowing ember, can ignite the resulting flammable mixture. This leads to an explosive or very rapid burning of the gases. Expansion due to the heat created by the combustion will expel the burning gases out through the opening. This phenomenon, termed *backdraft*, can be extremely hazardous, and many firefighters have lost their lives due to this very rapidly occurring event.

In Figure 2.4 this would be illustrated by a straight, vertical line, rising directly from Point C and reaching high temperatures. Usually, a backdraft will only last for a very short time, in the order of seconds (backdrafts lasting for minutes have, however, been observed). A backdraft will usually be followed by flashover, since the thermal insult will ignite all combustible fuel in the enclosure, leading to a fully developed enclosure fire.

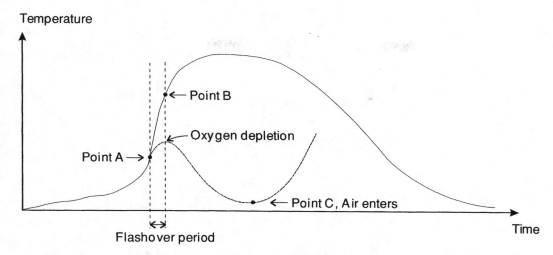

FIGURE 2.4 Enclosure fire development in terms of gas temperatures; some of the many possible paths a room fire may follow. (Adapted from Bengtsson[11].)

Smoke gas explosion: When unburnt gases from an underventilated fire flow through leakages into an closed space connected to the fire room, the gases there can mix very well with air to form a combustible gas mixture. A small spark is then enough to cause a smoke gas explosion, which can have very serious consequences. This phenomenon is, however, very seldom observed in enclosure fires.

2.3 STAGES IN ENCLOSURE FIRE DEVELOPMENT

In this section, we present an overview of how fire development in enclosures is commonly divided into stages. This can be done in terms of several environmental variables; we mainly discuss this in terms of enclosure temperatures and in terms of mass flows and pressure differences across the enclosure openings.

2.3.1 Fire Development in Terms of Enclosure Temperatures

Enclosure fires are often discussed in terms of the temperature development in the compartment and divided into different stages accordingly. Figure 2.5 shows an idealized variation of temperature with time, along with the growth stages, for the case where there is no attempt to control the fire. Walton and Thomas[5] list these stages as

- ignition
- growth
- flashover
- fully developed fire
- decay

Ignition: Ignition can be considered as a process that produces an exothermic reaction characterized by an increase in temperature greatly above the ambient. It can occur either by piloted ignition (by flaming match, spark, or other pilot source) or by spontaneous ignition (through accumulation of heat in the fuel). The accompanying combustion process can be either flaming combustion or smoldering combustion.

Growth: Following ignition, the fire may grow at a slow or a fast rate, depending on the type of combustion, the type of fuel, interaction with the surroundings, and access to oxygen. The fire

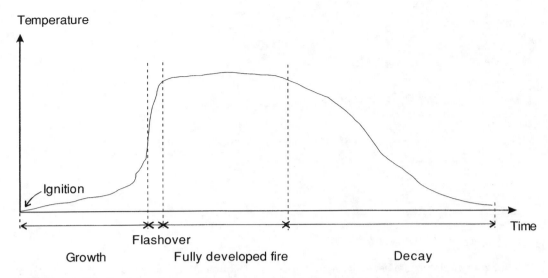

FIGURE 2.5   Idealized description of the temperature variation with time in an enclosure fire.

can be described in terms of the rate of energy released and the production of combustion gases. A smoldering fire can produce hazardous amounts of toxic gases while the energy release rate may be relatively low. The growth period of such a fire may be very long, and it may die out before subsequent stages are reached.

The growth stage can also occur very rapidly, especially with flaming combustion, where the fuel is flammable enough to allow rapid flame spread over its surface, where heat flux from the first burning fuel package is sufficient to ignite adjacent fuel packages, and where sufficient oxygen and fuel are available for rapid fire growth. Fires with sufficient oxygen available for combustion are said to be fuel-controlled.

Flashover: Flashover is the transition from the growth period to the fully developed stage in fire development. The formal definition from the International Standards Organization[6] is given as "the rapid transition to a state of total surface involvement in a fire of combustible material within an enclosure." In fire safety engineering, the word is used as the demarcation point between two stages of a compartment fire, i.e., pre-flashover and post-flashover.

Flashover is not a precise term: several variations in definition can be found in the literature. The criteria given usually demand that the temperature in the compartment has reached 500–600°C, or that the radiation to the floor of the compartment is 15 to 20 kW/m^2, or that flames appear from the enclosure openings. These occurrences may all be due to different mechanisms resulting from the fuel properties, fuel orientation, fuel position, enclosure geometry, and conditions in the upper layer. Flashover cannot be said to be a mechanism, but rather a phenomenon associated with a thermal instability.

Fully developed fire: At this stage the energy released in the enclosure is at its greatest and is very often limited by the availability of oxygen. This is called ventilation-controlled burning (as opposed to fuel-controlled burning), since the oxygen needed for the combustion is assumed to enter through the openings. In ventilation-controlled fires, unburnt gases can collect at the ceiling level, and as these gases leave through the openings they burn, causing flames to stick out through the openings. The average gas temperature in the enclosure during this stage is often very high, in the range of 700 to 1200°C.

Decay: As the fuel becomes consumed, the energy release rate diminishes and thus the average gas temperature in the compartment declines. The fire may go from ventilation-controlled to fuel-controlled in this period.

2.3.2 Fire Development in Terms of Flow through Openings

A second way of dividing the enclosure fire into a number of stages is to observe the mass flows in and out through the enclosure openings while the fire progresses. The mass flows in turn depend on the pressure differences across the opening. Figure 2.6 shows how the fire develops in an enclosure with a single opening in terms of the pressure profiles across the opening.

The pressure for the outside atmosphere is represented by a straight line, sloping because the weight of the column of air increases as we approach the ground level. Assuming that the temperature of the air in the lower layer of the compartment equals the temperature outside, we know that the pressure in the lower layer is given by a straight line with the same slope as that for the outside. There is a kink in the pressure line inside the compartment when we move into the upper layer; the hot air is lighter than the cold air.

The pressure of the outside atmosphere is the same for all four cases in Figure 2.6 and is denoted by the symbol P_o. The pressure inside the compartment is denoted P_i. We shall discuss these pressure profiles in greater detail in Chapter 5.

Stage A: In the first stage of the fire the pressure inside is higher than the pressure outside the compartment. This is due to the expansion of the hot gases, which have a greater volume than the cold gases. If the opening is not at ceiling level, the cold gases will be forced out through the opening due to the hot gas expansion. As a result, the pressure difference across the opening is positive (with respect to the compartment), and there will be no inflow through the opening, only outflow of cold gases.

Stage B: Stage B only lasts a few seconds and is often ignored. The smoke layer has just reached the top of the opening and the hot gases have started to flow out. The pressure inside is still higher than the pressure outside and both hot and cold gases flow out through the opening. There is no mass flow into the compartment.

Stage C: In this stage the hot gases flow out through the top of the opening and the mass balance demands that cold gases of equal mass flow in through the lower part of the opening. This stage can last for a considerable time, until the room is filled entirely with smoke or until flashover occurs. Stages A, B, and C are associated with the growth stage mentioned above and the pre-flashover stage.

Stage D: This stage is often termed the "well-mixed" stage, where the compartment is filled with smoke that is assumed to be well mixed, i.e., assumed to have some single, average temperature. This stage is associated with the fully developed fire mentioned above. In many cases flashover has occurred between stages C and D.

The pressure profiles are discussed further in Chapter 5, where we derive equations for calculating the mass flow rates out of and into the compartment, and in Chapter 8, where we derive methods for calculating the smoke filling rate in an enclosure.

2.3.3 Other Common Terms Describing Enclosure Fire Stages

In fire safety engineering design two distinctly different design situations often occur. One has to do with **the pre-flashover fire,** where the emphasis is on the safety of humans. The design load in this case is characterized by an energy release rate curve, where the growth phase of the fire is of most importance. We discuss this further in Chapter 3.

A very different design situation arises when the objective is to ensure structural stability and safety of firefighters. Here **the post-flashover fire** is of greatest concern, and the design load is characterized by a temperature–time curve where the fully developed fire stage is of greatest concern. We discuss this further in Chapter 6.

The fire development is therefore sometimes simply divided into the pre-flashover stage and the post-flashover stage.

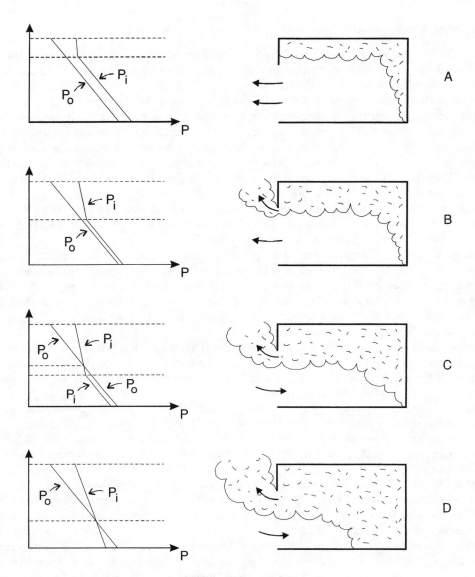

FIGURE 2.6   Pressure profile across the opening as the fire develops.

After ignition and during the initial fire growth stage, the fire is said to be **fuel-controlled,** since, in the initial stages, there is sufficient oxygen available for combustion and the growth of the fire entirely depends on the characteristics of the fuel and its geometry.

When the fire grows toward flashover it may become **ventilation-controlled,** when there is not enough oxygen available to combust most of the pyrolyzing fuel. The energy release rate of the fire is then determined by the amount of oxygen that enters the enclosure openings, and the fire is therefore termed ventilation-controlled. During the decay stage the fire will eventually return to being fuel-controlled.

The growth stage of the fire (and the pre-flashover fire) is therefore often associated with the fuel-controlled fire, and the fully developed stage (and the post-flashover fire) is often associated with the ventilation-controlled fire.

2.4 FACTORS INFLUENCING FIRE DEVELOPMENT IN AN ENCLOSURE

The factors that influence the development of a fire in an enclosure can be divided into two main categories: those that have to do with the enclosure itself, and those that have to do with the fuel. These factors are

- the size and location of the ignition source
- the type, amount, position, spacing, orientation, and surface area of the fuel packages
- the geometry of the enclosure
- the size and location of the compartment openings
- the material properties of the enclosure boundaries

Ignition source: An ignition source can consist of a spark with a very low energy content, a heated surface or a large pilot flame, for example. The source of energy is either chemical, electrical, or mechanical. The greater the energy of the source, the quicker the subsequent fire growth on the fuel source. A spark or a glowing cigarette may initiate smoldering combustion, which may continue for a long time before flaming occurs, often producing low heat but considerable amounts of toxic gases. A pilot flame usually produces flaming combustion directly, resulting in flame spread and fire growth.

The location of the ignition source is also of great importance. A pilot flame positioned at the lower end of, say, a window curtain may cause rapid upward flame spread and fire growth. The same pilot flame would cause much slower fire growth were it placed at the top of the curtain, resulting in slow, creeping, downward flame spread.

Fuel: The type and amount of combustible material is, of course, one of the main factors determining the fire development in an enclosure. In building fires the fuel usually consists of solid materials, such as the furniture and fittings normally seen in the enclosure interior; in some industrial applications the fuel source may also be in liquid state.

Heavy, wood-based furniture usually causes slow fire growth but can burn for a long time. Some modern interior materials include porous lightweight plastics, which cause more rapid fire growth but burn for a shorter time. A high fire load therefore does not necessarily constitute a greater hazard; a rapid fire growth is more hazardous in terms of human lives.

The position of the fuel package can have a marked effect on the fire development. If the fuel package is burning away from walls, the cool air is entrained into the plume from all directions. When placed close to a wall, the entrainment of cold air is limited. This not only causes higher temperatures but also higher flames since combustion must take place over a greater distance.

Figure 2.7 shows temperatures measured above fires in 1.22-m-high stacks of wood pallets.[7] Curve A shows temperatures as a function of height above pallets burning without the presence of walls. Curve B is for a similar stack near a wall, and curve C is for a stack in a corner.

The spacing and orientation of the fuel packages are also of importance. The spacing in the compartment determines to a considerable extent how quickly the fire spreads between the fuel packages. Upward flame spread on a vertically oriented fuel surface will occur much more rapidly than lateral spread along a horizontally oriented fuel surface.

Combustible lining materials mounted on the compartment walls and/or ceiling can cause very rapid fire growth. Figure 2.8 shows results from a small-scale room test where the lining material is mounted on the walls only (with noncombustible ceiling) and where the material is mounted on both walls and ceiling.[8] In both cases an initial flame was established on the lining material along one corner of the room. With material mounted on the ceiling, the flame spreads with the flow of gases (concurrent-flow flame spread), causing rapid growth. With a noncombustible ceiling the flame spreads horizontally (opposed-flow flame spread) across the material, a process that is much slower and requires the lining material to be heated considerably before the flame can spread

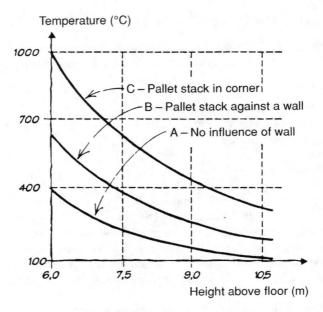

FIGURE 2.7 The temperature in the fire plume as a function of height above a burning stack of wood pallets. A, B, and C indicate burning away from walls, by a wall, and in a corner, respectively. (Adapted from Alpert and Ward[7]. With permission.)

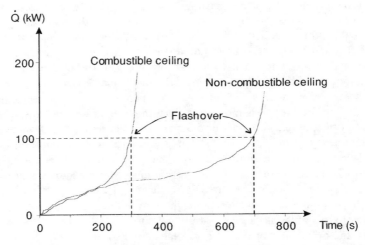

FIGURE 2.8 Energy release rate vs. time in a small room with particle board mounted on walls only and with particle board mounted on walls and ceiling. Flashover occurs when the energy release rate is 100 kW in this compartment.

rapidly over it. As a result, the time to flashover is 4 minutes in the former case and 12 minutes in the latter case.

A fuel package of a large surface area will burn more rapidly than an otherwise equivalent fuel package with a small surface area. A pile of wooden sticks, for example, will burn more rapidly than a single log of wood of the same mass.

Enclosure geometry: The hot smoke layer and the upper bounding surfaces of the enclosure will radiate toward the burning fuel and increase its burning rate. Other combustible items in the room will also be heated up. The temperature and thickness of the hot layer and the temperature of the upper bounding surfaces thus have a considerable impact on the fire growth.

A fuel package burning in a small room will cause relatively high temperatures and rapid fire growth. In a large compartment, the same burning fuel will cause lower gas temperatures, longer smoke-filling times, less feedback to the fuel, and slower fire growth.

The fire plume entrains cold air as the mixture of combustion products and air move upward toward the ceiling. The amount of cold air entrained depends on the distance between the fuel source and the hot layer interface. In an enclosure with a high ceiling this causes relatively low gas temperatures, but due to the large amount of air entrained, the smoke-filling process occurs relatively rapidly. The smaller the floor area, the faster the smoke-filling process.

With a low ceiling the heat transfer to the fuel will be greater. Additionally, the flames may reach the ceiling and spread horizontally under it. This results in a considerable increase in the feedback to the fuel and to other combustibles, and a very rapid fire growth is imminent.

For enclosures with a high ceiling and a large floor area, the flames may not reach the ceiling and the feedback to the fuel is modest. The fire growth rather occurs through direct radiation from the flame to nearby objects, where the spacing of the combustibles becomes important.

In buildings with a large floor area but a low ceiling height, the feedback from the hot layer and ceiling flames can be very intensive near the fire source. Further away, the hot layer has entrained cold air and has lost heat to the extensive ceiling surfaces, and the heat flux to the combustible materials, in the early stage of the fire, is therefore lower than the heat flux closer to the fire source.

We can conclude that the proximity of ceilings and walls can greatly enhance the fire growth. Even in large spaces, the hot gases trapped under the ceiling can heat up the combustibles beneath and result in extremely rapid fire spread over a large area. The fire that so tragically engulfed the Bradford City Football Stadium in England in 1985 is a vivid reminder of this.[9]

Compartment openings: Once flaming combustion is established the fire must have access to oxygen for continued development. In compartments of moderate volume that are closed or have very small leakage areas, the fire soon becomes oxygen-starved and may self-extinguish or continue to burn at a very slow rate depending on the availability of oxygen.

For compartments with ventilation openings, the size, shape, and position of such openings become important for the fire development under certain circumstances. During the growth phase of the fire, before it becomes ventilation-controlled, the opening may act as an exhaust for the hot gases, if its height or position is such that the hot gases are effectively removed from the enclosure. This will diminish the thermal feedback to the fuel and cause slower fire growth. For other circumstances the geometry of the opening does not have a very significant effect on fire growth during the fuel-controlled regime.

It is when the fire becomes controlled by the availability of oxygen that the opening size and shape first become all-important. Kawagoe found, mainly through experimental work, that the rate of burning depended very strongly on the "ventilation factor," defined as $A_o \sqrt{H_o}$, where A_o is the area of the opening and H_o is its height.[10] The importance of this factor can also be shown by theoretically analyzing the flow of gases in and out of a burning compartment. We shall derive this result in Chapter 5. It is thereby shown that the rate of burning is controlled by the rate at which air can flow into the compartment. An increase in the factor $A_o \sqrt{H_o}$ will lead to an equal increase in the burning rate. This is valid up to a certain limit when the burning rate becomes independent of the ventilation factor and the burning becomes fuel-controlled.

Properties of bounding surfaces: The material in the bounding surfaces of the enclosure can affect the hot gas temperature considerably, and thereby the heat flux to the burning fuel and other combustible objects. Certain bounding materials designed to conserve energy, such as mineral wool, will limit the amount of heat flow to the surfaces so that the hot gases will retain most of their energy.

The material properties controlling the heat flow through the construction are the conductivity (k), density (ρ), and heat capacity (c). They are commonly collected in a property called *thermal inertia* and given as the product $k\rho c$. Insulating materials have a low thermal inertia; materials with relatively high thermal inertia, such as brick and concrete, allow more heat to be conducted into the construction, thereby lowering the hot gas temperatures.

2.5   SUMMARY

The previous sections have attempted to give a qualitative description of enclosure fire development and an introduction to the terms that scientists, fire safety engineers, and fire service personnel most frequently use to describe this development.

A fire in a compartment can develop in an infinite number of ways; we have here attempted to give a general description of the most significant enclosure fire stages. However, no words can adequately describe the enormous power and destructiveness that result from an enclosure fire going through flashover or backdraft.

It is therefore important that the reader either attends enclosure fire experiments or views video films from such experiments or from real fires. An abundance of such video films exist, for example, "Countdown to Disaster," produced by the National Fire Protection Association in the U.S., and "Anatomy of a Fire," produced by the Building Research Establishment in the U.K.

Finally, it is necessary for the reader to have a clear understanding of the terms fire safety professionals use to describe fire development. It is important to be well acquainted with the meaning of such terms as flashover, the pre-flashover fire, the post-flashover fire, the fuel-controlled fire, the ventilation-controlled fire, and the fully developed fire.

REFERENCES

1. Friedman, R., "Status of Mathematical Modeling of Fires," FMRC Technical Report RC81-BT-5, Factory Mutual Research Corp., Boston, 1981.
2. Faraday, M., *The Chemical History of a Candle*, Thomas Y. Crowell, New York, 1957. (First published in London, 1861.)
3. Lyons, J.W., Ed., "Fire Research on Cellular Plastics: The Final Report of the Products Research Committee (PRC), Library of Congress Catalog Card Number 80-83306, Products Research Committee, 1980.
4. Cooper, L.Y., "Compartment Fire-Generated Environment and Smoke Filling," in *The SFPE Handbook of Fire Protection Engineering*, 2nd ed., National Fire Protection Association, Quincy, MA, 1995.
5. Walton, W.D. and Thomas, P.H., "Estimating Temperatures in Compartment Fires," in *The SFPE Handbook of Fire Protection Engineering*, 2nd ed., National Fire Protection Association, Quincy, MA, 1995.
6. ISO, "Glossary of Fire Terms and Definitions," ISO/CD 13943, International Standards Organization, Geneva, 1996.
7. Alpert, R.L. and Ward, E.J., "Evaluation of Unsprinklered Fire Hazards," *Fire Safety Journal*, Vol. 7, No. 2, pp. 127–143, 1984.
8. Andersson, B., "Model Scale Compartment Fire Tests with Wall Lining Materials," Report LUTVDG/(TVBB-3041), Department of Fire Safety Engineering, Lund University, Lund, Sweden, 1988.
9. FPA, "Fire tragedy at a football stadium," *Fire Prevention* No. 181, July/August, Fire Protection Association, London, 1985.
10. Kawagoe, K., "Fire Behaviour in Rooms," Report No. 27, Building Research Institute, Tokyo, 1958.
11. Bengtsson, L., "Flashover, Backdraft and Smoke Gas Explosion from a Fire Service Perspective" (in Swedish), Department of Fire Safety Engineering, Lund University, Lund, Sweden, 1998.

3 Energy Release Rates

The rate at which energy is released in a compartment fire controls to a considerable extent the environmental consequences of the fire, such as the plume flows, the hot gas temperatures, and the rate of descent of the hot gas layer. The following chapters in this textbook outline a number of calculational procedures to enable an estimation of the environmental consequences of a fire. Most of these procedures require some knowledge of the energy release rate. This chapter outlines the methods commonly used to estimate the energy release rate produced by a fire.

CONTENTS

3.1 Terminology 25
3.2 Introduction 26
3.3 Factors Controlling Energy Release Rates in Enclosure Fires 27
    3.3.1 Burning Rate and Energy Release Rate 27
    3.3.2 Enclosure Effects 28
3.4 Energy Release Rates Based on Free Burn Measurements...28
    3.4.1 Measurement Techniques and Parameters Measured..29
    3.4.2 Pool Fires 32
    3.4.3 Various Products 34
    3.4.4 The t-squared Fire..38
3.5 The Design Fire 39
    3.5.1 Background 39
    3.5.2 The Growth Phase...39
    3.5.3 The Steady Phase 43
    3.5.4 The Decay Phase...44
    3.5.5 A More Complex Design Fire 44
    3.5.6 Energy Release Rates Used in This Book...44
References 45
Problems and Suggested Answers 45

3.1 TERMINOLOGY

Burning rate or mass loss rate — The mass rate of solid or liquid fuel vaporized and burned. It is expressed as mass flow per unit time, typically in kg/s or g/s and is here denoted as $\dot{m}$. It can also be expressed as mass flux or mass burning rate per unit area, typically in kg/(m² s). In this case it is denoted as $\dot{m}''$. A distinction should be made between burning rate and mass loss rate (fuel supplied), since all of the fuel supplied may not be burned. For burning objects with unlimited air supply, the terms are synonymous.

Combustion efficiency — The ratio between the effective heat of combustion and the complete heat of combustion is termed the combustion efficiency and is denoted χ.

Energy release rate or heat release rate — When an object burns it releases a certain amount of energy per unit time, usually given in kW (= kJ/s) and denoted $\dot{Q}$. For most materials the energy release rate changes with time. This is also often termed heat release rate (sometimes shortened to HRR). The term energy release rate is, however, more appropriate, because heat is strictly speaking energy transported due to a temperature difference. But because of the general usage of the expressions, these terms will be taken as synonymous.

Heat of combustion — A measure of how much energy is released when a unit mass of material combusts, typically given in kJ/kg or kJ/g. It is important to distinguish between the complete heat of combustion, denoted ΔH_c, and the effective heat of combustion, denoted ΔH_{eff}. The former is a measure of the energy released when the combustion is complete, leaving no residual fuel and releasing all of the chemical energy of the material. The effective heat of combustion is more appropriate for fires, where some residue is left and the combustion is not necessarily complete. This is also sometimes termed the *chemical heat of combustion*.

Heat of gasification — A measure of how much energy is needed to gasify some unit mass of the fuel. It is typically given in kJ/kg and is denoted ΔH_g.

3.2   INTRODUCTION

Energy release rate (often termed heat release rate or HRR) is measured in watts, kilowatts, or megawatts. Table 3.1 gives some characteristic values of energy released by various burning fuel packages and heat output from various sources.

Fire development is generally characterized in terms of energy release rate vs. time. Once an engineer has arrived at the energy release rate vs. time relationship for a certain scenario, it is termed the *design fire*. Table 3.1 indicates that for many design purposes the design fire energy output could be in the range 100 kW to 50 MW.

There are basically two approaches available when determining the design fire for a given scenario. One is based on knowledge of the amount and type of combustible materials in the compartment of fire origin. The other is based on knowledge of the type of occupancy, where very little is known about the details of the fire load.

In the first case, an object is assumed to ignite and start to burn. The resulting energy release rate vs. time can in many cases be estimated using data from previous experiments where energy release rate has been measured. Such results have been summarized in a number of publications.[1,2,3]

However, in many design situations there is very little information available on the combustible content of the room of fire origin. In this case, knowledge of the type of occupancy, any available statistics, and engineering judgment must be used to arrive at a design fire.

In Section 3.3, we discuss the factors controlling energy release rates in enclosure fires. Section 3.4 provides examples of how experimental data can be used to estimate this for various fuels and products. Finally, Section 3.5 discusses how this information can be used to arrive at a design fire for a certain scenario.

TABLE 3.1
Rough Measure of Energy Released or Generated from Various Sources

A burning cigarette	5 W
A typical light bulb	60 W
A human being at normal exertion	100 W
A burning wastepaper basket	100 kW
A burning 1m² pool of gasoline	2.5 MW
Burning wood pallets, stacked to the height of 3 m	7 MW
Burning polystyrene jars, in cartons, 2 m², 4.9 m high	30–40 MW
Output from a typical reactor at a Nuclear Power Plant	500–1000 MW

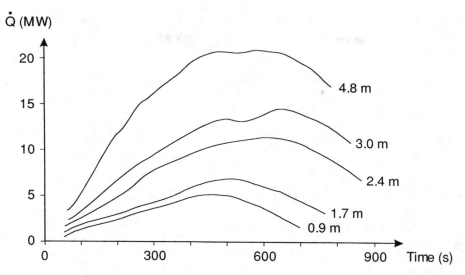

FIGURE 3.1 Energy release rate measured when burning 1.2 m by 1.2 m wood pallets, stacked to different heights.

3.3 FACTORS CONTROLLING ENERGY RELEASE RATES IN ENCLOSURE FIRES

The rate at which energy is released in a fire depends mainly on the type, quantity, and orientation of fuel and on the effects that an enclosure may have on the energy release rate.

The energy release rate will vary with time. Figure 3.1 shows a schematic graph of the energy release rate vs. time measured when wood pallet stacks of different heights burn. Such measurements are often termed "free burn" tests, indicating that the items are burning without any effects of the enclosure in which the fire takes place.

3.3.1 BURNING RATE AND ENERGY RELEASE RATE

Burning rate is the mass rate of solid or liquid fuel vaporized and burned. It is expressed as mass flow per unit time, typically in kilograms per second or grams per second and is denoted as $\dot{m}$. It can also be expressed as mass flux or mass burning rate per unit area, typically kilograms per square meter per second. In this case it is denoted as $\dot{m}''$.

A general predictive formula for steady burning mass flux is given by

$$\dot{m}'' = \frac{\dot{q}''}{\Delta H_g} \tag{3.1}$$

where $\dot{q}''$ is the net heat flux from the flame to the fuel and used to release the volatiles, and ΔH_g is called the heat of gasification. The net heat flux from the flame to the fuel is typically measured in kilowatts per square meter. The heat of gasification is a measure of how much energy is needed to gasify some unit mass of the fuel and is typically given in kiloJoules per kilogram. For liquid fuels this is the same as the heat of evaporation: the energy needed to evaporate the liquid once it has reached its boiling point.

The heat of gasification is not a constant property for solids, but it is a fundamental property for liquid fuels. Because of the difficulty in quantitatively determining the net heat flux to the fuel

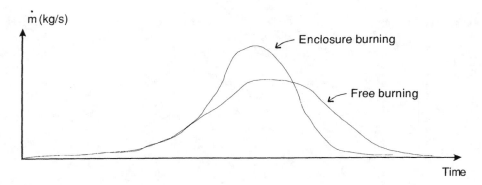

FIGURE 3.2   The enclosure effect on mass loss rate.

surface, Eq. (3.1) is seldom used in practice, and generally we need experimental results to estimate the burning rate.

3.3.2   ENCLOSURE EFFECTS

When an item burns inside an enclosure, two factors mainly influence the energy released and the burning rate. First, the hot gases will collect at the ceiling level and heat the ceiling and the walls. These surfaces and the hot gas layer will radiate heat toward the fuel surface, thus enhancing the burning rate. Second, the enclosure vents (doors, windows, leakage areas) may restrict the availability of oxygen needed for combustion. This causes a decrease in the amount of fuel burnt, leading to a decrease in energy release rate and an increase in the concentration of unburnt gases.

Figure 3.2 shows a schematic of an arbitrary item burning. One curve shows the burning rate when the object is burning in the open (free burn experiment). The other curve shows the effect of burning the item in an enclosure (with an opening), where the hot surfaces and gases transfer heat to the fuel bed, thus increasing the burning rate (compared to the somewhat slower burning rate expected from a free burn experiment).

If, however, the opening is relatively small, the limited availability of oxygen will cause incomplete combustion, resulting in a decrease in energy release rate, which in turn causes lower gas temperatures and less heat transfer to the fuel. The fuel will continue to release volatile gases at a similar or somewhat lower rate. Only a part of the gases combust, releasing energy, and unburnt gases will be collected at ceiling level. The unburnt gases can release energy when flowing out through an opening and mixing with oxygen, causing flames to appear at the opening.

In summary, compartment heat transfer can increase the mass loss rate of the fuel, while compartment vitiation of the available air near the floor will decrease the mass loss rate.

3.4   ENERGY RELEASE RATES BASED ON FREE BURN MEASUREMENTS

The only practical way to determine the burning rate or energy release rate of an item is by direct measurement. Such measurements are termed *free burn measurements*, meaning that the enclosure effects are minimized; the hot gases are vented away from the fuel and there is no limitation on air supply to the fuel. The results can then be used by engineers as guidelines when determining the design fire for a certain scenario. In the case of liquid fuels, such measurements have resulted in expressions that allow the energy release rate to be calculated if the liquid pool diameter is known. Below, we briefly discuss the most common measurement techniques, discuss methods for calculating energy release rate from pool fires, and show experimentally determined energy release rate curves for various residential and industrial items.

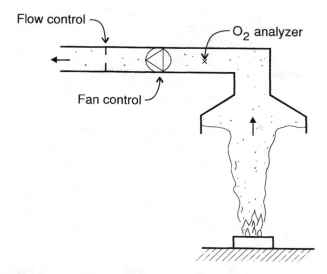

FIGURE 3.3 A schematic of an oxygen consumption calorimeter.

3.4.1 MEASUREMENT TECHNIQUES AND PARAMETERS MEASURED

Energy release rate: The most common method to measure energy release rate is known as *oxygen consumption calorimetry*. The basis of this method is that for most gases, liquids, and solids, a more or less constant amount of energy is released per unit mass of oxygen consumed. This constant has been found to be 13,100 kJ per kilogram oxygen consumed and is considered to be accurate with very few exceptions to about ±5% for many hydrocarbon materials (see Huggett[9]).

A schematic of an oxygen consumption calorimeter is shown in Figure 3.3. After ignition, all the combustion products are collected in a hood and removed through an exhaust duct. The flow rate and the composition of the gases is measured in the duct. This gives a measure of how much oxygen has been used for combustion. Using the above constant, the energy release rate can be computed.

EXAMPLE 3.1

The gases from a burning slab of polymethylmethacrylate (PMMA) are collected in a hood and removed through an exhaust duct. The volumetric flow of gases and the gas temperature are measured, and thus the mass flow of gases is calculated to be 0.05 kg/s. The mass fraction of oxygen is found to settle at an average value of around 15%. Give a rough estimate of the average energy released by the object.

SUGGESTED SOLUTION

The mass fraction of oxygen in air is ~23%. The mass rate of oxygen used for combustion is therefore $(0.23 - 0.15) \cdot 0.05$ kg/s. Each kilogram of oxygen used gives 13,100 kJ of energy. The energy released, or the effect from the burning object, is therefore

$$\dot{Q} = 13,100 \text{ kJ/kg} \cdot (0.23 - 0.15) \cdot 0.05 \text{ kg/s} = 52.4 \text{ kJ/s} = 52.4 \text{ kW}$$

PMMA happens to burn at a roughly constant rate, but usually solid objects burn at a rate that varies with time, and the energy release rate will therefore vary with time. Examples of the energy release rate vs. time of various products are given in Section 3.4.2.

Burning rate: Another common method of assessing the energy release rate is by measuring the burning rate, or the mass loss rate. This is done by weighing the fuel package as it burns, using weighing devices or load cells. An estimation of the energy release rate requires that the effective heat of combustion is known. The energy release is then calculated using the equation

$$\dot{Q} = \dot{m}\Delta H_{eff} \qquad (3.2)$$

where $\dot{m}$ is given in kilograms per second or grams per second and ΔH_{eff} is the effective heat of combustion (see below), often given in kiloJoules per kilogram or kiloJoules per gram. The resulting energy release rate will then be in kiloJoules per second, which is the same as kilowatts.

It should be noted that the term "burning rate" is really a misnomer for the mass loss rate, because all of the fuel volatilized may not burn when there is insufficient oxygen available. However, most textbooks equate the two terms and for our purposes we shall assume them to be identical. Note, however, that in a compartment fire where there is insufficient oxygen, all of the mass loss from the fuel will not burn.

The average burning rates for very many products and materials have been experimentally determined in free burn tests. For many materials the burning rate is reported per horizontal burning area and given in kilograms per square meter per second. If the area of the fuel is known, as well as the effective heat of combustion, Eq. (3.2) becomes

$$\dot{Q} = A_f \dot{m}''\Delta H_{eff} \qquad (3.3)$$

where A_f is the horizontal burning area of the fuel. Note that when the burning rate is given per unit area, it is denoted by the symbol $\dot{m}''$.

The average burning rate per unit area of various materials, measured in free burn tests, is given in Table 3.2. Other sources give such results for many more products and materials (see, for example, Babrauskas,[1] Särdqvist,[2] and Babrauskas and Greyson[3]).

Heat of combustion: The effective heat of combustion (sometimes called the chemical heat of combustion) can be estimated from Eq. (3.2) if both the energy release rate and the mass loss rate have been measured as described above. It is important to distinguish between the effective heat of combustion, ΔH_{eff}, and the complete heat of combustion, ΔH_c. The complete heat of combustion can be measured in a device called the bomb calorimeter, where the sample is completely combusted under high pressure in pure oxygen, leaving almost no residue and releasing almost all of its potential energy. This is not representative of real fires, where typically only 70 to 80% of the mass is converted to volatiles that burn almost completely, leaving some char or residue. Additionally, some of the volatiles do not combust completely, leaving some combustible components such as CO, soot, and unburnt hydrocarbons in the products of combustion.

To exemplify the distinction between the effective heat of combustion and the complete heat of combustion we shall look at the case of PMMA burning where there is not much oxygen. When the products burn completely with the reaction

$$C_5H_8O_2 \text{ (g)} + 6 \text{ } O_2 \text{ (g)} \rightarrow 5 \text{ } CO_2 \text{ (g)} + 4 \text{ } H_2O \text{ (g)}$$

the complete heat of combustion, ΔH_c, is 24.9 kJ/g. When less oxygen is available some carbon monoxide will be formed, and some of the carbon may be used to form soot particles. If we assume that the reaction instead occurs as

$$C_5H_8O_2 \text{ (g)} + 4.5 \text{ } O_2 \text{ (g)} \rightarrow 3 \text{ } CO_2 \text{ (g)} + 4 \text{ } H_2O \text{ (g)} + CO \text{ (g)} + C \text{ (s)}$$

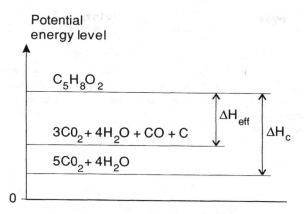

Potential energy level

$C_5H_8O_2$

$3CO_2 + 4H_2O + CO + C$ $\quad \Delta H_{eff}$

$\quad \Delta H_c$

$5CO_2 + 4H_2O$

0

FIGURE 3.4 Heats of combustion for PMMA at two different reaction stoichiometries.

the heat of combustion will be 18.2 kJ/g. If we assume that this latter reaction occurs in a specific room fire, we can say that the effective heat of combustion for PMMA, in exactly that scenario, is 18.2 kJ/g. The combustion efficiency (see below) in this case would be $\frac{18.2}{24.9} = 0.73$. Figure 3.4 shows this schematically.

In enclosure fires the exact reaction stoichiometry is not known and therefore one does not usually tabulate values of ΔH_{eff}. However, in some free burn experiments this value is recorded and tabulated, and it is then valid for well-ventilated enclosure fires.

Values of ΔH_c for many different materials are known and are available in the literature. The work of Tewarson[4,5] is especially useful for this purpose. Note that there is some variability in both notation and terms; the effective heat of combustion is sometimes termed the chemical heat of combustion and given the symbol ΔH_{ch}. The complete heat of combustion is sometimes termed the net heat of complete combustion and given the symbol ΔH_T, where the subscript T stands for total. When using values from various sources, one must carefully observe the definition of terms and symbols.

Table 3.2 lists some common materials, their burning rates, and the complete heat of combustion, ΔH_c. The data is collected from Tewarson,[5] where more information on how the data was arrived at is found.

Combustion efficiency: The ratio between the effective heat of combustion and the complete heat of combustion is termed the *combustion efficiency* and is given by the relation

$$\chi = \frac{\Delta H_{eff}}{\Delta H_c} \tag{3.4}$$

Alcohols, such as methanol and ethanol, burn with a flame that is hardly visible, indicating that very little soot is produced; hence, it has a combustion efficiency close to unity. This is also true for many gaseous fuels such as methane. For fuels that produce sooty flames, such as oil, the combustion efficiency is significantly lower, typically around 60 to 70%. Note that in the above discussion we have assumed that there is plenty of available oxygen for combustion; in small enclosures this is often not the case, and the combustion efficiency decreases markedly when oxygen concentration decreases.

The values of burning rate (for relatively large diameters) and complete heat of combustion given in Table 3.2 can now be used to calculate the energy release rate from various burning materials. Equation (3.3) becomes

$$\dot{Q} = A_f \dot{m}'' \chi \Delta H_c \tag{3.5}$$

EXAMPLE 3.2

Assume that the burning PMMA in Example 3.1 was weighed as it burnt, giving the average mass loss rate of 3 g/s. The complete heat of combustion for PMMA has been found to be roughly 25 kJ/g (see Tables 3.2 and 3.3). Assuming that the combustion efficiency is 70%, estimate the resulting energy release rate. Also, give a rough estimate of the size of the burning PMMA slab.

SUGGESTED SOLUTION

Using Eq. (3.3) we find

$$\dot{Q} = 3 \text{ g/s} \cdot 25 \text{ kJ/g} \cdot 0.7 = 52.5 \text{ kW}$$

From Table 3.2 we find that PMMA has an average burning rate per unit area of 30 g/(m²s), while Table 3.3 gives 20 g/(m²s). This is partly due to experimental differences. Assuming an average value of 25 g/(m²s), the size of the burning slab is roughly 3/25 = 0.12 m².

Note that the measured value of $\dot{m}''$ from Table 3.2 is valid for larger diameter than that considered in Example 3.2; we can expect a lower value for this small diameter. This will be addressed below.

3.4.2 POOL FIRES

Accidental spills of liquid fuels in industrial process and power plant systems can pose a serious fire hazard. Some liquids are highly volatile at ambient temperatures; they can evaporate and form a flammable mixture with air, leading to a possible explosion in a confined space. Other liquids have a high flashpoint and require localized heating to achieve ignition. Once ignited, however, very rapid flame spread will occur over the liquid spill surface. In free burn conditions, the burning rate will quickly reach a constant value, depending on the diameter of the spill.

Diameter dependence: Fairly extensive pool fire experiments have been carried out for a wide range of liquids. It has been found that, for diameters larger than 0.2 m, the burning rate increases with diameter up to a certain value, which we shall call the asymptotic diameter mass loss rate, denoted $\dot{m}''_\infty$, usually given in kilograms per square meter per second. The values of burning rate given in Table 3.3 are generally for large diameters, but for smaller diameters the burning rate reduces. A typical relationship can be seen in Figure 3.5, where the mass loss rate for gasoline pools of various diameters is shown.

The mass loss rate from free burning pools depends not only on diameter but also on two empirical constants that characterize the particular fuel used and are a function of the radiative heat flux from the flame toward the fuel surface. One of these is the extinction–absorption coefficient of the flame, denoted k, and the other is the mean beam length corrector, denoted β. For pool fire calculation purposes it is not necessary to determine these two constants separately, only their product.

Using a large number of free burn experiments with liquids of various diameters, the following correlation equation was arrived at:

$$\dot{m}'' = \dot{m}''_\infty \cdot \left(1 - e^{-k\beta D}\right) \tag{3.6}$$

where $\dot{m}''$ is the free burn mass loss rate and $\dot{m}''_\infty$ and $k\beta$ depend on the liquid type and are given in Table 3.3. The diameter, denoted D, is assumed to be circular. Square and similar configurations

TABLE 3.2
Burning Rate per Unit Area and Complete Heat of Combustion for Various Materials

Material (values in brackets indicate pool diameters tested)	$\dot{m}''$ (kg/m²s)	ΔH_c (MJ/kg)
Aliphatic Carbon–Hydrogen Atoms		
Polyethylene	0.026	43.6
Polypropylene	0.024	43.4
Heavy fuel oil (2.6–23 m)	0.036	—
Kerosene (30–80 m)	0.065	44.1
Crude oil (6.5–31 m)	0.056	—
n-Dodecane (0.94 m)	0.036	44.2
Gasoline (1.5–223 m)	0.062	—
JP-4 (1–5.3 m)	0.067	—
JP-5 (0.6–1.7 m)	0.055	—
n-Heptane (1.2–10 m)	0.075	44.6
n-Hexane (0.75–10 m)	0.077	44.8
Transformer fluids (2.37 m)	0.025–0.030	—
Aromatic Carbon–Hydrogen Atoms		
Polystyrene (0.93 m)	0.034	39.2
Xylene (1.22 m)	0.067	39.4
Benzene (0.75–6.0 m)	0.081	40.1
Aliphatic Carbon–Hydrogen–Oxygen Atoms		
Polyoxymethylene	0.016	15.4
Polymethylmethacrylate, PMMA (2.37 m)	0.030	25.2
Methanol (1.2–2.4 m)	0.025	20
Acetone (1.52 m)	0.038	29.7
Aliphatic Carbon-Hydrogen–Oxygen–Nitrogen Atoms		
Flexible polyurethane foams	0.021–0.027	23.2–27.2
Rigid polyurethane foams	0.022–0.025	25.0–28.0
Aliphatic Carbon–Hydrogen–Halogen Atoms		
Polyvinylchloride	0.016	16.4
Tefzel™ (ETFE)	0.014	12.6
Teflon™ (FEP)	0.007	4.8

Source: Tewarson, A., in *SFPE Handbook of Fire Protection Engineering*, 2nd ed., National Fire Protection Association, Quincy, MA, 1995. With permission.

can be treated as a pool of equivalent circular area. Some solid thermoplastic materials behave as liquid pools when they burn; Table 3.3 contains examples of these as well. Note that there is some difference in the values given in Tables 3.2 and 3.3; this must be attributed to different test conditions.

Note that the heat of combustion given in Table 3.3 is the complete heat of combustion. To calculate the energy release rate one should therefore use Eq. (3.5), which takes account of the combustion efficiency.

For alcohols, the diameter dependence is negligible. The flames contain little soot and are nearly invisible to the human eye, so radiation to the surface is much less than that for sootier flames. The mass loss rate is therefore a relatively constant value for almost all diameters larger than 0.2 m. Also, the combustion efficiency for alcohols is close to unity for the same reason.

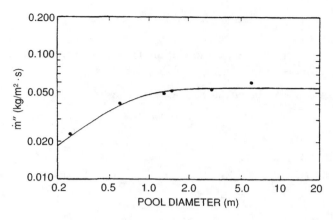

FIGURE 3.5   Mass loss rate for gasoline pools of various diameters. (From Babrauskas[1]. With permission.)

For sootier flames, such as those that appear when transformer oil burns, the diameter dependence is considerable and the combustion efficiency is lower, perhaps around 60 to 70% for free burn experiments.

EXAMPLE 3.3

A pump breakdown causes 20 l of transformer oil to spill into a sump of 2 m² floor area. The oil is warm and ignites. Estimate the energy released and the duration of the fire.

SUGGESTED SOLUTION

Table 3.3 gives $\dot{m}''_{\infty} = 0.039$ kg/(m²s), $\Delta H_c = 46.4$ MJ/kg, and $k\beta = 0.7$ (m⁻¹). The diameter is approximated assuming a circular spill of area 2 m², so $D = \sqrt{\dfrac{4 \cdot 2}{\pi}} = 1.6$ m. Using Eq. (3.6), we find $\dot{m}'' = 0.039(1 - \exp(-0.7 \cdot 1.6)) = 0.026$ kg/(m²s). Using Eq. (3.5) and assuming the combustion efficiency to be ~70%, we find $\dot{Q} = 2 \cdot 0.026 \cdot 0.7 \cdot 46.4 = 1.69$ MW. Table 3.3 gives the density of the oil as 760 kg/m³. The total mass of oil is therefore $M = \dfrac{20}{1000} \cdot 760 = 15.2$ kg. The total mass loss rate is $\dot{m} = 2 \cdot 0.026 = 0.052$ kg/s. The duration of the fire is therefore $\dfrac{15.2}{0.052} = 290$ seconds, or roughly 5 minutes.

3.4.3   VARIOUS PRODUCTS

In the previous sections we concentrated on materials that, when burning, will exhibit a roughly constant burning rate. Most products and solid materials, however, exhibit a varying burning rate with time. Starting at ignition, the burning rate typically increases to a maximum value and then decreases until most of the combustible material is burnt. For the majority of products the burning can in most instances be divided into the stages of ignition, growth, and decay.

In this section we give some typical energy release rate results from free burn tests of various products and commodities. These can be used by the engineer as benchmarks or guidelines but they cannot be used as precise estimates because within a given item class there can be wide

TABLE 3.3

Data for Large Pool ($D > 0.2$ m) Burning Rate Estimates

Material	Density (kg/m³)	$\dot{m}''_\infty$ (kg/m²s)	ΔH_c (MJ/kg)	$k\beta$ (m⁻¹)
Cryogenics				
Liquid H₂	70	0.017	120.0	6.1
LNG (mostly CH₄)	415	0.078	50.0	1.1
LPG (mostly C₃H₈)	585	0.099	46.0	1.4
Alcohols				
Methanol (CH₃OH)	796	0.017	20.0	a
Ethanol (C₂H₅OH)	794	0.015	26.8	b
Simple organic fuels				
Butane (C₄H₁₀)	573	0.078	45.7	2.7
Benzene (C₆H₆)	874	0.085	40.1	2.7
Hexane (C₆H₁₄)	650	0.074	44.7	1.9
Heptane (C₇H₁₆)	675	0.101	44.6	1.1
Xylene (C₈H₁₀)	870	0.09	40.8	1.4
Acetone (C₃H₆O)	791	0.041	25.8	1.9
Dioxane (C₄H₈O₂)	1035	0.018	26.2	5.4ᵇ
Diethyl ether (C₄H₁₀O)	714	0.085	34.2	0.7
Petroleum products				
Benzine	740	0.048	44.7	3.6
Gasoline	740	0.055	43.7	2.1
Kerosine	820	0.039	43.2	3.5
JP-4	760	0.051	43.5	3.6
JP-5	810	0.054	43.0	1.6
Transformer oil, hydrocarbon	760	0.039ᵇ	46.4	0.7ᵇ
Fuel oil, heavy	940–1000	0.035	39.7	1.7
Crude oil	830–880	0.022–0.045	42.5–42.7	2.8
Solids				
Polymethylmethacrylate (C₅H₈O₂)ₙ	1184	0.020	24.9	3.3
Polypropylene (C₃H₆)ₙ	905	0.018	43.2	
Polystyrene (C₈H₈)ₙ	1050	0.034	39.7	

[a] Value independent of diameter in turbulent regime.

[b] Estimate uncertain, since only two points available.

Source: Babrauskas, V., in *SFPE Handbook of Fire Protection Engineering*, 2nd ed., National Fire Protection Association, Quincy, MA, 1995. With permission.

variations. Also, the mode of ignition in each case will have some influence on the initial fire growth in time. Most of the experimental data shown here is taken from the *SFPE Handbook*.[1]

Wood Pallets: A fairly common fuel source at industrial locations is a stack of wood pallets. Figure 3.6 shows energy release rate vs. time of a typical pallet stack, with a height of 1.22 m, burning in the open. Each pallet is 1.22 m by 1.22 m and is 0.14 m high, so the stack in Figure 3.6 consists of roughly nine pallets stacked on top of each other.

Further experiments have shown that the peak energy release rate increases with the stack height, as can be seen in Figure 3.7. Figure 3.1 also gave examples of wood pallets burning at different stack heights.

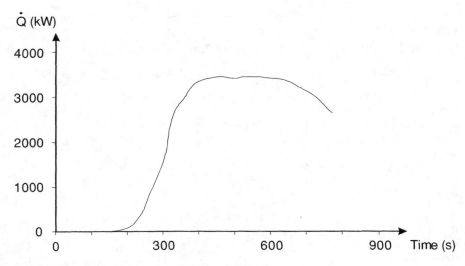

FIGURE 3.6   Typical energy release rate from a wood pallet stack.

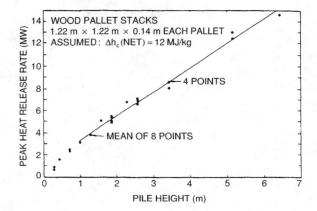

FIGURE 3.7   Dependence of pallet stack height on peak energy release rate. (From Babrauskas[1]. With permission.)

Upholstered furniture and mattresses: A significant amount of data on energy release rate from upholstered furniture is available in the literature (see further references in Babrauskas[1] and Särdqvist[2]). Figure 3.8 shows the energy release rate of a typical upholstered sofa, a two-seat sofa (loveseat), and an upholstered chair. Figure 3.9 shows similar results for typical mattresses.

Various other items: Figure 3.10 shows the energy release rates from trash bags of various weights. Figure 3.11 shows the results from two experiments with television sets, and Figure 3.12 shows results from three experiments with Christmas trees.

Summary: We can here recommend two publications that include collections of a wide range of energy release rate data: the *SFPE Handbook of Fire Protection Engineering* (Babrauskas[1]) and *Initial Fires* (Särdqvist[2]). Additionally, numerous publications have published energy release rate data on a wide range of specific items. Tewarson[4,5] gives data on the heat of combustion and species generations of various chemicals and products.

Again, it should be noted that the data on energy release rates should be used by the engineer as guidelines only and cannot be used as precise estimates. This is because there is a wide variation within each class of item. Also, the mode of ignition in each case will have some influence on the

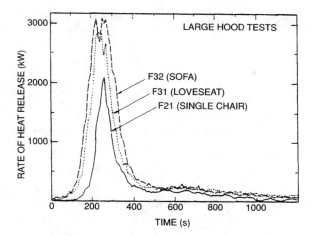

FIGURE 3.8   Typical upholstered furniture energy release rates. (From Babrauskas[1]. With permission.)

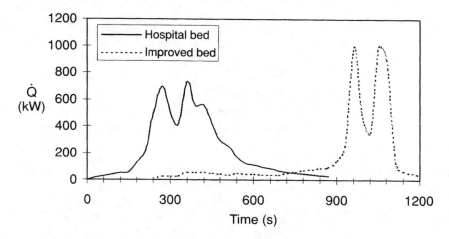

FIGURE 3.9   Typical mattress energy release rate. (From Särdqvist[2].)

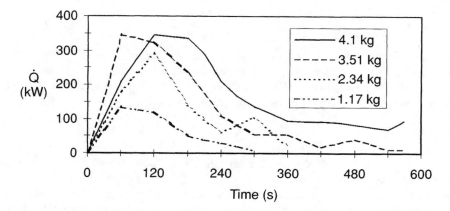

FIGURE 3.10   Energy release rates for trash bags. (Adapted from Babrauskas[1]. With permission.)

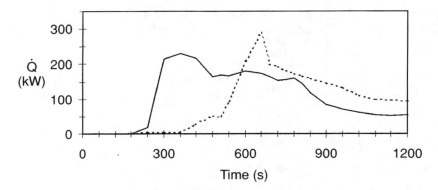

FIGURE 3.11 Energy release rates from two experiments with television sets. (Adapted from Babrauskas[1]. With permission.)

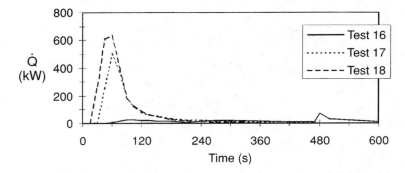

FIGURE 3.12 Energy release rates from three experiments with Christmas trees. (Adapted from Babrauskas[1]. With permission.)

initial fire growth in time. However, tests have shown that identical specimens ignited with different procedures (i.e., a cigarette vs. a wastebasket) show a similar peak energy release rate and a similar burning time. The time to ignition, or the time until the rapid growth period starts, can vary considerably.

3.4.4 THE T-SQUARED FIRE

In real fires the initial growth period is nearly always accelerating. A simple way to describe the accelerating growth is to assume that the energy release rate increases as the square of the time. By multiplying time squared by a factor α, various growth velocities can be simulated, and the energy release rate as a function of time could be expressed as

$$\dot{Q} = \alpha \cdot t^2 \tag{3.7}$$

where α is a growth factor (often given in kilowatts per second squared (kW/s²)) and t is the time from established ignition, in seconds.

This relationship has been found to fit well with the growth rates exhibited by various different commodities, but only after ignition has been well established and the fire has started to grow. This starting time, denoted t_0, depends both on the commodity in question and the manner in which it is ignited and is generally taken to be the time from ignition until flaming occurs, when significant energy starts to be released.

Table 3.4 lists a large number of furniture calorimeter tests and the growth rate factor, α, exhibited in each test. It also lists the starting time, t_0, for each test. An example of one of the tests is given in Figure 3.13, showing how the experimental data is fitted to arrive at a growth rate factor, α.

The t-squared fire has been used extensively in the U.S. for the design of detection systems, and guidance on selecting values for the growth time associated with various materials is available in NFPA 204M.[6] This includes typical values for α corresponding to fires that grow "ultra fast," "fast," "medium," and "slow." These correspond to the time it takes to reach 1000 Btu/s (= 1055 kW). Figure 3.14 shows how the energy release rate increases with these growth velocities, and Table 3.5 gives the corresponding values of α and the time it takes to reach 1055 kW.

Nelson[7] has collected and summarized energy release rate data for various commodities and materials, giving rough indications of growth rate and peak energy release rates. His data is given in Table 3.6.

3.5 THE DESIGN FIRE

3.5.1 BACKGROUND

Building fire regulations commonly require that two main objectives be met: life safety of the occupants and structural stability of the building. Two distinctly different design procedures are applied in each case.

In the case of structural stability, the objective is to protect property and ensure that firefighters can gain entry to the building without the risk of a structural collapse. Here, the time frame is relatively long (often 0.5 to 3 hours), the fire is assumed to have caused flashover, and the design fire is usually given as a temperature–time curve. We shall not discuss this design case here. However, Chapter 6 will address the issue of post-flashover temperatures in compartments.

In the case where the objective is to facilitate escape for the occupants, the time frame is usually relatively short (most often less than 30 minutes) and the design fire is specified as energy release rate vs. time; this case is the focus of our book.

When an engineer is to design a beam, a design load must be chosen. The beam cannot be designed for all possible load combinations that can occur during the building life span. Similarly, when designing for fire safety, the purpose of the design is not to simulate accurately all possible fires that could occur in the building. The engineer must first decide upon how much material can be expected to burn and how rapidly it will burn, and thus arrive at a design fire. The purpose of the design work is mainly to help the engineer form an opinion of the possible consequences and establish a basis for appraising the risks due to a building fire. The engineer can then compare the effectiveness of different safety measures; for example, how smoke ventilation through the roof would compare with installing early warning devices with regard to allowing means of escape to the occupants.

There is no exact methodology or procedure available to the engineer for defining the design fire. The engineer must use all available information on the building contents and building type and use engineering judgment to arrive at a design fire. The engineer can then perform a simple sensitivity analysis by changing the design fire to check the reliability of the design solution.

An example of the simplest way to construct a design fire curve is given in Figure 3.15. The curve is divided into the growth phase, the steady phase, and the decay phase. We shall, in the following, examine crude ways in which the engineer can estimate the duration and magnitude of these phases.

3.5.2 THE GROWTH PHASE

In real fires the initial fire development is nearly always accelerating. A suitable way to describe this mathematically is to use the t-squared fire described in Section 3.4.4. The choice of the growth

TABLE 3.4
Fire Growth Rates for Various Commodities

Test No.	Description	α (kW/s^2)	t_0 (s)
15	Metal wardrobe 41.4 kg (total)	0.4220	10
18	Chair F33 (trial loveseat) 39.2 kg	0.0066	140
19	Chair F21, 28.15 kg (initial stage of fire growth)	0.0344	110
19	Chair F21, 28.15 kg (later stage of fire growth)	0.04220	190
21	Metal wardrobe 40.8 kg (total) (average growth)	0.0169	10
21	Metal wardrobe 40.8 kg (total) (later growth)	0.0733	60
21	Metal wardrobe 40.8 kg (total) (initial growth)	0.1055	30
22	Chair F24, 28.3 kg	0.0086	400
23	Chair F23, 31.2 kg	0.0066	100
24	Chair F22, 31.9 kg	0.0003	150
25	Chair F26, 19.2 kg	0.0264	90
26	Chair F27, 29.0 kg	0.0264	360
27	Chair F29, 14.0 kg	0.1055	70
28	Chair F28, 29.2 kg	0.0058	90
29	Chair F25, 27.8 kg (later stage of fire growth)	0.2931	175
29	Chair F25, 27.8 kg (initial stage of fire growth)	0.1055	100
30	Chair F30, 25.2 kg	0.2931	70
31	Chair F31, (loveseat) 39.6 kg	0.2931	145
37	Chair F31, (loveseat) 40.4 kg	0.1648	100
38	Chair F32, (sofa) 51.5 kg	0.1055	50
39	1/2 inch plywood wardrobe w/ fabrics 68.8 kg	0.8612	20
40	1/2 inch plywood wardrobe w/ fabrics 68.32 kg	0.8612	40
41	1/8 inch plywood wardrobe w/ fabrics 36.0 kg	0.6594	40
42	1/8 inch plywood wardrobe w/ fire-ret. (int. fin. initial)	0.2153	50
42	1/8 inch plywood wardrobe w/ fire-ret. (int. fin. later)	1.1722	100
43	Repeat of 1/2 inch plywood wardrobe 67.62 kg	1.1722	50
44	1/8 inch plywood wardrobe w/ fire ret., latex paint 37.26 kg	0.1302	30
45	Chair F21, 28.34 kg (large hood)	0.1055	120
46	Chair F21, 28.34 kg	0.5210	130
47	Chair, adjustable back metal frame, foam cushion, 20.8 kg	0.0365	30
48	Easychair CO7 11.52 kg	0.0344	90
49	Easychair 15.68 kg (F-34)	0.0264	50
50	Chair metal frame minimum cushion, 16.52 kg	0.0264	120
51	Chair molded fiberglass no cushion, 5.28 kg	0.0733	20
52	Molded plastic patient chair, 11.26 kg	0.0140	2090
53	Chair metal frame w/padded seat and back 15.5 kg	0.0086	50
54	Loveseat metal frame w/foam cushions 27.26 kg	0.0042	210
55	Group chair metal frame w/foam cushion, 6.08 kg	Never exceeded 50 kW	
56	Chair wood frame w/latex foam cushions, 11.2 kg	0.0042	50
57	Loveseat wood frame w/foam cushions, 54.60 kg	0.0086	500
61	Wardrobe 3/4 inch particle board, 120.33 kg	0.0469	0
62	Bookcase plywood w/aluminum frame, 30.39 kg	0.2497	40
64	Easychair molded flexible urethane frame, 15.98 kg	0.0011	750
66	Easychair, 23.02 kg	0.1876	3700
67	Mattress and boxspring, 62.36 kg (initial fire growth)	0.0086	400
67	Mattress and boxspring, 62.36 kg (initial fire growth)	0.0009	90

Source: Adapted from Schifility, R.P., et al., in *SFPE Handbook of Fire Protection Engineering*, 2nd ed., National Fire Protection Association, Quincy, MA, 1995.

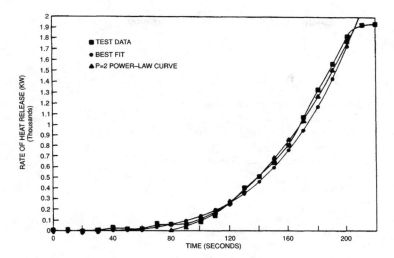

FIGURE 3.13   An example of one of the tests reported in Table 3.4. (From Schifility et al.[8] With permission.)

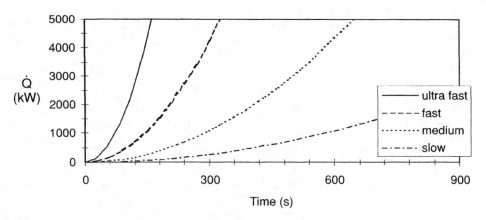

FIGURE 3.14   Energy release rates for different growth rates.

TABLE 3.5
Values of α for Different Growth Rates
According to NFPA 204M

Growth Rate	α (kW/s²)	Time (s) to reach 1055 kW
ultra fast	0.19	75
fast	0.047	150
medium	0.012	300
slow	0.003	600

Source: NFPA, *Guide for Smoke and Heat Venting*, NFPA
204M, National Fire Protection Association, Quincy, MA, 1985.

TABLE 3.6
Energy Release Rate Data

Description	Growth Rate	kW/m² of floor area
Fire retarded treated mattress (including normal bedding)	S	17
Lightweight type C upholstered furniture[b]	M	170[a]
Moderate-weight type C upholstered furniture[b]	S	400[a]
Mail bags (full) stored 5 ft high	F	400
Cotton/polyester innerspring mattress (including bedding)	M	565[a]
Lightweight type B upholstered furniture[b]	M	680[a]
Medium-weight type C upholstered furniture[b]	S	680[a]
Methyl alcohol pool fire	UF	740
Heavyweight type C upholstered furniture[b]	S	795[a]
Polyurethane innerspring mattress (including bedding)	F	910[a]
Moderate-weight type B upholstered furniture[b]	M	1020[a]
Wooden pallets 1 1/2 feet high	M	1420
Medium-weight type B upholstered furniture[b]	M	1645[a]
Lightweight type A upholstered furniture[b]	F	1700[a]
Empty cartons 15 ft high	F	1700
Diesel oil pool fire (>about 3 ft dia.)	F	1985
Cartons containing polyethylene bottles 15 ft high	UF	1985
Moderate-weight type A upholstered furniture[b]	F	2500[a]
Particle board wardrobe/chest of drawers	F	2550[a]
Gasoline pool fire (>about 3 ft dia.)	UF	3290
Thin plywood wardrobe with fire-retardant paint on all surfaces	UF	3855[a]
Wooden pallets 5 ft high	F	3970
Medium-weight type A upholstered furniture[b]	F	4080[a]
Heavyweight type A upholstered furniture[b]	F	5100[a]
Thin plywood wardrobe (50 in. × 24 in. × 72 in. high)	UF	6800[a]
Wooden pallets 10 ft high	F	6800
Wooden pallets 16 ft high	F	10200

[a] Peak rates of energy release were of short duration. These fuels typically showed a rapid rise to the peak and a corresponding rapid decline. In each case the fuel package tested consisted of a single item.

[b] The classification system used to describe upholstered furniture is as follows:

Lightweight = Less than about 5 lbs/ft² of floor area. A typical 6-ft long couch would weigh under 75 lbs.
Moderate-weight = About 5–10 lbs/ft² of floor area. A 6-ft long couch would weigh between 75 and 150 lbs.
Medium-weight = About 10–15 lbs/ft² of floor area. A 6-ft long couch would weigh between 150 and 300 lbs.
Heavyweight = More than about 15 lbs/ft² of floor area. A typical 6-ft long couch would weigh over 300 lbs.
Type A = Furniture with untreated or lightly treated foam plastic padding and nylon or other melting fabric.
Type B = Furniture with lightly or untreated foam plastic padding or nylon or other melting fabric, but not both.
Type C = Furniture with cotton or treated foam plastic padding, having cotton or other fabric that resists melting.

Source: From Nelson, H.E., "FPETOOL: Fire Protection Engineering Tools for Hazard Estimation," National Institute of Standards and Technology Internal Report 4380, pp. 93–100, Gaithersburg, MD, 1990.

rate (the choice of the factor α) depends on how much is known about the building contents and the building type.

If considerable knowledge of the building contents are available, a suitable ignition scenario can be assumed, and experimental data on materials (such as the data given in Table 3.4) can be used to determine the growth rate factor α.

In many instances, there is very scarce information available on the building contents, and the engineer must use information on the building type and use. There is currently considerable ongoing

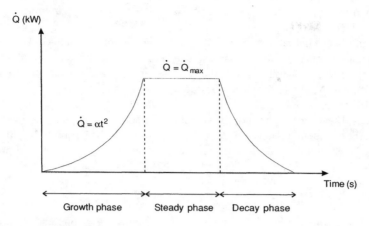

FIGURE 3.15   A simple design fire curve.

TABLE 3.7
Typical Growth Rates Recommended for
Various Types of Occupancies

Type of Occupancy	Growth Rate α
Dwellings, etc.	medium
Hotels, nursing homes, etc.	fast
Shopping centers, entertainment centers	ultra fast
Schools, offices	fast
Hazardous industries	Not specified

activity internationally to give recommendations on such growth factors for various building types. As an example, Table 3.7 gives the growth rates that have been suggested as recommendations in Swedish occupancies (not approved yet). Note that these values are only suggested; the designer must use engineering judgment and carry out sensitivity analysis to check the reliability of the design solution.

It should also be noted that the initial ignition time, before the fire starts the growth phase, has not been taken into account here. The fire is assumed to start growing at zero time. In all instances there is a shorter or longer time to ignition, where little energy is being evolved but the smoldering fuel may produce smoke that can trigger detection devices even before the growth phase starts. This ignition phase should be accounted for in the final design solution. Examples of experimentally measured such times, for a given ignition scenario, are given as t_0 in Table 3.4.

3.5.3   THE STEADY PHASE

The growth phase of the fire can lead either to a stage where the fuel reaches a maximum burning rate (where the fire is termed *fuel-bed controlled*) or to a stage where there is insufficient oxygen for continued combustion (where the fire is termed *ventilation-controlled*). In the latter case this may lead to flashover, where excess fuel is combusted in an adjacent room or in the openings leading to the outside.

In both cases, some assumption on the fuel must be made in order to estimate both the magnitude, $\dot{Q}_{max}$, and the duration, t_s, of the steady phase. Where there is considerable knowledge of the building contents, the experimental results discussed in Section 3.4 can be used directly. If no such information is available, some assumptions on the scenario at hand must be made to arrive at values for both $\dot{Q}_{max}$ and t_s.

For the simple case where the fire compartment has no openings to adjacent rooms and openings only to the outside air, $\dot{Q}_{max}$ can be assessed by considering the amount of air available for combustion. Any excess fuel will burn in the outside air and will not influence the conditions in the fire compartment. Methods for assessing the amount of air entering the compartment will be discussed in Chapter 5.

In most cases, however, the fire compartment has openings to adjacent rooms and excess fuel may burn there, thus influencing the environmental conditions in the building. It is therefore essential to base the steady phase of the design fire on the amount and type of fuel, and not solely on the availability of oxygen.

3.5.4 THE DECAY PHASE

In most practical design situations, the first 10 to 30 minutes are of interest when the objective is to ensure human safety and escape, since the Fire Department is assumed to start its rescue and firefighting operations within that time interval. The steady phase is therefore most often assumed to continue, and no decay phase is specified.

If, however, considerable knowledge is available on the building contents, and only a limited amount of fuel can be burnt, some assumptions on the decay phase can be made. The engineer must then rely on experimental data (such as are given in Section 3.4) to assess when and how rapidly the energy release rate decreases.

3.5.5 A MORE COMPLEX DESIGN FIRE

When there is very exact information on the material content of the fire compartment, a more realistic design fire curve can be constructed. This requires that a number of material parameters for the fuel packages are known, and that their relative positions in the compartment are known as well as the mode of ignition. Figure 3.16 gives a schematic example of such a design fire where one of the fuel packages is ignited, burns, and causes ignition of the other fuel package.

The energy release rate from the first fuel package can be estimated using experimental results. Once the energy release rate from this fuel package is known, the resulting flame height can be calculated using methods that will be discussed in Chapter 4. The heat transfer toward the second fuel package can be calculated using methods that will be discussed in Chapter 8, and classical heat conduction theory can be used to calculate the time to ignition. Thus, the first fuel package is ignited at time t_0 and reaches steady burning at time t_1. The second fuel package ignites at time t_2, reaches steady burning at time t_3. At time t_4 the first fuel package reaches the decay period, and at time t_5 only the second fuel package is burning. It should be noted that this procedure will only give very rough estimates and that very detailed information on the fuel packages and room geometry are required.

3.5.6 ENERGY RELEASE RATES USED IN THIS BOOK

In Section 3.5 we summarily outlined some of the simplest ways in which an engineer arrives at a design fire. The design fire is most commonly used as input to computer fire models to predict certain environmental conditions in a building, such as the temperatures and smoke layer heights. This information is then used to assess whether the occupants are likely to escape. If not, some measures are taken, such as installing early warning devices or smoke vents, and the computer simulation is repeated with new input data, again checking the hazard to the occupants.

In this textbook, however, we shall be studying much simpler methods to assess the environmental consequences of a fire in an enclosure. We shall state the fundamental laws governing the dominant physical processes and from there derive simple analytical solutions that can be used for

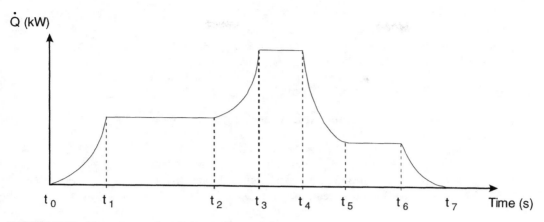

FIGURE 3.16   A more complex design fire.

hand calculations. The purpose is more to facilitate an understanding of these processes than to outline a complete methodology for design purposes.

In order to arrive at analytical solutions describing the physical processes, some simplifying assumptions will be made. In many cases these assumptions will require that the energy release rate is taken to be a constant value. In some cases the energy release rate can be assumed to be quasi-steady (stepwise-steady).

As a consequence, many of the examples and problems contained in this book will require the growth and decay phases of the design fire to be ignored and the energy release rate to be described by a constant value, corresponding to the value of $\dot{Q}_{max}$, as described in Section 3.5.3 above.

REFERENCES

1. Babrauskas, V., "Burning Rates," *SFPE Handbook of Fire Protection Engineering*, 2nd ed., National Fire Protection Association, Quincy, MA, 1995.
2. Särdqvist, S., "Initial Fires: RHR, Smoke Production and CO Generation from Single Items and Room Fire Tests," ISRN LUTVDG/TVBB--3070--SE, Department of Fire Safety Engineering, Lund University, Lund, Sweden, 1993.
3. Babrauskas, V., Greyson, S., Eds., *Heat Release in Fires*, E. & F. N. Spon, London, 1992.
4. Tewarson, A., "Generation of Heat and Chemical Compounds in Fires," *SFPE Handbook of Fire Protection Engineering*, 2nd ed., National Fire Protection Association, Quincy, MA, 1995.
5. Tewarson, A., "Flammability," *Physical Properties of Polymers Handbook*, Ed. Mark, J.A., 1996.
6. NFPA, *Guide for Smoke and Heat Venting*, NFPA 204M, National Fire Protection Association, Quincy, MA, 1985.
7. Nelson, H.E., "FPETOOL: Fire Protection Engineering Tools for Hazard Estimation," National Institute of Standards and Technology Internal Report 4380, Gaithersburg, MD, pp. 93–100, 1990.
8. Schifility, R.P., Meacham, B.J., and Custer, L.P., "Design of Detection Systems," *SFPE Handbook of Fire Protection Engineering*, 2nd ed., National Fire Protection Association, Quincy, MA, 1995.
9. Huggett, C., "Estimation of Rate of Heat Release by Means of Oxygen Consumption Measurements," *Fire and Materials*, Vol. 4, pp. 61–65, 1980.

PROBLEMS AND SUGGESTED ANSWERS

3.1 Estimate the potential energy release rate from a circular heptane pool with diameter 1.2 m.

Suggested answer: 2.6 MW

3.2 Estimate the potential energy release rate from a rectangular pool of transformer oil where the dimensions are
   (a) 2.0 m by 1.5 m
   (b) 2.5 m by 1.8 m

   Suggested answer: (a) 2.8 MW; (b) 4.6 MW

3.3 Estimate a value for the constant maximum energy release rates from burning pools of
   (a) 2 m² transformer oil fire
   (b) 2 m² wood pallets, pallet stack height is 1.5 m
   (c) mail bags, full, stored 1.5 m high, floor area of fuel is 2 m²
   (d) the Christmas tree fire in Figure 3.12 (test 17), diameter 1 m
   (e) sofa in Figure 3.8, projected floor area of fuel is 2 m²

   Suggested answers: (a) 1.7 MW; (b) 8 MW; (c) 0.8 MW; (d) 0.5 MW; (e) 2.8 MW

3.4 Suggest the form of a design fire curve for a 1 m² wood pallet fire, stored 1.5 m high.

   Suggested answer: $\alpha = 0.047$ kW/s² till $t = 290$ s, then constant effect of 3970 kW

3.5 A heptane pool with a diameter $D = 1.2$ m is burning in a room 3 m by 4 m and 2.4 m high. The room has an opening that is 1 m wide and 2 m high. The mass flow rate of air through the opening can be approximated by $\dot{m}_a = 0.5A_o\sqrt{H_o}$ kg/s, where A_0 is the area of the opening and H_0 is its height.
   (a) Calculate the maximum energy release rate that can be attained within the compartment.
   (b) Discuss the effect the enclosure will have on the heptane fire.

   Suggested answer: (a) Free burning heptane pool: 2.6 MW. Inflow of oxygen is roughly 0.33 kg/s, which allows a maximum energy release rate of 4.3 MW if all the oxygen is used for combustion. (b) See discussion on p. 28.

4 Fire Plumes and Flame Heights

When a mass of hot gases is surrounded by colder gases, the hotter and less dense mass will rise upward due to the density difference, or rather, due to buoyancy. This is what happens above a burning fuel source, and the buoyant flow, including any flames, is referred to as a *fire plume*. As the hot gases rise, cold air will be entrained into the plume, causing a layer of hot gases to be formed. Many applications in fire safety engineering have to do with estimating the properties of the hot layer and the rate of its descent. This depends directly on how much mass and energy is transported by the plume to the upper layer. This chapter will explain some of the most fundamental properties of fire plumes and provide analytical expressions for their properties. Further, the size and geometry of the flames due to a burning object are of great interest to the fire safety engineer. An estimation of the flame height can facilitate calculations on heat transfer to distant objects, secondary fuel and fire detection, and suppression equipment. This chapter discusses flame heights in general and gives expressions for calculation of the flame heights for certain given scenarios.

CONTENTS

4.1   Terminology 48
4.2   Introduction 48
    4.2.1   Flame Characteristics..48
    4.2.2   Turbulent Fire Plume Characteristics 52
4.3   The Ideal Plume...54
    4.3.1   Assumptions 54
    4.3.2   Initial Considerations 55
    4.3.3   The Continuity Equation for Mass 57
    4.3.4   The Momentum and Buoyancy Equation.......................................57
    4.3.5   Solution of the Two Differential Equations 58
    4.3.7   Inserting the Constants and Concluding..60
4.4   Plume Equations Based on Experiments..62
    4.4.1   The Zukoski Plume..62
    4.4.2   The Heskestad Plume...63
    4.4.3   The McCaffrey Plume..67
    4.4.4   The Thomas Plume 69
4.5   Line Plumes and Bounded Plumes...71
    4.5.1   Wall and Corner Interactions with Plumes...................................71
    4.5.2   Line Source Plumes 72
4.6   Ceiling Jets..73
    4.6.1   Ceiling Jet Temperatures and Velocities.......................................74
    4.6.2   Flame Extensions under Ceilings 76
References 77
Problems and Suggested Answers 78

4.1   TERMINOLOGY

Axisymmetric plume — The buoyant axisymmetric plume, caused by a diffusion flame formed above the burning fuel, is the most commonly used plume in fire safety engineering. An axis of symmetry is assumed to exist along the vertical centerline of the plume, and air is entrained horizontally from all directions. Other fire plume categories include, for example, line plumes, which may be formed above a long and narrow burner, allowing air to be entrained from two sides only.

Plume — When a mass of hot gases is surrounded by colder gases, the hotter and less dense mass will rise upward due to the density difference, or rather, due to buoyancy. This is what happens above a burning fuel source, and the buoyant flow, including any flames, is referred to as a fire plume.

Plume mass flow rate — The total mass flowing upward, at a certain height above the fuel source, within the plume boundaries. The plume mass flow rate is most often given in kg/s and is denoted $\dot{m}_p$.

Plume radius — The axisymmetric plume radius at a some height above the fuel source. The plume radius is given in m and is denoted b.

Plume temperature — The temperature of the gases within the plume boundaries, at a certain height above the fuel source. At any given height the highest temperature is at the plume centerline (for axisymmetric plumes), decreasing toward the edge of the plume. The centerline temperature changes with height is given in °C or K and denoted T_0, where the subscript "0" refers to the centerline.

Plume velocity — The velocity at which the gases within the plume boundaries move upward, at a certain height above the fuel source. In an axisymmetric plume, the highest velocity (at a given height) is at the plume centerline. This centerline velocity is given in m/s and is denoted u_0, where the subscript "0" refers to the centerline.

4.2   INTRODUCTION

This introduction discusses some of the general characteristics of fire plumes and flames most commonly occurring in building fires. Definitions are given of a number of concepts that will be used in later sections. The discussion is divided into two parts, one on flames and the other on fire plumes.

4.2.1   FLAME CHARACTERISTICS

In most fire safety engineering applications we are concerned with the so-called buoyant, turbulent diffusion flame. Here we discuss some of the processes that characterize such flames, show results from flame height measurements, and give correlation equations that can be used for estimating flame heights.

Diffusion: Diffusion flames refer to the case where fuel and oxygen are initially separated, and mix through the process of diffusion. Burning and flaming occur where the concentration of the mixture is favorable to combustion. Although the fuel and the oxidant may come together through turbulent mixing, the underlying mechanism is *molecular diffusion*. This is the process in which molecules are transported from a high to low concentration. Opposite to this is the *premixed flame*, as is the case with a welder's torch, where the fuel and the oxidant are mixed before ignited. Flames in accidental fires are nearly always characterized as diffusion flames.

Buoyancy: When a mass of hot gases is surrounded by colder gases, the hotter and less dense mass will rise upward due to the density difference, or rather, due to *buoyancy*. The upward velocity of the flow within a flame will be dominated by the buoyancy force if the velocity at which the fuel is injected is not exceptionally high. Opposite to the buoyancy-dominated flames are the flows

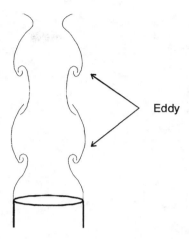

FIGURE 4.1 Flame fluctuations due to periodic eddy shedding.

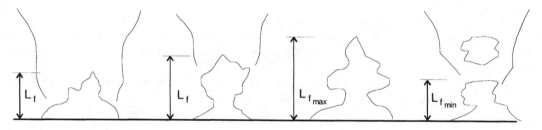

FIGURE 4.2 Characteristic sketch of flame height fluctuations.

formed above a high-pressure gaseous fuel source (for example, a ruptured pipeline under high-pressure) where the flow is not buoyancy dominated, but momentum dominated; these are termed *jet flames*.

Turbulence: Very small diffusion flames can be laminar, such as the flame on a candle. Larger diffusion flames are turbulent and will fluctuate with periodic oscillations with large eddies shedding at the flame edge (see Figure 4.1). The eddies, which are visible in turbulent plumes (more so in momentum-driven plumes than in buoyancy-driven ones), roll up along the outside of the plume and are a result of the instability between the hot flame and the cold air.

These random fluctuations, which are characteristic of turbulence, will give rise to periodic flame height (and shape) fluctuations. The fluctuations normally have a frequency of the order of 1–3 Hz, i.e., will occur between one and three times per second; in general, this shedding depends on fire diameter. Figure 4.2 shows a characteristic sketch of this phenomenon where L_f is the visible flame height as a function of time. In our treatment we consider only time–mean results, i.e., the mean flame height, denoted L, given in meters.

Definition of mean flame height: In order to provide engineering equations allowing calculation of the flame height, we must first define the mean flame height. This is most conveniently done by averaging the visible flame height over time. The luminosity of the lower part of the flaming region appears fairly steady. The upper part fluctuates or, in other words, is intermittent. The graph in Figure 4.3 is generally used to define the mean flame height.

The intermittency, denoted I, is shown on the vertical axis, where a value of 1 indicates the appearance of a flame at all times. The horizontal axis shows the distance above the fire source, z. The height at which the intermittency is 0.5, i.e., the height above which flame appears half the time, is defined as the mean flame height, L. The experimental procedures for measuring flame

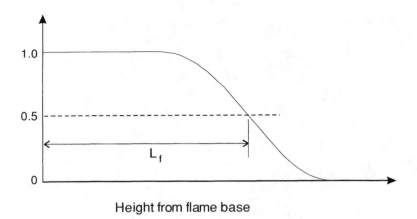

FIGURE 4.3   Definition of mean flame height.

usually involve video equipment. The data achieved is fairly consistent with flame heights that are averaged by the human eye.

Flame height correlations: Due to the turbulent nature of the flames, we cannot provide engineering equations for flame heights that are derived from first principles. We must therefore investigate which properties influence the flame height and use experimental data to express the flame height in terms of the dominating properties.

The nondimensional Froude number, denoted Fr, is used in hydraulics when describing liquid flows, also approximately applicable to the high-temperature gas in the flames.

$$Fr = \frac{u^2}{g \cdot D}$$

where u is the flow velocity, g is the acceleration due to gravity, and D is the diameter of the flow source. The numerator is in proportion to the momentum and the denominator in proportion to the gravity or buoyancy. The Froude number can be expressed in terms of energy release rate by noting that $Q = \dot{m}\Delta H_c$, where $\dot{m}$ is the burning rate and ΔH_c is the heat of combustion. Further, the burning rate can be expressed as $\dot{m} = u\rho A$ where u is the velocity of the gas, ρ is the density of the gas, and A is the area of the fuel source (directly related to D^2). The relationship between the Froude number, the energy release rate, and the diameter of the source can therefore be said to be of the form

$$Fr \propto \frac{\dot{Q}^2}{D^5}$$

The geometry of turbulent diffusion flames has been found to scale with the square root of the Froude number. Representing the flame geometry as the flame height normalized by the source diameter, L/D, we can write

$$\sqrt{Fr} \propto \frac{L}{D} \propto \sqrt{\frac{\dot{Q}^2}{D^5}} \propto \frac{\dot{Q}}{D^{5/2}}$$

A vast number of experiments have been carried out that relate flame heights to energy release rate and source diameter. Experimenters have found it convenient to express the data in terms of a nondimensional energy release rate, denoted Q^*, and given by the expression

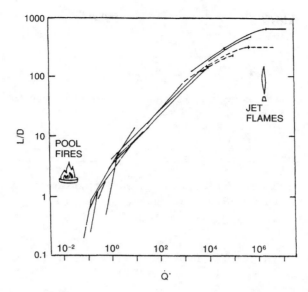

FIGURE 4.4 Normalized flame height vs. dimensionless energy release rate. (Adapted from McCaffrey[10]. With permission.)

$$\dot{Q}^* = \frac{\dot{Q}}{\rho_\infty c_p T_\infty \sqrt{gD} D^2} \tag{4.1}$$

where ρ_∞, c_p, and T_∞ refer to the properties of ambient air. This dimensionless energy release rate parameter, which can be said to represent the square root of the Froude number, has been found to be very important in controlling the geometry of fire plumes, and we shall return to it in later sections.

From the above equations we see that we should be able to express flame heights as a function of diameter and energy release rate for a wide range of Froude numbers. Indeed, Figure 4.4 shows a representation of a great number of experimental results, where the mean flame height normalized by source diameter, L/D, has been plotted against the dimensionless energy release rate, $\dot{Q}^*$. Each line in the plot represents the results of separate experiments carried out by different investigators.

The left side of the plot shows fires where the diameter is of the same order of magnitude as the flame height and the Froude number is low, indicating buoyancy-dominated flows. Buoyancy dominates at intermediate Froude numbers as well. The upper right-hand corner represents the high Froude number, high momentum jet flame regime.

In most fire situations $\dot{Q}^*$ will be less than 10, and for most large fires less than 2. We are therefore interested primarily in the left-hand side of Figure 4.4. Experimenters have found that the normalized mean flame height, L/D, correlates well with $\dot{Q}^{*2/5}$ over a wide range of values (roughly $1 < \dot{Q}^* < 1000$, as seen by the straight line in Figure 4.4). A great number of correlation equations for flame heights have been presented in the literature, representing various of the regimes shown in Figure 4.4.

One of the most useful such equations, presented by Heskestad,[1] gives good results for the different regimes in Figure 4.4, except for the jet flame regime. The equation expresses mean flame height divided by diameter as

$$\frac{L}{D} = 3.7\dot{Q}^{*2/5} - 1.02 \tag{4.2}$$

This relationship maintains the 2/5 power of $\dot{Q}^*$ over the large intermediate regime while exhibiting an increasing slope at small $\dot{Q}^*$, as is seen in Figure 4.4.

A more convenient form of Eq. (4.2) gives the mean flame height as a function of energy release rate and diameter:

$$L = 0.235\dot{Q}^{2/5} - 1.02D \qquad (4.3)$$

where the energy release rate is given in kilowatts (kW) and the diameter given in meters (m), resulting in the mean flame height in meters.

EXAMPLE 4.1

Example 3.3 discussed a pump breakdown where 20 l of transformer oil were spilled into a sump of 2 m^2 and ignited. Calculate the resulting flame height.

SUGGESTED SOLUTION

Example 3.3 found the energy release rate to be 1.69 MW and an equivalent circular diameter to be 1.6 m. Using Eq. (4.3), $L = 0.235\dot{Q}^{2/5} - 1.02D$, we find

$$L = 0.235 \cdot (1690)^{2/5} - 1.02 \cdot 1.6 = 2.96 \text{ m}$$

Therefore, the flame will be roughly 3 m high.

Concluding remarks: The above results give only mean flame heights. Experimental results have been used to arrive at engineering expressions for flame heights due to the fluctuating nature of turbulent diffusion flames. These experiments are most often carried out with simple gaseous fluids such as methane, natural gas, or propane, which do not model all of the interesting properties of real fuels. Further, they do not take into account more complex geometries, such as furniture, where the solid fuel is distributed in the vertical and horizontal directions. Nevertheless, Eq. (4.3) has given good approximations of flame heights from real fuels, but it is important to keep these limitations in mind.

In this section we have discussed only unbounded flames, where there is no influence of nearby walls, ceilings, or openings. We have not discussed the influence other environmental factors, such as wind, may have on the flame height. Some of these issues will be discussed in Section 4.5.

4.2.2 TURBULENT FIRE PLUME CHARACTERISTICS

Fire plumes can be characterized into various groups depending on the scenario under investigation. In this section we shall concentrate on the plume most commonly used in fire safety engineering, the so-called *buoyant axisymmetric plume*, caused by a diffusion flame formed above the burning fuel. An axis of symmetry is assumed to exist along the vertical centerline of the plume. Other fire plume categories include, for example, line plumes, which may be formed above a long and narrow burner, allowing air to be entrained from two sides only as the hot gases rise (see Section 4.5).

The axisymmetric fire plume is conventionally divided into the three zones, as shown in Figure 4.5. In the *continuous flame* zone the upward velocity is near zero at the base and increases with height. In the *intermittent flame* zone the velocity is relatively constant, and in the *far field* zone the velocity decreases with height. Figure 4.6 shows some of the characteristics of a buoyant axisymmetric plume.

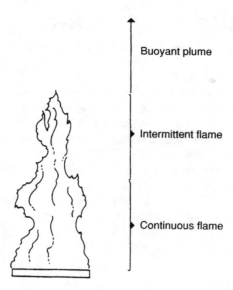

FIGURE 4.5   The three zones of the axisymmetric buoyant plume. (Adapted from McCaffrey[10].)

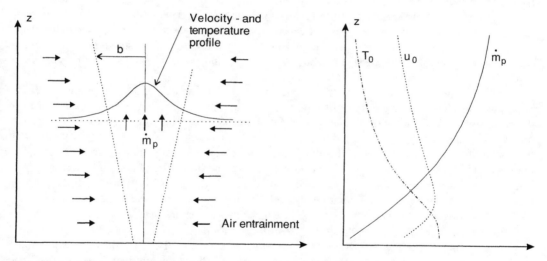

FIGURE 4.6   Some of the characteristics of a buoyant axisymmetric plume.

Plume velocity: The highest velocity is at the centerline of the plume, as shown by the velocity profile in Figure 4.6. This centerline velocity, denoted u_0, changes with height. The right-hand side of Figure 4.6 shows that the centerline velocity is close to zero at the fuel bed. The only mass flowing upward at the fuel bed is the mass loss rate of the fuel bed. The centerline velocity increases with height in the continuous flame region. Above the flames the velocity decreases with height, since more ambient air is entrained, which cools the plume.

Plume temperature: The highest temperature is at the plume centerline, decreasing toward the edge of the plume with a similar profile as the velocity profile. The centerline temperature, denoted T_0, changes with height. It is roughly constant in the continuous flame region and represents the mean flame temperature. The temperature decreases sharply above the flames as an increasing amount of ambient air is entrained into the plume. The symbol ΔT_0 is used for the centerline plume temperature rise above the ambient temperature, T_∞. Thus, $\Delta T_0 = T_0 - T_\infty$.

Plume mass flow rate: The mass flow rate in the plume, denoted $\dot{m}_p$ and most often given in kilograms per second (kg/s), is the total mass flowing upward, at a certain height, within the plume boundaries. The plume mass flow increases steadily with height, since ambient air is continually entrained over the plume height.

This mass consists of a mixture of combustion products and ambient air entrained into the plume, most of the mass stemming from the ambient air entrained and only a small portion stemming from the combustion products. For a burning wastebasket, with a total effect of around 100 kW, the plume mass flow rate 2 m above it is roughly 1 kg/s. At a height of 4 m the plume mass flow rate is roughly 3 kg/s. The plume mass flow rate should not be confused with the mass burning rates released from the fuel bed, as discussed in Chapter 3. The above-mentioned wastebasket would release fuel gases at roughly the rate of 0.005 kg/s. This is insignificant in relation to the mass plume rate of 1 kg/s at the height of 2 m.

Plume radius: The point source plume radius, denoted b, spreads roughly with an angle of 15° to the vertical. This corresponds approximately to the point where the centerline temperature rise has declined to $0.5\ \Delta T_0$. Similar values are achieved when the plume radius is defined as the point where the centerline velocity has declined to $0.5\ u_0$.

4.3   THE IDEAL PLUME

In this section we consider a very simple type of a fire plume, sometimes referred to as the *ideal plume* or the *point-source plume*. We shall set up the fundamental equations for continuity, momentum, and buoyancy and thus arrive at analytical solutions for the mass flow, velocity, and temperature of the gases in this simplified plume. The purpose is to facilitate an understanding of how plume equations are generally arrived at. In Section 4.4 we expand upon the subject and provide more generally valid expressions, based on experiments.

Consider Figure 4.7. We assume the simple case of a point source of heat at height $z = 0$. The energy is considered to be totally transported in the plume, and no radiative heat is emitted from the point source. The force driving the system can therefore be assumed to arise due to the density difference of the hot air above the point and the cold surrounding air. We shall further assume that the flow profile across the section of the plume, at any height, is a so-called *top hat* profile. Therefore, the upward velocity is assumed to be constant across the width of the plume and zero outside it. The plume temperature is similarly assumed to be constant across any section of the plume. Note that in Section 4.2 the plume temperature and velocity were denoted by the symbols T_0 and u_0, the suffix "0" referring to the centerline axis. Since these properties are here assumed to be constant across the plume, we shall drop the suffix "0" in the symbols.

Here, T_∞ and ρ_∞ are the temperature and density, respectively, of the surrounding ambient air, the upward velocity is denoted u, the temperature increase above the ambient is denoted ΔT, and the height above the point source is denoted z. We shall also assume that there is a relationship between the upward velocity in the plume and the horizontal entrainment velocity into the plume, v, such that $v = \alpha \cdot u$, where α is called the entrainment coefficient. Section 4.3.1 summarizes the main assumptions.

4.3.1   ASSUMPTIONS

To arrive at simple analytical solutions expressing the plume properties, we must make many restricting assumptions:

1. We assume that all the energy is injected at the point source of origin and that energy remains in the plume, i.e., that there are no heat losses in the system due to radiative losses. In real fire plumes the radiative part is typically 20 to 40% of the total energy released from many common fuel sources.

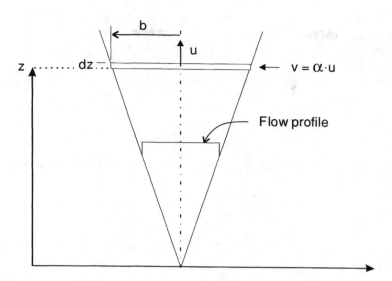

FIGURE 4.7 The ideal plume.

2. We assume that the density variations throughout the plume height are small and only need to be considered when the difference $(\rho_\infty - \rho)$ appears directly. The ideal plume theory is therefore sometimes referred to as the weak plume theory, where, due to mixing (entrainment) of air, the plume temperature is only slightly higher than the ambient. At certain points in the derivation we shall therefore assume that $\rho_\infty \approx \rho$. However, when expressing the buoyancy force, which is caused by the density difference, $(\rho_\infty - \rho)$, this assumption does not apply. This approximation is sometimes referred to as the Boussinesq approximation. In practical terms this means that the equations cannot be used at heights close to the fire source but will give reasonable answers at heights further above the source.

3. We assume that the velocity, temperature, and force profiles are of similar form independent of the height, z. We further assume that these profiles are so-called top hat profiles, so that the velocity and temperature are constant over the horizontal section at height z along the radius b, and that $u = 0$, and $T = T_\infty$ outside the plume radius.

4. We assume that the air entrainment at the edge of the plume is proportional to the local gas velocity in the plume, so that the entrainment velocity can be written as $v = \alpha \cdot u$, where α is a constant and is taken to be ≈ 0.15 (applies to top hat profile). In other words, the horizontal entrainment velocity is assumed to be 15% of the upward plume velocity. This value is difficult to measure but has been found to correspond reasonably with experimentally measured values.

4.3.2 INITIAL CONSIDERATIONS

We seek analytical expressions for the following variables as a function of height z:

The temperature difference at height z	$\Delta T(z)$ given in [°C] or [K]
The radius of the plume at height z	$b(z)$ given in [m]
The upward gas velocity at height z	$u(z)$ given in [m/s]
The plume mass flow at height z	$\dot{m}_p(z)$ given in [kg/s]

We start by setting up general expressions for the mass flow, the momentum, the buoyancy force, and the energy release rate. These expressions will then be incorporated in the differential equations for continuity of mass and for momentum/buoyancy, and we will examine solutions to the differential equations. This will allow us, through dimensional analysis, to identify the constants arising from the assumed solutions to the differential equations and thereby set up expressions for the above-mentioned variables.

The expression for the *mass flow rate* at some height z, where the plume radius equals b and the cross-sectional area equals πb^2, can be written as

$$\dot{m}_p = \pi b^2 \rho u \tag{4.4}$$

The differential *buoyancy force* acting on the mass within the small differential segment dz can be expressed as

$$dF = g(\rho_\infty - \rho) \cdot dz \cdot \pi b^2 \tag{4.5}$$

where g is the acceleration due to gravity. The time rate of *momentum* (mass flow times velocity) at height z can be written as

$$\dot{m}_p u = \pi b^2 \rho u^2 \tag{4.6}$$

These three equations will be used in Sections 4.3.3 and 4.3.4 to set up two differential equations, one for continuity and one for momentum and buoyancy.

At a later stage in the derivation we will wish to express the density difference $(\rho_\infty - \rho)$, or $\Delta\rho$, in terms of energy release rate, $\dot{Q}$. We therefore seek a relationship between the density difference and the energy release rate.

Assuming that there are no radiative heat losses in the plume, the energy flow rate of the gases at height z can be written as

$$\dot{Q} = \dot{m}_p c_p \Delta T = \pi b^2 \rho u \cdot c_p \Delta T$$

where c_p is the specific heat at constant pressure of the gases. Using the ideal gas law and rewriting the temperature difference as $\Delta T = \dfrac{\Delta\rho}{\rho} \cdot T_\infty$, (since $T \cdot \rho = T_\infty \cdot \rho_\infty$, $\Delta T = T - T_\infty$, $\Delta\rho = \rho_\infty - \rho$), the expression becomes

$$\dot{Q} = \pi b^2 \rho u \cdot c_p \frac{\Delta\rho}{\rho} T_\infty$$

We can therefore write the following relationship between density difference and energy release rate, which will become useful to us in Section 4.3.4:

$$\Delta\rho b^2 = \frac{\dot{Q}}{\pi u c_p T_\infty} \tag{4.7}$$

We now write the differential equations for mass and momentum/buoyancy with respect to height $\left(\dfrac{d}{dz}\right)$. This will give us two differential equations. Assuming solutions for these and using dimensional analysis will give us the coefficients we need to arrive at analytical solutions for the variables we wish to express.

4.3.3   THE CONTINUITY EQUATION FOR MASS

The continuity equation for mass simply states that the increase in mass flowing up through the differential element dz must be entrained through the sides of element dz. We shall first express the rate of change of mass flowing through the element and then express the mass entrained through the sides. Equating the two will lead to the differential equation that we seek.

The *rate of change of mass* over height dz can be written (using Eq. (4.4))

$$\frac{d\dot{m}_p}{dz} = \frac{d(\pi b^2 \rho u)}{dz} \tag{4.8}$$

The rate at which air is entrained in through the sides of element dz equals the area of the sides ($= 2\pi b\, dz$) times the horizontal entrainment velocity, v, times the density of the ambient air, ρ_∞. Earlier we assumed that the horizontal entrainment velocity could be expressed in terms of the upward velocity as $v = \alpha \cdot u$. To get the *rate of entrained air* per unit height we must also divide by dz, which gives

$$2\pi b \cdot dz \cdot \alpha \cdot u \cdot \rho / dz \tag{4.9}$$

We now make use of the weak plume assumption, i.e., that there are only small density differences with height. We can therefore assume, in Eq. (4.8), that ρ does not change with respect to height. As a result, both ρ and π can be moved outside the brackets in Eq. (4.8).

We now equate (4.8) and (4.9), since the rate of change of mass over dz must equal the rate of mass entrained through the sides of dz, resulting in the continuity equation

$$\frac{d}{dz}(b^2 u) = 2\alpha u b \tag{4.10}$$

We will use this equation in Section 4.3.5, but first we will set up the differential equation for momentum and buoyancy.

4.3.4   THE MOMENTUM AND BUOYANCY EQUATION

The rate of change of momentum over height dz must equal the buoyancy forces per unit height acting on the element dz. Differentiating Eq. (4.6) with respect to height, the *rate of change of momentum* per unit height z becomes

$$\frac{d}{dz}(\dot{m}_p u) = \frac{d}{dz}(\pi \rho b^2 u^2)$$

The differential *buoyancy force* acting on the mass within the small differential segment dz was given by Eq. (4.5). To get the buoyancy force acting on element dz per unit height, we divide Eq. (4.5) by dz, getting

Buoyancy force per unit height, $dF/dz = g(\rho_\infty - \rho) \cdot \pi b^2$

Equating the rate of change of momentum with the buoyancy force per unit height, we get

$$\frac{d}{dz}\left(\pi\rho b^2 u^2\right) = g\Delta\rho\pi b^2 \tag{4.11}$$

Again we make use of the weak plume assumption, so the ρ is assumed constant with height. Further, we wish to express the density difference, $\Delta\rho$, in terms of the energy release rate. Using Eq. (4.7) and treating ρ and π as constants, we arrive at the differential equation

$$\frac{d}{dz}(b^2 u^2) = \frac{\dot{Q}g}{\pi u c_p T_\infty \rho} \tag{4.12}$$

We will need to make one further assumption to arrive at an easily manageable differential equation for the momentum and buoyancy. We assume that ρ in the above equation is roughly equivalent to ρ_∞, which can be justified by the fact that we are considering the weak plume. This is the Boussinesq approximation mentioned in Section 4.3.1. Equation (4.12) therefore becomes what we shall call the momentum–buoyancy equation

$$\frac{d}{dz}(b^2 u^2) = \frac{\dot{Q}g}{\pi u c_p T_\infty \rho_\infty} \tag{4.13}$$

The main results from the two last sections are two differential equations, the continuity equation (4.10) and the momentum–buoyancy equation (4.13).

4.3.5   SOLUTION OF THE TWO DIFFERENTIAL EQUATIONS

To facilitate an analytical solution of the two differential equations (4.10) and (4.13), we shall assume that the radius, b, and the velocity, u, change as some power of the height, z, and write

$$b = C_1 \cdot z^m$$

$$u = C_2 \cdot z^n$$

We insert this into Eq. (4.10) and (4.13) and seek the constants C_1 and C_2 as well as the constants m and n.

Equation (4.10) becomes

$$\frac{d}{dz}(C_1^2 z^{2m} \cdot C_2 z^n) = 2\alpha C_1 z^m \cdot C_2 z^n$$

Differentiating the left-hand side with respect to z, we get

$$C_1^2 C_2 (2m + n) z^{2m+n-1} = 2\alpha C_1 C_2 z^{m+n} \tag{4.14}$$

Similarly, Eq. (4.13) becomes

$$\frac{d}{dz}(C_1^2 z^{2m} \cdot C_2^2 z^{2n}) = \frac{\dot{Q}g}{\pi c_p T_\infty \rho_\infty \cdot C_2 z^n}$$

Differentiating the left-hand side with respect to z, we get

$$C_1^2 C_2^2 (2m + 2n) z^{2m+2n-1} = \frac{\dot{Q}g}{\pi c_p T_\infty \rho_\infty \cdot C_2} \cdot z^{-n} \qquad (4.15)$$

We can now determine the constants m and n as well as the constants C_1 and C_2 by simple dimensional analysis.

Constants m and n: The power of z must be the same on both sides of the above equations. We can therefore set up the following equations for m and n:

From Eq. (4.14)

$$2m + n - 1 \quad = m + n$$

$$2m - 1 \quad = m$$

$$\Rightarrow \quad m \quad = 1$$

From Eq. (4.15)

$$2m + 2n - 1 \quad = -n$$

$$2n + 1 \quad = -n$$

$$\Rightarrow \quad n \quad = -1/3$$

Constants C_1 and C_2: Similarly, the constants C_1 and C_2 must result in identical expressions for both sides of Eq. (4.14) and (4.15). We divide Eq. (4.14) by $z^{2/3}$ and Eq. (4.15) by $z^{1/3}$. The constants on both sides can then be equated.

From Eq. (4.14)

$$C_1^2 C_2 \left(2 - \frac{1}{3}\right) = 2C_1 C_2 \alpha$$

$$C_1 \cdot \frac{5}{3} = 2\alpha$$

$$\Rightarrow \quad C_1 = \frac{6}{5}\alpha$$

From Eq. (4.15)

$$C_1^2 C_2^2 (2 - 2/3) = \frac{\dot{Q}g}{\pi c_p T_\infty \rho_\infty} \cdot \frac{1}{C_2}$$

$$\left(\frac{6}{5}\alpha\right)^2 \cdot \frac{4}{3} \cdot C_2^3 = \frac{\dot{Q}g}{\pi c_p T_\infty \rho_\infty}$$

$$\frac{48}{25}\alpha^2 C_2^3 = \frac{\dot{Q}g}{\pi c_p T_\infty \rho_\infty}$$

$$\Rightarrow \quad C_2 = \left(\frac{25}{48\alpha^2} \cdot \frac{\dot{Q}g}{\pi c_p T_\infty \rho_\infty}\right)^{1/3}$$

4.3.7 INSERTING THE CONSTANTS AND CONCLUDING

We now summarize our findings and write down expressions for b, u, $\dot{m}_p$, and ΔT by substituting the constants C_1, C_2, m, and n into the equations arrived at above.

In Section 4.3.5 we assumed the solution for plume radius to be of the form $b = C_1 z^m$ and the solution for the upward velocity to be of the form $u = C_2 z^n$. Inserting the constants, we find

$$b = \frac{6}{5}\alpha \cdot z \tag{4.16}$$

and

$$u = \left(\frac{25}{48\alpha^2} \cdot \frac{\dot{Q}g}{\pi c_p T_\infty \rho_\infty}\right)^{1/3} \cdot z^{-1/3} \tag{4.17}$$

The entrainment coefficient, α, has from experiments been found to be roughly 0.15 for a top hat velocity profile. Inserting this into Eq. (4.17) we get a useful engineering relationship for the gas velocity in the plume:

$$u = 1.94\left[\frac{g}{c_p T_\infty \rho_\infty}\right]^{1/3} \dot{Q}^{1/3} z^{-1/3} \tag{4.18}$$

From Eq. (4.4) and using the expressions for b and u in Eq. (4.16) and (4.17) above, we arrive at an expression for the plume mass flow:

$$\dot{m}_p = \pi b^2 \rho u = \pi\rho\left(\frac{6}{5}\alpha z\right)^2\left(\frac{25}{48\alpha^2} \cdot \frac{\dot{Q}g}{\pi c_p T_\infty \rho_\infty}\right)^{1/3} z^{-1/3}$$

We see that there are two densities appearing in the above equation, ρ and ρ_∞. Once more we use the weak plume assumption and set $\rho \approx \rho_\infty$. Substituting this into the above equation, and setting $\alpha = 0.15$, we get a simple engineering expression for the mass plume flow at height z:

$$\dot{m}_p = 0.20\left(\frac{\rho_\infty^2 g}{c_p T_\infty}\right)^{1/3} \dot{Q}^{1/3} \cdot z^{5/3} \tag{4.19}$$

The form of the above expression is well known in fire safety engineering, and we shall return to it when discussing the so-called Zukoski plume equation in Section 4.4.1.

ΔT is arrived at from the energy contents of the plume gases; $\dot{Q} = \dot{m}_p c_p \Delta T$, assuming there are no radiative heat losses to the environment. Solving this for ΔT and inserting $\dot{m}_p$ from Eq. (4.19) leads to an equation for the plume temperature difference at height z:

$$\Delta T = 5.0\left(\frac{T_\infty}{g c_p^2 \rho_\infty^2}\right)^{1/3} \dot{Q}^{2/3} \cdot z^{-5/3} \tag{4.20}$$

It is useful to note the form of the relationships in Eq. (4.19) and (4.20). The plume mass flow increases with the energy release rate to the 1/3 power and with the height to the 5/3 power. The plume temperature increases with the energy release rate to the 2/3 power and decreases with the height to the –5/3 power.

A note on units and gas properties: Equations (4.16)–(4.20) have been arrived at from fundamental principles, and the variables are given in SI units. However, as we shall see in Section 4.4, some plume equations are arrived at through experiments and include constants that assume that the energy release rate, $\dot{Q}$, is given in [kW]. When this is the case the heat capacity of the gases, c_p, must be given in [kJ/(kg K)] to ensure correct dimensions. It is a tradition in fire safety engineering to use [kW] when discussing plumes, and we shall therefore do so throughout this chapter. Also, most of the gases in the plume stem from ambient entrained air. Therefore, the gas properties in the above equations are taken to be those of air, and the value for c_p is traditionally taken at ambient temperature. It is important to note that T_∞ in the above equations must be given in degrees Kelvin.

EXAMPLE 4.2

A wastepaper basket, standing on the floor, burns with a total energy release rate of ≈100 kW. Calculate the mass flow in the plume at heights 2 m and 4 m above the floor. Also calculate the plume temperatures at these heights.

SUGGESTED SOLUTION

Although we have assumed a point source of heat when deriving Eq. (4.19) and (4.20), a wastepaper basket is of small enough dimensions, in relation to the heights at which we wish to investigate plume properties, to give us approximate answers. However, realistically we must take into account the radiant part of the energy release rate, assuming that roughly 30% of the total energy is lost by radiation from the flames. The rest of the energy (the convective energy release rate) creates the buoyancy, which drives the plume flows. We shall therefore use $\dot{Q} = 0.7 \cdot 100 = 70$ kW in our calculations.

We shall use the following approximate ambient air data:

$T_\infty = 293$ K

$\rho_\infty = 1.2$ kg/m³

$c_p = 1.0$ kJ/(kg K)

$g = 9.81$ m/s²

Mass flow at the height of 2 m: Using Eq. (4.19) we find

$$\dot{m}_p = 0.20\left(\frac{1.2^2 \cdot 9.81}{1.0 \cdot 293}\right)^{1/3} 70^{1/3} \cdot 2^{5/3} = 0.95 \text{ kg/s}$$

Mass flow at the height of 4 m: Choosing $z = 4$ in the above equation gives a mass flow of 3 kg/s. So, from the height of 2 m to the height of 4 m, the plume mass flow has increased threefold.

Plume temperature at the height of 2 m: Using Eq. (4.20) we find

$$\Delta T = 5.0\left(\frac{293}{9.81 \cdot 1.0^2 \cdot 1.2^2}\right)^{1/3} 70^{2/3} \cdot 2^{-5/3} = 73 \text{ K}$$

So, at 2 m height the plume temperature is 73 + 20 = 93°C, or 73 + 293 = 366 K.

Plume temperature at the height of 4 m: Choosing $z = 4$ in the above equation gives a temperature difference of 23°. The plume temperature is then 43°C or 316 K. So, the plume temperature has decreased by 50° from the height of 2 m to 4 m, since a large amount of ambient air has entrained and cooled the plume over this height.

When using the above equations it is important to keep in mind the assumptions made and listed in Section 4.3.1. The effects produced by fuel sources with complex geometries, where the solid fuel is distributed in the vertical and horizontal directions, are not taken into account by the above equations due to the point source assumption. The weak plume assumption means that the equations can only be used for plume properties at a reasonable distance above the flames.

In the following section we relax some of these assumptions, give somewhat different expressions, and show how these compare to experiments.

4.4   PLUME EQUATIONS BASED ON EXPERIMENTS

In the previous section we derived expressions for the plume properties from fundamental principles, incorporating several restricting assumptions in order to achieve analytical solutions. In this section we start by discussing the so-called Zukoski plume, which is largely identical to the ideal plume, and compare calculated plume mass flows with measured plume mass flows. We then move on to the so-called Heskestad plume and the McCaffrey plume, where the point source and the weak plume assumptions are relaxed. Finally, we consider the Thomas plume, which is intended for entrainment in the near field or flame region.

4.4.1   THE ZUKOSKI PLUME

Zukoski[2] carried out experiments where the plume gases were collected in a hood. By controlling the flow rate in the hood exhaust, the hot gas layer height could be kept constant (see Figure 4.8). The flow rate in the hood exhaust was therefore equal to the plume mass flow at the layer interface.

Several experimental measurements on the plume mass flow rate as a function of height and energy release rate were made possible by adjusting the fuel height and energy release rate. Zukoski used the ideal plume theory and adjusted Eq. (4.19) very slightly to get a best fit with the experiments. The resulting plume mass flow equation became

$$\dot{m}_p = 0.21 \left(\frac{\rho_\infty^2 g}{c_p T_\infty} \right)^{1/3} \dot{Q}^{1/3} \cdot z^{5/3} \tag{4.21}$$

which is identical to Eq. (4.19) except for the very small adjustment of the constant from 0.20 to 0.21. The experimental mass flows were compared to the calculated ones for three different burner sizes and different energy release rates. Figure 4.9 gives an example of the results for a relatively large burner diameter (0.5 m). The vertical axis gives the experimental values, $\dot{m}_{exp}$, divided by the ones calculated using Eq. (4.21), $\dot{m}_p$, and the horizontal gives the ratio of the height, z, to the flame height, L. The agreement between calculations and experiments is reasonable, but Eq. (4.21) underestimates the plume mass flow rate somewhat. For smaller burners, the plume mass flow rate is slightly overestimated by the equation.

Equation (4.21) is also commonly shown in the form

$$\dot{m}_p = 0.071 \dot{Q}^{1/3} \cdot z^{5/3} \tag{4.22}$$

where the ambient air properties are assumed to be $T_\infty = 293$ K, $\rho_\infty = 1.1$ kg/m^3, $c_p = 1.0$ kJ/(kg K) and $g = 9.81$ m/s^2.

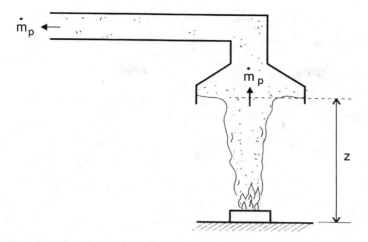

FIGURE 4.8 Plume flow rate experiments.

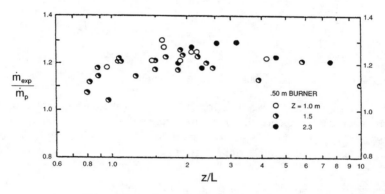

FIGURE 4.9 Experimentally measured and calculated plume mass flows. (From Zukoski[2]. With permission.)

The expressions for plume velocity and plume temperature associated with the Zukoski mass flow equation can be assumed to be identical to Eq. (4.18) and (4.20), since there is hardly any difference in Eq. (4.19) and (4.21).

4.4.2 THE HESKESTAD PLUME

In this section we discuss a set of equations proposed by Heskestad[1] describing plume properties. Three of the main assumptions for the ideal plume will be removed or limited:

1. The point source assumption is relaxed by introducing a "virtual origin" at height z_0 (see Figure 4.10). Also, account will be taken of the fact that some plume properties depend on the convective energy release rate, Q_c.
2. The "top hat" profiles across the plume for velocity and temperature will be replaced by the more realistic Gaussian profiles, as shown in Figure 4.6. We therefore reintroduce the symbols ΔT_0 for the centerline plume temperature and u_0 for the centerline velocity, as described in Section 4.2 and characterized in Figure 4.6.
3. The Boussinesq approximation will be removed so that large density differences can be taken into account. This means that we shall not assume that $\rho_\infty \approx \rho$ in certain equations. Because of the Boussinesq approximation, the ideal plume theory is said to describe weak plumes; the equations discussed in this section are said to describe strong plumes.

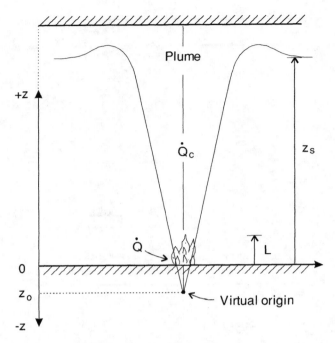

FIGURE 4.10 Some plume properties discussed in this section.

Section 4.2 and Figure 4.6 gave a description of some of the basic plume properties; we now introduce some additional concepts, illustrated in Figure 4.10.

Virtual origin: The virtual origin, denoted z_0, depends on the diameter of the fire source and the total energy released. The virtual origin is given by

$$z_0 = 0.083 \dot{Q}^{2/5} - 1.02D \qquad (4.23)$$

where D is the diameter of the fuel source in [m] and $\dot{Q}$ is the total energy release rate in [kW]. This expression has been derived from experimental data, and represents a "best fit" for pool fires. The value of z_0 may be negative and lie beneath the fuel source, indicating that the area of the fuel source is large compared to the energy being released over that area. For fire sources where the fuel releases high energy over a small area, z_0 may be positive and therefore lie above the fuel source.

Mean flame height: Some of the equations introduced in this section describe plume properties in two regions: above the mean flame height and below the mean flame height. This height, denoted L, was given by Eq. (4.3) in Section 4.2.1 and is reproduced here for completeness:

$$L = 0.235 \dot{Q}^{2/5} - 1.02D \qquad (4.3)$$

using the same notation as in Eq. (4.23).

Energy release rate: The total energy release rate, $\dot{Q}$, is used when calculating the mean flame height and the position of the virtual origin. However, when estimating other plume properties we shall use the convective energy release rate, $\dot{Q}_c$, since this is the part of the energy release rate that causes buoyancy. The energy losses due to radiation from the flames are typically in the order of 20 to 40% of the total energy release rate. The higher of these values are valid for the sootier and more luminous flames, often from fuels that burn with a low combustion efficiency. The convective energy release rate is therefore often in the range $\dot{Q}_c = 0.6\dot{Q}$ to $\dot{Q}_c = 0.8\dot{Q}$.

Plume radius and centerline temperature and velocity: Heskestad took into accou relaxations of the assumptions mentioned above and modified the form of the ideal plume equa He then examined experimental data and found that the plume radius, centerline temperature, and centerline velocity obey the following relations, valid above the flame height:

$$b = 0.12(T_0/T_\infty)^{1/2}(z - z_0) \tag{4.24}$$

$$\Delta T_0 = 9.1 \left(\frac{T_\infty}{gc_p^2\rho_\infty^2} \right)^{1/3} \dot{Q}_c^{2/3}(z - z_0)^{-5/3} \tag{4.25}$$

$$u_0 = 3.4 \left(\frac{g}{c_p T_\infty \rho_\infty} \right)^{1/3} \dot{Q}_c^{1/3}(z - z_0)^{-1/3} \tag{4.26}$$

The convective energy release rate is given in [kW], the temperatures in [K], the heat capacity (of ambient air) in [kJ/(kg K)], and other properties of ambient air are given in SI units.

Note that these plume properties depend on the same powers of energy release rate and height as did the ideal plume. Also, observe that the above equations are valid only above the mean flame height.

It is useful for later comparisons with other plume equations to note that when using the approximate properties of ambient air ($T_\bullet = 2\backslash 93K$, $r_\bullet = 1.2$ kg/m³, $c_p = 1.0$ kJ/(kg K), and $g = 9.81$ m/s²), Eq. (4.25) can be rewritten as

$$\Delta T_0 = 25 \cdot \left(\frac{\dot{Q}_c^{2/5}}{(z - z_0)} \right)^{5/3} \tag{4.25a}$$

and Eq. (4.26) can be rewritten as

$$u_0 = 1.0 \cdot \left(\frac{\dot{Q}_c}{(z - z_0)} \right)^{1/3} \tag{4.26a}$$

Plume mass flow rates: Entrainment into the flame region and into the plume above the flames scale differently, and distinguishing between these is necessary. Heskestad gave the following equations for the plume mass flow rates above and below the mean flame height: For $z > L$, i.e., plume mass flow rate above the flame height,

$$\dot{m}_p = 0.071\dot{Q}_c^{1/3} \cdot (z - z_0)^{5/3} + 1.92 \cdot 10^{-3} \cdot \dot{Q}_c \tag{4.27}$$

For $z < L$, i.e., plume mass flow rate at or below the flame height,

$$\dot{m}_p = 0.0056\dot{Q}_c \frac{z}{L} \tag{4.28}$$

The convective energy release rate is given 5in [kW], the resulting plume mass flow rate in [kg/s].

Calculational procedure:

1. Calculate the position of the virtual origin, z_0, using Eq. (4.23). Observe the use of the total energy release rate, $\dot{Q}$, given in [kW].
2. Calculate the mean flame height, L, using Eq. (4.3). Again, use the total energy release rate, $\dot{Q}$, given in [kW].
3. Determine z, the height at which the plume properties are to be evaluated.
4. Calculate the convective energy release rate, $\dot{Q}_c$, often taken to be 0.7 $\dot{Q}$, and given in [kW]. Select the appropriate equation, depending on which property is sought. Use SI units for the ambient air properties. For c_p, use [kJ/(kg K)] (since the energy release is given in [kW]).

EXAMPLE 4.3

Examples 3.3 and 4.1 discussed a pump breakdown where 20 l of transformer oil were spilled into a sump of 2 m² and ignited. Calculate the plume mass flow rate and plume centerline temperature at heights 2.5 m and 6 m.

SUGGESTED SOLUTION

In Example 3.3 we found $\dot{Q}$ = 1.69 MW and D = 1.6 m.

1. We calculate z_0 from Eq. (4.23), finding $z_0 = 0.083 \cdot (1690)^{2/5} - 1.02 \cdot 1.6 = -0.01$ m. This means that the virtual origin is 1 cm below the floor level. The value is insignificantly different from zero, but we shall carry the number through the calculations to demonstrate its use in the equations.

2. The mean flame height is calculated from Eq. (4.3) as $L = 0.235 \cdot (1690)^{2/5} - 1.02 \cdot 1.6$ = 2.96 m (as we found in Example 4.1).

3. The heights at which we wish to evaluate the plume properties are (a) z = 2.5 m and (b) z = 6 m.

4a. The height z = 2.5 m is within the flame region, since the mean flame height L = 2.96 m. To evaluate the plume flow at z = 2.5 m, we therefore use Eq. (4.28). We must determine the convective part of the energy release rate and assume that it is 70% of the total energy release rate, so that $\dot{Q}_c = 0.7 \cdot 1690 = 1183$ kW. Equation (4.28) gives

$$\dot{m}_p = 0.0056 \cdot 1183 \cdot \frac{2.5}{2.96} = 5.6 \text{ kg/s}$$

We are also asked to evaluate the centerline plume temperature at z = 2.5 m. However, we know that Eq. (4.24)–(4.26) are valid only above the flames so we cannot use them at the height z = 2.5 m. We shall return to this problem in the next section.

4b. The height z = 6 m is well above the flames. To calculate the plume mass flow rate we use Eq. (4.27):

$$\dot{m}_p = 0.071(1183)^{1/3} \cdot (6 + 0.01)^{5/3} + 1.92 \cdot 10^{-3} \cdot 1183 = 17 \text{ kg/s}$$

Note the plus sign in the brackets for heights, since z_0 is a negative value.

The centerline plume temperature can be evaluated using Eq. (4.25) where we shall use the same ambient air data as in Example 4.2, i.e., $T_\infty = 293$ K (= 20°C), $\rho_\infty = 1.2$ kg/m³, $c_p = 1.0$ kJ/(kg K), and $g = 9.81$ m/s², so that

$$\Delta T_0 = 9.1 \left(\frac{293}{9.81 \cdot 1.0^2 \cdot 1.2^2} \right)^{1/3} 1183^{2/3} (6 + 0.01)^{-5/3} = 140°C \text{ or } 140 \text{ K}.$$

The centerline plume temperature is $T_0 = \Delta T_0 + T_\infty$, which gives 160°C or 433 K.

4.4.3 THE MCCAFFREY PLUME

McCaffrey[3] used experimental data and dimensional analysis to arrive at plume relationships for upward velocity and temperature. He divided the plume into three regions, as shown in Figure 4.5: the continuous flame region, the intermittent region, and the plume.

These relationships were of the form

$$\Delta T_0 = \left(\frac{\kappa}{0.9 \cdot \sqrt{2g}} \right)^2 \left(\frac{z}{\dot{Q}^{2/5}} \right)^{2\eta - 1} \cdot T_\infty \tag{4.29}$$

$$u_0 = \kappa \left(\frac{z}{\dot{Q}^{2/5}} \right)^{\eta} \dot{Q}^{1/5} \tag{4.30}$$

The constants η and κ vary depending on the three regions. They are given in Table 4.1.

Figure 4.11 shows results from McCaffrey's experiments using a methane flame, where the centerline plume temperature rise is plotted against $z/\dot{Q}^{2/5}$. Results using Eq. (4.29) are given by the solid line. Equation (4.29) shows a reasonably good agreement with the experiments over all regions. The constants are arrived at by correlations using the total energy release rate, and therefore $\dot{Q}$ is used, and not $\dot{Q}_c$, when calculating the plume properties.

Figure 4.12 shows data from the same experiments, now showing the variation of centerline plume velocity with height, plotted as $u_0/\dot{Q}^{1/5}$ vs. $z/\dot{Q}^{2/5}$. Equation (4.30) is represented by the solid line through the data.

It should be noted that in McCaffrey's plume equations, as in all the plume equations presented in this text, the plume properties are assumed to be independent of fuel, and only dependent on energy release rate, $\dot{Q}$ or $\dot{Q}_c$. Because flames from different fuels have different luminosity, radiation losses vary considerably between fuels, and therefore flame temperature varies from fuel to fuel. These fuel effects are not as marked above the continuous flame region.

Also, for very large fires the flame temperature can be as high as 1200°C, whereas Eq. (4.29) predicts a maximum temperature rise in the flame region as being 800°C. The higher temperatures in the continuous flame region for very large fires are believed to be due to blockage of soot, which hinders radiation losses.

TABLE 4.1
Constants in McCaffrey's Plume Equations

Region	$z/\dot{Q}^{2/5}$ [m/kW²/⁵]	η	κ
Continuous	< 0.08	1/2	6.8 [m¹/²/s]
Intermittent	0.08–0.2	0	1.9 [m/(kW¹/⁵s)]
Plume	> 0.2	−1/3	1.1 [m⁴/⁴/(kW¹/³s)]

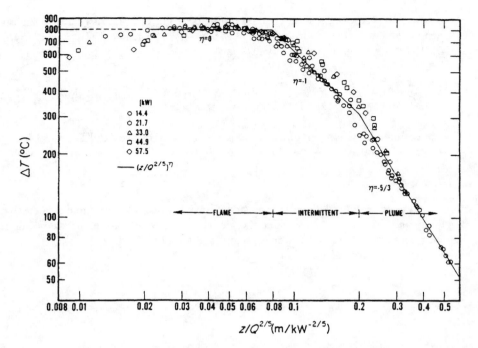

FIGURE 4.11 Variation of centerline plume temperature rise with height and energy release rate. (From Drysdale[11]. With permission.)

EXAMPLE 4.4

In Example 4.3 we could not estimate the flame temperatures since Eq. (4.24)–(4.26) are not valid in the flame region. Use Eq. (4.29) to calculate the centerline plume temperatures at heights $z = 2.5$ m and $z = 6$ m, as specified in Example 4.3.

SUGGESTED SOLUTION

The constants arrived at experimentally by McCaffrey assume the use of the total energy release rate, which in our case is $\dot{Q} = 1690$ kW. First we determine which region the height $z = 2.5$ m falls into by calculating $z/\dot{Q}^{2/5} = 2.5/1690^{2/5} = 0.13$, which indicates the intermittent region of the flame. This gives the values $\eta = 0$ and $\kappa = 1.9$. The centerline plume temperature rise thus becomes

$$\Delta T_0 = \left(\frac{1.9}{0.9 \cdot \sqrt{2 \cdot 9.81}}\right)^2 \left(\frac{2.5}{1690^{2/5}}\right) \cdot 293 = 520 \text{ K or } 520°C .$$

The centerline plume temperature in the intermittent region is therefore $T_0 = 540°C$. Setting $z = 6$ m in the above calculations, we find that $z/\dot{Q}^{2/5} = 6/1690^{2/5} = 0.3$, indicating the plume region. This gives the values $\eta = -1/3$ and $\kappa = 1.1$. The centerline plume temperature rise becomes

$$\Delta T_0 = \left(\frac{1.1}{0.9 \cdot \sqrt{2 \cdot 9.81}}\right)^2 \left(\frac{6}{1690^{2/5}}\right)^{-5/3} \cdot 293 = 160 \text{ K}$$

This gives $T_0 = 180°C$, which compares reasonably with the earlier calculated value of $160°C$ in Example 4.3.

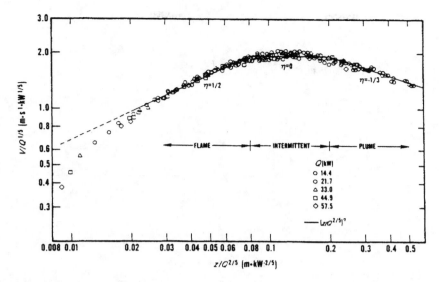

FIGURE 4.12 Variation of centerline plume velocity with height and energy release rate. (From Drysdale[11]. With permission.)

Comparison of the McCaffrey and Heskestad plume equations: To compare the McCaffrey and Heskestad plume equations we note that, for the plume region, Eq. (4.29) can be rewritten (using $T_\infty = 293K$ and $g = 9.81$ m/s^2) as

$$\text{Plume region } \Delta T_0 = 22.3 \cdot \left(\frac{\dot{Q}^{2/5}}{z} \right)^{5/3} \tag{4.29a}$$

Similarly, Eq. (4.30) can be rewritten, for the plume region, as

$$\text{Plume region } u_0 = 1.1 \cdot \left(\frac{\dot{Q}}{z} \right)^{1/3} \tag{4.30a}$$

We note that the exponential constants are the same for the plume temperature equations (4.25a) and (4.29a) and for the plume velocity equations (4.26a) and (4.30a). To allow a more direct comparison, we assume that $z_0 = 0$ and that $\dot{Q}_c = 0.7 \cdot \dot{Q}$ in the Heskestad equations. The constant in Eq. (4.25a) should then be multiplied by a factor $0.7^{2/3}$ ($= 0.79$), reducing this constant from 25 to 19.7. For Eq. (4.26a), the factor of 1.0 should be multiplied by $0.7^{1/3}$ ($= 0.89$), reducing this constant from 1.0 to 0.89.

So, for both plume temperatures and plume velocities, the McCaffrey equations will result in values roughly 10% higher than those given by the Heskestad equations.

4.4.4 THE THOMAS PLUME

The experimental data on which the above plume equations are based did not include experiments where the mean flame height, L, was significantly less than the fuel source diameter, D. Thomas et al. found that in the continuous flame region, or in the near field, the plume mass flow rate was more or less independent of the energy release rate and more a function of the perimeter of the fire, P, and the height above the fire source, z.[4] This has been found to be particularly valid for fires where the mean flame height is considerably smaller than the diameter. The Thomas plume mass flow rate equation is written as

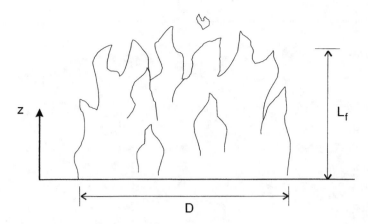

FIGURE 4.13 Characteristic sketch of the Thomas plume.

$$\dot{m}_p = 0.188 \cdot P \cdot z^{3/2} \qquad (4.31)$$

where P is the fire perimeter ($= \pi D$) in [m] and z is the height, in [m], at which the plume mass flow rate, in [kg/s], is to be evaluated. Figure 4.13 shows a characteristic sketch of the Thomas plume.

Note that the plume shape is no longer assumed to be conical, but rather cylindrical. This is typical for larger fires, where the flame height tends to be lower than the fire diameter. Figure 4.14 shows experimental data where the burner diameter is large compared to the energy release rate or, in other words, where diameter is large compared to the mean flame height. Using Eq. (4.3) we find that the flame height to diameter ratio, in Figure 4.14, is in the interval $0.28 < L/D < 1.44$.

Equation (4.31) is represented by the straight line through the data. Noting that, for circular sources, the fire perimeter can be written in terms of the diameter as $P = \pi D$, Eq. (4.31) can then be written

$$\dot{m}_p = 0.59 \cdot D \cdot z^{3/2} \qquad (4.31a)$$

Observe that the mass flow rates predicted by Eq. (4.31) are only valid up to the flame tip, but prediction of mass flow rates above this height have also been found to agree well with data. The equation is especially useful for cases where $L/D < 1$ and for cases where the fire source is noncircular and P is the perimeter of the source.

EXAMPLE 4.5

Consider a 3-m diameter pool of burning methanol. What is the plume mass flow at the mean flame height?

SUGGESTED SOLUTION

Using the data in Table 3.3 and Eq. (3.5), we find that the energy release rate is $\dot{Q} = 0.017$ $\cdot \pi \dfrac{3^2}{4} \cdot 20 \cdot 0.9 = 2.16$ MW, assuming high combustion efficiency ($= 0.9$) for this relatively nonsooty alcohol flame. The mean flame height, given by Eq. (4.3), is $L = 0.235 \cdot (2160)^{2/5} - 1.02 \cdot 3 = 2$ m. The plume mass flow rate from Eq. (4.31a) is $\dot{m}_p = 0.59 \cdot 3 \cdot 2^{3/2} = 5$ kg/s.

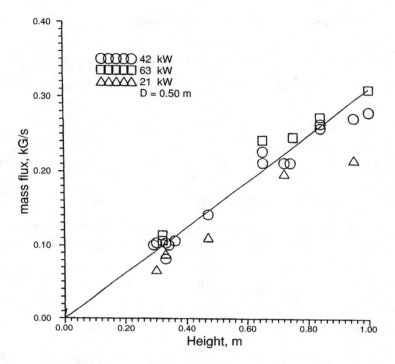

FIGURE 4.14 Plume mass flow rate for low values of *L/D*. (Adapted from Zukoski[5]. With permission.)

4.5 LINE PLUMES AND BOUNDED PLUMES

In the above sections we discussed the axisymmetric, unbounded fire plume, where the fuel source has been assumed to be circular and the plume has been assumed to be free from the interference of walls and other surfaces. Here we discuss line plumes and bounded plumes. Little data is available on such plumes, but we shall in the following give a few expressions for plumes of this type. The special case where the plume flow impinges on a ceiling will be dealt with in Section 4.6.

4.5.1 WALL AND CORNER INTERACTIONS WITH PLUMES

Zukoski discusses studies made where fire sources are placed near or flush with walls and corners.[5] Figure 4.15 shows a characteristic sketch of three cases studied. Experimenters reported that when a circular burner was placed with one edge tangent to a vertical wall (Figure 4.15a), there was very little influence on plume geometry and plume entrainment up to a height of three times the burner diameter. However, when a semicircular burner was placed with its straight edge against a wall (Figure 4.15b), the plume was attached to the wall and developed as a half plume with plume properties closely approximating these for a full circular burner of twice the energy release rate.

The plume mass flow can therefore be calculated to be half of the plume mass flow of a fire with twice the energy release rate. We can use the simple Zukoski plume mass flow equation (4.22) to develop a relationship for the case in Figure 4.15b by writing

$$\dot{m}_{p,wall} = \frac{1}{2} 0.071 (2\dot{Q})^{1/3} z^{5/3}$$

which simplifies to

$$\dot{m}_{p,wall} = 0.045 \dot{Q}^{1/3} z^{5/3} \tag{4.32}$$

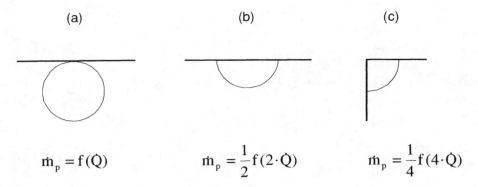

$$\dot{m}_p = f(\dot{Q}) \qquad\qquad \dot{m}_p = \frac{1}{2}f(2 \cdot \dot{Q}) \qquad\qquad \dot{m}_p = \frac{1}{4}f(4 \cdot \dot{Q})$$

FIGURE 4.15 Fire sources near walls and corners.

Similarly, for the case of the corner (Figure 4.15c), the plume mass flow is roughly one quarter of the flow from an unbounded fire with four times the energy release rate. Again, using Eq. (4.22) we find

$$\dot{m}_{p,corner} = \frac{1}{4}0.071\left(4\dot{Q}\right)^{1/3} z^{5/3}$$

which simplifies to

$$\dot{m}_{p,corner} = 0.028\dot{Q}^{1/3}z^{5/3} \qquad\qquad\qquad (4.33)$$

It should be noted that Eqs. (4.32) and (4.33) are approximations. Further, as shown for the cases in Figures 4.15b and 4.15c, the fuel source must be placed flush with the wall or in the corner for the equations to be valid. Williamson et al. carried out experiments where a 0.3 by 0.3 m burner was placed flush in a corner, 5 cm from the corner and 10 cm from the corner.[12] They found that with those relatively small stand-off distances, the flame did not attach to the wall. This indicates that in the flame region, the fuel source does not have to be moved far from the wall or corners to exhibit plume properties similar to the plume equations reported in Section 4.3.

Mean flame heights in the cases above will be longer than for an unbound axisymmetrical plume, since less air is entrained and the fuel must travel a longer distance to become fully combusted.

4.5.2 LINE SOURCE PLUMES

Figure 4.16 shows results, by Hasemi and Nishihata,[6] from flame height measurements vs. energy release rate. The burner dimensions range from a square burner to a line burner where the ratio of the longer side to the shorter side is 10. The burner dimensions are given in [m] where the shorter side length is denoted A and the longer side length is denoted B.

The following equation has been suggested for line sources, where the longer side, B, is more than three times the shorter side:

$$L = 0.035\left(\frac{\dot{Q}}{B}\right)^{2/3} \qquad\qquad\qquad (4.34)$$

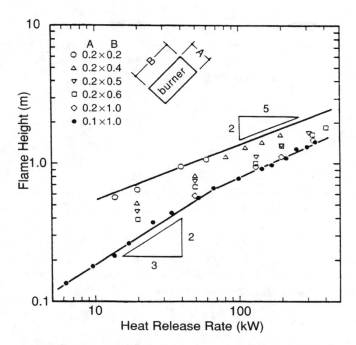

FIGURE 4.16 Flame heights vs. energy release rate for various burner geometries. (Adapted from Hasemi and Nishihata[6].)

The topmost solid line in Figure 4.16 represents the rectangular source flame heights as calculated using Eq. (4.3). The lower solid line in Figure 4.16 represents the line burner case where $A = 0.1$ m and $B = 1.0$ m, as calculated by Eq. (4.34). The calculated line represents the data well except for the case of very high energy release rates. The dotted line between these represents the case where $B = 0.6$ m as calculated by Eq. (4.34). Again, the calculation fits the appropriate data ($A = 0.2$, $B = 0.6$) well, except for the case of very high energy release rates.

Experimental data on plume mass flow rates for line plumes is scarce. The following equation has been suggested for line sources, where the longer side, B, is more than three times the shorter side:

$$\dot{m}_p = 0.21 \left(\frac{\dot{Q}}{B} \right)^{2/3} z \tag{4.35}$$

Equation (4.35) will give approximate plume mass flows above the flames, in the range $L < z < 5B$. For heights larger than $5B$, the plume properties will resemble the axisymmetric plumes, and the equations given in Section 4.4 will be valid.

4.6 CEILING JETS

When a plume impinges on a ceiling, hot gases spread out radially to form a so-called *ceiling jet*. Most fire detection and fire suppression devices are placed near the ceiling surfaces. The magnitude and temperature of the ceiling jet flow determine to a great degree the response time of this equipment, and it is therefore important to quantify these ceiling jet parameters. We will, in this section, seek relationships for the velocities and temperatures in the ceiling jet.

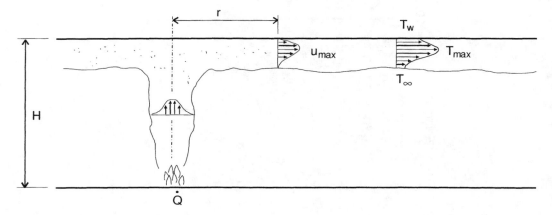

FIGURE 4.17 An idealization of the ceiling jet flow beneath a ceiling.

Further, if the flame itself impinges on the ceiling, we wish to calculate the length of the flame extension along the ceiling. Such information can be useful when estimating radiative heat transfer to objects, as well as for other engineering calculations.

4.6.1 CEILING JET TEMPERATURES AND VELOCITIES

Figure 4.17 shows an idealization of the ceiling jet flow beneath an unconfined (large area) ceiling. The distance from the plume centerline axis to the point at which we wish to examine the ceiling jet properties is denoted r and is given in [m]. The height from the fuel source to the ceiling is denoted H, given in [m].

Initially, the depth of the hot layer under the ceiling will typically be in the range 5 to 12% of the distance from the fuel source to the ceiling. The ceiling jet velocity profile will be of the form given in Figure 4.17, with zero velocity at both boundaries. The maximum flow velocity and the maximum temperature occur relatively close to the ceiling, typically around 1% of H, the distance from the source to the ceiling.

The temperature profile will be bounded by the ceiling temperature, denoted T_w, on one side and by the ambient temperature, denoted T_∞, on the other, as seen in Figure 4.17. As the ceiling jet moves radially outward, ambient air is entrained at the layer interface, cooling the flow. The ceiling jet is also cooled by heat transfer to the ceiling.

Alpert provided simple correlation equations, based on a generalized theory and experimental data, for predicting the maximum temperature and velocity at a given position, r, in a ceiling jet.[7] The experimental data were collected during test burns of various types of solid and liquid fuels with energy release rates ranging from roughly 500 kW to 100 MW under ceiling heights ranging from 4.6 to 15.5 m. The equations are given in two regions: a region close to the plume where the properties are independent of r, and a region further away from the plume where account must be taken of r.

The equation for ceiling jet maximum temperature for $r/H < 0.18$ is

$$T_{max} - T_\infty = \frac{16.9\dot{Q}^{2/3}}{H^{5/3}} \qquad (4.36)$$

and for $r/H > 0.18$ is

$$T_{max} - T_\infty = \frac{5.38(\dot{Q}/r)^{2/3}}{H} \qquad (4.37)$$

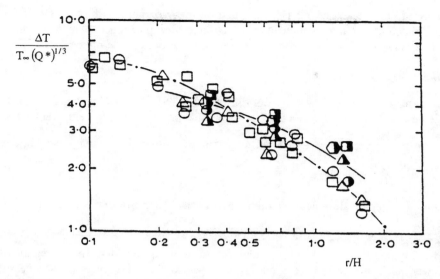

FIGURE 4.18   Dimensionless maximum ceiling jet temperature increase. (Adapted from Alpert[7].)

The equation for ceiling jet maximum velocity is for $r/H < 0.15$ is

$$u_{max} = 0.96 \left(\frac{\dot{Q}}{H} \right)^{1/3} \tag{4.38}$$

and for $r/H > 0.15$ is

$$u_{max} = \frac{0.195 \dot{Q}^{1/3} H^{1/2}}{r^{5/6}} \tag{4.39}$$

Figure 4.18 gives some of the data for the temperature correlation, plotted as a rather complex dimensionless temperature vs. r/H. The solid line through the data points represents Eq. (4.36). Observe that in the near plume region ($r/H < 0.18$), the ceiling jet temperatures and velocities are really a measure of the plume properties at ceiling level. Equation (4.36) can therefore be compared to the Heskestad and McCaffrey equations (4.25a) and (4.29a). These equations are of exactly the same form, but have a slightly higher pre-exponential constant, giving somewhat higher values of maximum plume temperature at ceiling level. Similarly, Eq. (4.38) can be compared to Eq. (4.26a) and (4.30a), which again are of exactly the same form, but give slightly higher values of maximum plume velocity at ceiling level.

It should be noted that, even though it is the convective energy release rate, $\dot{Q}_c$, that causes upward movement, Eq. (4.36)–(4.39) use the total energy release rate, $\dot{Q}$, since the experimental data were correlated using the total energy release rate.

All the units in Eq. (4.36)–(4.39) are SI units; note that the energy release rate is in [kW], as is customary when discussing plume flows.

EXAMPLE 4.6

We turn once again to Example 3.3, where an oilspill fire has $\dot{Q} = 1.69$ MW and $D = 1.6$ m. Assume the fire occurs in an industrial building of 6 m ceiling height and that detection devices at ceiling level are distributed in such a way that the maximum distance from the plume axis to the nearest device is 7 m. Calculate the maximum ceiling jet velocity and the maximum ceiling jet temperature at a detection device positioned 6 m from the plume centerline. Also, calculate the ceiling jet temperature at or near the plume axis.

SUGGESTED SOLUTION

For the case where $r = 7$ m, we use Eq. (4.37) and (4.39), since r/H is larger than 0.18. We assume an ambient temperature of 20°C, so the maximum ceiling jet temperature becomes $T_{max} = 20 + \dfrac{5.38 \cdot (1690/7)^{2/3}}{6} = 54°C$. The maximum ceiling jet velocity becomes $u_{max} = \dfrac{0.195 \cdot 1690^{1/3} \cdot 6^{1/2}}{7^{5/6}} = 1.1$ m/s.

The ceiling jet temperature near the plume axis is calculated from Eq. (4.36) as $T_{max} = 20 + \dfrac{16.9 \cdot 1690^{2/3}}{6^{5/3}} = 141°C$. This can be compared to the value of 180°C calculated using the Heskestad plume Eq. (4.25) and the value of 160°C using the McCaffrey equation (4.29). The lower value arrived at here is due to increased mixing of ambient air and heat losses when the plume impinges on the ceiling.

The above correlations apply only during times after fire ignition when the ceiling flow can be considered unconfined. When the flows have hit the side walls and a considerable hot gas layer has started to form and descend, the validity of the equations starts to diminish.

It should also be noted that the above equations are valid for steady fires only, where the energy release rate is constant. Calculational procedures for transient fire growth are also available through the use of computer programs. The special transient case, where the energy release rate is assumed to follow a t^2 growth curve, can also be solved by hand calculation methods; for details see Evans.[8]

The properties calculated here are the maximum temperature and maximum velocity in the ceiling jet. These occur very close to the ceiling, or at around 1% of the distance from the fuel to the ceiling. Detection and suppression devices placed outside of this range will experience cooler temperatures and lower velocities than predicted by the above equations, so that response time could be drastically increased.

4.6.2 FLAME EXTENSIONS UNDER CEILINGS

When a flame impinges on an unconfined ceiling, the unburnt gases will spread out radially and entrain air for combustion, and a circular flame will be established under the ceiling. We wish to calculate the length of the flame extension along the ceiling and use this information when estimating radiative heat transfer to objects in the enclosure.

Figure 4.19 shows a sketch of the scenario under consideration. Here, r_f is the radial flame extension in [m], L is the free-burning mean flame height as calculated by Eq. (4.3), D is the diameter of the fire source in [m], and H is the height from the fuel source to the ceiling in [m].

You and Faeth carried out experiments for this scenario and analyzed the data, resulting in the following approximate expression for the radial flame extension divided by fuel source diameter:[9]

$$\frac{r_f}{D} = 0.5\left(\frac{L - H}{D}\right)^{0.96} \tag{4.40}$$

This is roughly equivalent to saying that the radial flame extension is equal to half the free flame height part that would extend above the ceiling, or $r_f < 0.5(L - H)$. Equation (4.40) is intended

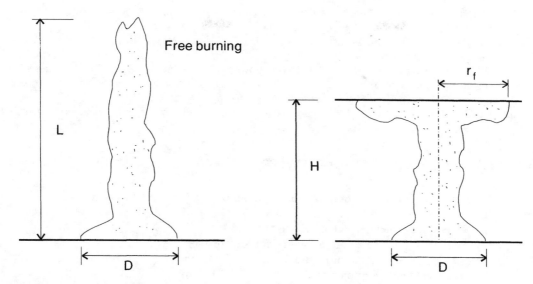

FIGURE 4.19 Flame extensions under ceilings.

only for very rough estimates, since the experiments it is based on were carried out with small flames, small heights, and low energy release rates.

Heskestad and Hamada[13] carried out six experiments for larger energy release rates (93–760 kW), which gave an average flame extension

$$r_f = 0.95(L - H) \tag{4.41}$$

The constant 0.95 was an average of values in the range 0.88–1.05. Equation (4.41) may provide more realistic estimates than Eq. (4.40) for larger flames, but more experimental evidence is needed.

REFERENCES

1. Heskestad, G., "Fire Plumes," *SFPE Handbook of Fire Protection Engineering*, 2nd ed., National Fire Protection Association, Quincy, MA, 1995.
2. Zukoski, E.E., Kubota, T., and Cetegen, B., "Entrainment in Fire Plumes," *Fire Safety Journal*, Vol. 3, pp. 107–121, 1980.
3. McCaffrey, B.J., *Purely Buoyant Diffusion Flames: Some Experimental Results*, NBSIR 79-1910, National Bureau of Standards, 1979.
4. Thomas, P.H., Hinkley, P.L., Theobald, C.R., and Simms, D.L., "Investigations into the Flow of Hot Gases in Roof Venting," Fire Research Technical Paper No. 7, HMSO, London, 1963.
5. Zukoski, E.E., "Properties of Fire Plumes," in *Combustion Fundamentals of Fire*, Cox, G., Ed., Academic Press, London, 1995.
6. Hasemi, Y. and Nishihata, M., "Fuel Shape Effects on the Deterministic Properties of Turbulent Diffusion flames," International Association of Fire Safety Science, Proceedings of the Second International Symposium, Hemisphere Publishing, pp. 275–284, Washington, D.C., 1989.
7. Alpert, R.L., "Calculation of response time of ceiling-mounted fire detectors," *Fire Technology*, Vol. 8, pp. 181–195, 1972.
8. Evans, D.D., "Ceiling Jet Flows," *SFPE Handbook of Fire Protection Engineering*, 2nd ed., National Fire Protection Association, Quincy, MA, 1995.
9. You, H.Z. and Faeth, G.M., "An Investigation on Fire Impingement on a Horizontal Ceiling," NBS-GCR-81-304, National Bureau of Standards, CFR, Gaithersburg, MD, 1981.

10. McCaffrey, B., "Flame Height," *SFPE Handbook of Fire Protection Engineering*, 2nd ed., National Fire Protection Association, Quincy, MA, 1995.
11. Drysdale, D., "Diffusion Flames and Fire Plumes," *An Introduction to Fire Dynamics*, Wiley-Interscience, New York, 1992.
12. Williamson, R.B., Revenaugh, A., and Mowrer, F.W., "Ignition Sources in Room Fire Tests and Some Implications for Flame Spread Evaluation," International Association of Fire Safety Science, Proceedings of the Third International Symposium, New York, pp. 657–666, 1991.
13. Heskestad, G. and Hamada, T., "Ceiling Jets of Strong Fire Plumes," *Fire Safety Journal*, Vol. 21, pp. 69–82, 1993.

PROBLEMS AND SUGGESTED ANSWERS

4.1 Estimate the flame height from the following burning objects (see Problem 3.3):
   (a) 2 m² transformer oil fire
   (b) 2 m² wood pallets, pallet stack height is 1.5 m
   (c) mail bags, full, stored 1.5 m high, floor area of fuel is 2 m²
   (d) the Christmas tree fire in Figure 3.12 (test 17), diameter 1 m
   (e) sofa in Figure 3.8, projected floor area of fuel is 2 m²

   Suggested answer: (a) 3.0 m; (b) 6.9 m; (c) 1.8 m; (d) 1.8 m; (e) 4.1 m

4.2 Calculate the mass flow rate in the plume at the height 5 m from the base of the fire for the fires in Problem 4.1, using the Heskestad plume equations. Assume for all of these that $\dot{Q}_c = 0.7\dot{Q}$.

   Suggested answer: (a) 13.2 kg/s; (b) 22.6 kg/s; (c) 10.8 kg/s; (d) 8.0 kg/s; (e) 15.4 kg/s.

4.3 In a high atrium a fire occurs with an energy release rate of 1500 kW on a floor area of 2 m². The mechanical smoke ventilation starts by means of a heat detector at the ceiling which is activated if the temperature rise is greater than $\Delta T = 40°C$. How high can the atrium be if
   (a) ΔT is taken to be the value at the central axis of the plume?
   (b) ΔT is taken to be the average temperature in the plume?

   Suggested answer: (a) Using Heskestad's plume equations, $z = 12.1$ m. (b) Using a simple energy balance, $\dot{Q}_c = \dot{m}_p c_p \Delta T$, we find that $\dot{m}_p = 26.3$ kg/s at this height. Using Heskestad's plume equation, $z = 8.2$ m.

4.4 Show, using Eq. (4.10) and (4.13), that the value of C_2 in the expression $u = C_2 z^{-1/3}$ is

$$C_2 = \left(\frac{25}{48\alpha^2} \cdot \frac{\dot{Q}g}{\pi c_p T_\infty \rho_\infty} \right)^{1/3}$$

4.5 Show, using the results from Problem 4.4 and assuming that $\alpha = 0.15$, that the mass flow rate in the plume can be written as

$$\dot{m}_p = 0.20 \left(\frac{\rho_\infty^2 g}{c_p T_\infty} \right)^{1/3} \dot{Q}^{1/3} \cdot z^{5/3}$$

4.6 A pool of methanol with area 0.1 m² burns. Assume 100% combustion efficiency and no heat loss due to radiation from the flame. Using the ideal plume, draw a graph of following variables as a function of z:
(a) plume radius
(b) plume velocity
(c) mass flow rate
(d) temperature rise

Suggested answer: (a) $b = 0.18\ z$; (b) $u = 1.9\ z^{-1/3}$; (c) $\dot{m}_p = 0.24\ z^{5/3}$; (d) $DT = 144\ z^{-5/3}$

4.7 Solve Problem 4.6 again, now using the Heskestad plume equations. Assume 100% combustion efficiency and 10% losses due to radiation from the flame.

Suggested answer: Using $\dot{Q} = 34$ kW and $\dot{Q}_c = 30.6$ kW, we get
(a) $b = 0.12[(245(z + 0.024)^{-5/3} + 293)/293]^{1/2}(z + 0.024)$
(b) $u_0 = 3.2(z + 0.024)^{-1/3}$,
(c) $\dot{m}_p = 0.29z$ (for $z < 0.6$), $\dot{m}_p = 0.22(z + 0.024)^{5/3} + 0.059$ (for $z > 0.6$)
(d) $DT_0 = 245(z + 0.024)^{-5/3}$

Observe that the equations for u_0 and DT_0 are valid only above the flame height.

4.8 A 2 m² liquid pool of gasoline burns on the floor in a large industrial hall. By ventilating the smoke through the roof, a smoke layer interface is established 6 m above the floor.
(a) Estimate the plume mass flow rate into the smoke layer.
(b) Estimate the maximum plume temperature at the interface height.
(c) Estimate the average plume temperature at the interface height. Assume the combustion to be 70% efficient and the flame radiative losses to be 30%.

Suggested answer: With $\dot{Q} = 3240$ kW and $\dot{Q} = 2270$ kW we get
(a) $\dot{m}_p = 20$ kg/s; (b) $DT_0 = 250°C$; $T_{max} = 270°C$; (c) $T_{av} = 130°C$

4.9 A pump in a large industrial hall containing 30 l gasoline breaks down, causing the gasoline to spill into a 3 m² sump surrounding the pump. The liquid pool is ignited by sparks. Assume the combustion to be 70% effective and the heat loss due to radiation from the flame to be 30%. The ambient temperature in the hall is 27°C. A pipeline of diameter $d = 0.15$ m runs horizontally at a 6 m height above the pump. To calculate the heat transfer to the cylindrical pipe one needs to know the convective heat transfer coefficient at the cable surface.
(a) Calculate the plume centerline temperature at 6 m height.
(b) Calculate the plume centerline velocity at this height.
(c) Calculate the convective heat transfer coefficient $h_c = Nu \cdot k/d$, where $Nu = 0.024\ Re^{0.805}$ and $Re = u_0 \cdot d/n$. The values of k and n for air are evaluated at an average film temperature $T_f = 450K$ giving $k = 0.037$ W/(m K) and $n = 31.7 \cdot 10^{-6}$ m²/s.
(d) Estimate the duration of the fire.

Suggested answer: (a) $T_0 = 370°C$; (b) $u_0 = 8.8$ m/s; (c) $h_c = 31$ W/(m² K); (d) about 140 s

4.10 A 2 m diameter pool of transformer oil burns. Calculate the plume mass flow rate at a height of 3 m using the Thomas plume. Compare the answer to the Heskestad plume.

Suggested answer: Assuming 60% combustion efficiency and 30% radiation losses from the flame we get: Thomas plume, $\dot{m}_p = 6.1$ kg/s, Heskestad plume $\dot{m}_p = 8.9$ kg/s

4.11 A wastepaper basket positioned flush into a corner of a room releases 100 kW. Give a rough estimate of the mass flow rate in the plume at a height of 2 m above the fuel source.

Suggested answer: 0.4 kg/s

4.12 In a 6 m high room a fire of 2 m² floor area releases 2 MW. Calculate the ceiling jet temperatures and ceiling jet velocities at a distance 2.5 m from the plume central axis. Assume $T_\infty = 20°C$.

Suggested answer: $T_{max} = 97°C$, $u_{max} = 2.8$ m/s

5 Pressure Profiles and Vent Flows for Well-Ventilated Enclosures

The classical representation of a fire in a room or building represents the structure with an opening, such as a door or a window, to the ambient surroundings or to an adjacent compartment. We shall call such an opening a *vent*. The flow of gases in and out through a vent is controlled by the pressure difference across the vent. In order to estimate any of the environmental consequences of a fire in an enclosure, it is necessary to quantify the mass of hot gases exiting and the mass of colder gases (air) entering the enclosure. This gives information on the mass and energy balance and therefore allows calculation of such environmental consequences as hot gas temperature and smoke layer height. This chapter derives engineering equations used to calculate pressure differences across vents, as well as equations for calculating the mass flow of gases in and out through vents, for several common enclosure fire scenarios. Vent flows for nearly closed compartments or those with very small vents and leakages, where the pressure build-up is mainly due to dynamic pressures, will be dealt with in Chapter 8.

CONTENTS

5.1   Terminology 81
5.2   Introduction 82
    5.2.1   Some Characteristics of Pressure 83
    5.2.2   Application to a Simple Example..86
    5.2.3   Mass Flow Rate through Vents...89
    5.2.4   Summary 91
5.3   Examples of Pressure Profiles in a Fire Room with a Vent 92
5.4   The Well-Mixed Case 95
    5.4.1   Mass Flow Rates and the Height of the Neutral Plane.......................................95
    5.4.2   A Simplified Expression for the Mass Flow Rate in through an Opening 99
    5.4.3   Taking into Account the Mass Produced in the Room (the Burning Rate) 101
    5.4.4   Summary 103
5.5   The Stratified Case...103
    5.5.1   Mass Flow Rates into and out of the Vent..103
    5.5.2   Special Case: Mass Flow out through a Ceiling Vent 107
References 111
Problems and Suggested Answers 111

5.1   TERMINOLOGY

Flow coefficient — When a pressure difference exists across an opening, fluid will be pushed through. In practical applications the fluid is not ideal (it is not friction-free, incompressible, and isothermal) and there will be some resistance to the flow. This resistance is taken into account by a flow coefficient, which basically states that only a part of the opening will allow the fluid to flow effectively through it. The flow coefficient is denoted C_d and takes the value 0.7 for most of the applications in this text.

Hydrodynamic pressure difference — The pressure difference caused by a static pressure head across an opening at a given height, where a volume of fluid goes from being at rest to having a velocity *v*.

Hydrostatic pressure difference — The pressure difference across an opening, at a given height, due to the weight of a column of gas or liquid.

Neutral plane — When hot and cold gases are at either side of an opening, the hot gases will flow out through the upper part of the opening and the cold gases will flow in through its lower part. The flow is caused by the pressure difference across the opening, which in turn is caused by temperature difference of the gases. At some height above a given reference height, the pressure difference will be zero, and this height is called the *height of the neutral plane*. Hot gases will flow out above the neutral plane and cold gases flow in below it.

Stratified case — When a fire compartment is assumed to be divided into two zones, an upper hot zone and a lower cold zone (assumed to have the same properties as the outside ambient air), a pressure profile will be established across the opening to the outside. The pressure profile will be identified with the "stratified" case.

Well-mixed case — An enclosure has an opening to the outside, ambient air. When the gas in the enclosure is assumed to have a uniform temperature over its entire volume (to be well mixed), the pressure profile across the opening is identified with "the well-mixed" case.

5.2   INTRODUCTION

The flow of fluids always occurs from a place of high pressure to a place of lower pressure. The flow of gases from an enclosure containing fire depends on the pressure differences between the enclosure and the surroundings. A quantification of the pressure differences is therefore of fundamental importance.

It is convenient to split the driving pressure differences into two categories. In the first category the pressure differences are caused by the fire. In the second category the pressure differences are caused by normal conditions that are either always present in a building or are created by differences in conditions inside and outside of the building.

Pressure differences caused by the fire are mainly of two types:

1. Pressure differences caused by thermal expansion of the enclosure gases. When the gases heat up they expand. In a very tightly closed room this will cause a rise in pressure. However, in nearly all buildings there are small leakage areas that cause such pressure build-up to be negligible, and this phenomenon is usually ignored in engineering calculations. In certain types of tightly closed rooms, such as an engine room in a ship, this pressure build-up may be very substantial and must be accounted for. In Chapter 8, when discussing smoke filling a hermetically closed room, we shall give expressions for calculating pressures due to prevented thermal expansion.

2. Pressure differences due to buoyancy of the hot gases, or rather, due to density differences between the hot and the cold gases. This is the most common cause of smoke flow in a building containing fire. In this chapter we shall use classical laws of hydraulics to derive equations for calculating the pressure differences and the mass flow caused by these.

Normal pressure differences in a building are mainly of three types:

1. Pressure differences due to density differences inside and outside of the building, or due to the difference in inside and outside temperatures. The pressure differences created are governed by the same laws of hydraulics as those mentioned in type 2 above. The equations that we shall derive and use in this chapter are therefore applicable to this type as well.

TABLE 5.1
Frequently Used Units of Pressure

Pa (= N/m²)	atm	bar	torr (= mmHg at 0°C)
1	$9.869 \cdot 10^{-6}$	$10 \cdot 10^{-6}$	$7.501 \cdot 10^{-3}$
133.3	$1.316 \cdot 10^{-3}$	$1.333 \cdot 10^{-3}$	1
$100 \cdot 10^3$	0.987	1	7750.1
$101.3 \cdot 10^3$	1	1.013	760

2. Pressure differences caused by atmospheric wind flows and wind loading on a building. This can have considerable impact on smoke flow and buoyancy-driven roof venting design. We do not discuss this here; a summary discussion on the topic is given by Kandola in the *SFPE Handbook*.[1]
3. Pressure differences due to mechanical ventilation. In smoke control design, mechanical ventilation is often used to remove smoke from a building. In Chapter 8 we discuss the smoke-filling process and derive equations that take into account mechanical ventilation for smoke removal. Ordinary air-conditioning systems usually provide a flow volume that is low enough for such effects to be ignored, compared to buoyancy-driven flows. However, in certain circumstances, the flow in an air-conditioning system should be accounted for, and special computer models are available for such calculations.

Contents of this Chapter: As seen from the above, the most common cause of smoke flow in a building is density differences between hot and cold gases. We therefore concentrate on types 2 and 3; these cases have a solid foundation in engineering hydraulics. In the remaining introductory section we discuss some of the concepts to be used in this chapter and give a brief revision of the main principles of engineering hydraulics. In order to enhance understanding of the main principles, we apply these equations to a very simple case: an enclosure with relatively narrow openings at the top and bottom. In this case the pressure differences can be assumed to be constant over the opening height; this simplifies the equations and serves as a suitable introductory example.

In Section 5.2 we look closely at types 2 and 3 above and discuss the various pressure profiles that are likely to occur for other types of openings. In Sections 5.3 and 5.4 we look at two of the most common cases in detail and derive engineering equations for flow calculations in each case.

5.2.1 SOME CHARACTERISTICS OF PRESSURE

Units of pressure: The pressure caused by the force 1 Newton [N] on an area of 1 m² is 1 Pascal [Pa]. An ordinary piece of paper resting on a table top causes the pressure of 1 Pa [= N/m²] on the table. A 1 m high column of water (of any surface area) will cause a pressure of 9810 Pa on the same table top. An ordinary door will be difficult to open against a pressure of 100 Pa.

Pressure is measured in many units, here we shall use Pa, which is the same as N/m². Table 5.1 gives conversions to some other frequently used units.

Atmospheric pressure: The standard atmospheric pressure at the earth's surface is assumed to be equal to 1 atm or $101.3 \cdot 10^3$ Pa. This pressure is due to the weight that the air's atmosphere exerts on the earth's surface. The pressure decreases at the rate of roughly 10 Pa/m as the height above the earth's surface increases. The magnitude of the atmospheric pressure varies with respect to where on the earth's surface one is situated, but for the engineering calculations we will perform it is common to assume the reference pressure at some chosen height to be $101.3 \cdot 10^3$ Pa. Since the calculations are based on a pressure difference across a vent, the magnitude of the standard atmospheric pressure does not appear directly in the equations.

Hydrostatic pressure: Consider the container of water depicted in Figure 5.1. At the water level, the pressure is assumed to be 1 atm. The pressure increases as the height decreases, due to the weight of the water, at the rate of 9810 Pa per meter height. The pressure that the water exerts on the table top can be expressed as a pressure difference, in relation to the atmospheric pressure, P_0. This pressure difference, $P - P_0$ or ΔP, can be expressed as the force exerted by the water, $F = Ah\rho g$ on the surface A, so

$$\Delta P = \frac{F}{A} = h\rho g \qquad (5.1)$$

where h represents the height of the water surface above the table top, ρ is the water density, g is the gravitational constant, and A is the water surface area. Equation (5.1) is said to state the *hydrostatic pressure difference*.

In enclosures it is often the difference in temperatures, or densities, between the inside and the outside air that causes air flow in and out of the enclosure. Consider an enclosure with openings at the top and the bottom of an enclosure, as depicted in Figure 5.2. The ambient outside density and temperature are denoted ρ_a and T_a, and the inside gas density and temperature are denoted ρ_g and T_g, where the inside temperature is higher than the outside. We assume that there are heating sources inside the enclosure that keep the temperature constant. There will then be a flow of cold air (T_a) entering the enclosure through the lower opening and a flow of hot air (T_g) exiting through the upper opening. The reason for this is the hydrostatic pressure difference.

The hydrostatic pressure due to the column of cold air is represented by a straight, negatively sloping line, indicating increased pressure as height decreases. Inside the enclosure the hydrostatic pressure is given by a somewhat differently sloping line, since the air there is hotter and less dense. Plotting these two on top of each other results in a diagram such as the one given in Figure 5.2.

Relative to the enclosure there will be a positive pressure difference at the upper opening, ΔP_u, causing air to flow out, and a negative pressure difference at the lower opening, ΔP_l, causing air to be drawn in. Other names for this flow phenomenon in heated buildings are "stack effect" or "chimney effect."

Neutral plane: Since there are no discontinuities possible in the pressure profiles inside or outside the enclosure, the profiles will cross at a certain height, called the *height of the neutral plane*. At this height the pressure differences are zero. The neutral plane can lie anywhere between the two openings in Figure 5.2, and its height is determined by the flow resistance at the openings and the magnitude of the temperature difference. If, for example, the lower opening is large, or has low flow resistance, the neutral plane will be close to the lower opening.

Hydrodynamic pressure: When discussing hydrostatic pressure, we were considering a fluid at rest. In the example given in Figure 5.1 the water in the container was at rest; if we drill a hole at the bottom of the container, the hydro*static* pressure will be converted to a hydro*dynamic* pressure, expressed in terms of the velocity of the water exiting the container. In the example given in Figure 5.2 we considered the hot air inside the enclosure at rest and the cold air outside also at rest. However, in the opening itself, the air is not at rest, and the hydrostatic pressure is converted to hydrodynamic pressure. This hydrodynamic pressure must equal the hydrostatic pressure. By setting up the equations for these two, we will arrive at a relationship between pressure difference, height, density differences, and velocity. This will allow us to calculate the mass flow into and out of the vents. In order to achieve this we shall use the Bernoulli equation.

The Bernoulli equation: By applying the principle of conservation of energy to a flowing fluid, an expression can be derived that gives the theoretical net energy balance of an incompressible fluid at any point along its flow path. This is known as the *Bernoulli equation*. We shall be using the Bernoulli equation to expresses the relationship between the pressure and velocity within a fluid flow.

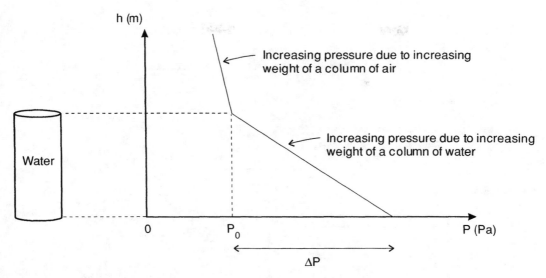

FIGURE 5.1 Hydrostatic pressure difference exerted by a fluid on a surface.

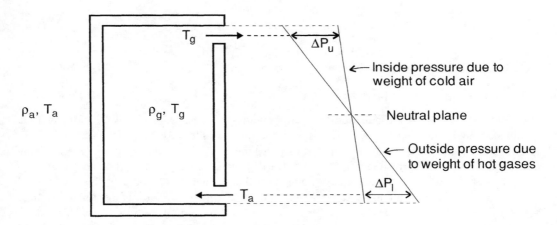

FIGURE 5.2 Hydrostatic pressure differences for a heated enclosure.

Consider Figure 5.3. The Bernoulli equation for the two points in the figure can be expressed as

$$P_1 + \frac{1}{2}v_1^2\rho_1 + h_1\rho_1 g = P_2 + \frac{1}{2}v_2^2\rho_2 + h_2\rho_2 g \tag{5.2}$$

assuming that there are no friction losses within the system. Since we will mainly be looking at the laminar flow of gases through vents where losses due to viscosity are low, we will accept the assumption of no losses and apply the equation to such cases. Using Eq. (5.2) also implies that the fluid is incompressible, which is a reasonable assumption for the applications we will develop.

The terms P_1 and P_2 [Pa] are the static pressure head, the velocity term on both sides is the hydrodynamic pressure head, and the last term on both sides is the hydrostatic pressure head. Velocity is denoted v [m/s], density r [kg/m³], height above some reference point is denoted h [m], and g is the gravitational constant [m/s²].

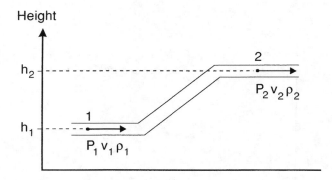

FIGURE 5.3   The Bernoulli principle.

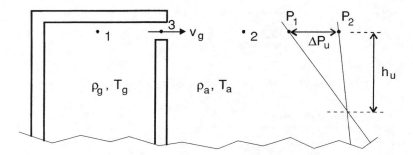

FIGURE 5.4   A close-up of the top opening in Figure 5.2.

5.2.2   APPLICATION TO A SIMPLE EXAMPLE

We now apply the Bernoulli equation to the top opening of the enclosure given in Figure 5.2 to facilitate an understanding of the basic terms and to quantify the pressure differences.

Figure 5.4 is a close-up of the top opening in Figure 5.2. The temperature inside is T_g and outside is T_a, with the corresponding densities ρ_g and ρ_a, respectively. We shall be considering the pressure at point 1 inside the enclosure, point 2 on the outside, and point 3 at the opening. All points are at the same height above some reference level. Geometric variables referring to the upper vent have the subscript "u," and those referring to the lower vent the subscript "l". We shall adopt the neutral plane as a reference height, and the points considered are at a height h_u above that level.

Expression for the hydrostatic pressure difference at the top vent: We are first interested in deriving an equation for the pressure difference between points 1 and 2, which is denoted ΔP_u and is equal to $P_1 - P_2$. Since the velocities at points 1 and 2 are considered to be zero, we can rewrite Eq. (5.2) to get

$$P_1 - P_2 = h_2 \rho_2 g - h_1 \rho_1 g$$

Written in the notation given in Figure 5.4, and noting that $h_1 = h_2 = h_u$, the expression for ΔP_u becomes

$$\Delta P_u = h_u (\rho_a - \rho_g) g \tag{5.3}$$

which is the hydrostatic pressure difference across the vent.

Expression for the hydrodynamic pressure difference at the top vent: We now wish to link the hydrostatic pressure difference to the velocity of flow in the vent opening, in order to be able to calculate mass flow rate through it. We therefore consider the Bernoulli equation at point 1 and point 3, and Eq. (5.2) is rewritten as

$$P_1 + \frac{1}{2}v_1^2\rho_1 + h_1\rho_1g = P_3 + \frac{1}{2}v_3^2\rho_3 + h_3\rho_3g$$

We note that the gas velocity at point 1 is zero, so $v_1 = 0$. The density of the gas inside the enclosure is the same as that exiting through the vent, so $r_1 = r_3 = r_g$. Finally, the heights of points 1 and 3 from the neutral plane are the same, so $h_1 = h_3 = h_u$. This leaves us with the simple expression

$$P_1 - P_3 = \frac{1}{2}v_3^2\rho_3$$

which in the notation of Figure 5.4 becomes

$$\Delta P_u = \frac{1}{2}v_g^2\rho_g \tag{5.4}$$

Expression for the flow velocity at the top vent: In order to arrive at an expression for the velocity of the gas exiting the vent, we isolate v_g in Eq. (5.4) to get

$$v_g = \sqrt{\frac{2\Delta P_u}{\rho_g}}$$

We then substitute DP_u from Eq. (5.3) to get

$$v_g = \sqrt{\frac{2h_u(\rho_a - \rho_g)g}{\rho_g}} \tag{5.5}$$

Note that we have expressed the velocity at a certain height h_u, which is at the center of the vent. The velocity will be slightly higher at the top of the vent and slightly lower at the bottom of the vent, since the pressure at these locations is slightly higher and lower, respectively. Since the vent in this example is relatively narrow with respect to the height h_u, we can say that Eq. (5.5) expresses the mean flow velocity in the vent.

Corresponding expressions for the lower vent: In a similar manner as above, by applying the Bernoulli equation, we can derive expressions for the lower vent. Figure 5.5 is a reproduction of Figure 5.2, with some additional notation. Variables referring to the upper vent have the subscript "u," as before, and those referring to the lower vent have the subscript "l."

Our reference height is the neutral plane (where pressure difference is zero), so the height to the center of the lower opening is negative, with the magnitude h_l. This leads to a negative hydrostatic pressure difference (indicating that air is being sucked in rather than pushed out) and a negative velocity out of the enclosure (indicating flow into the enclosure). As an alternative to the negative signs, we can choose h_l to be positive and define the positive direction of velocity to be into the enclosure, as we have done in Figure 5.5. This is what we shall do in the following equations; negative pressure difference and negative velocity are replaced by a coordinate system that defines h_l to be positive downward from the neutral plane and flow into the enclosure to be positive.

We now apply Eq. (5.2) to the lower vent at points 1, 2, and 3, corresponding to the notation given for the upper vent in Figure 5.4.

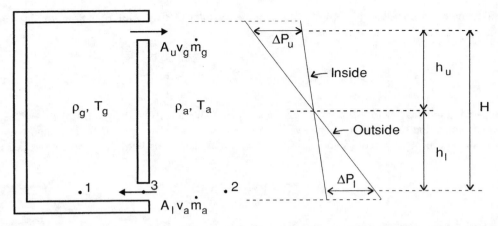

FIGURE 5.5 Notation describing the vent flows for a heated enclosure with relatively narrow openings at the top and bottom.

The hydrostatic pressure difference across the lower vent can be written (using the Bernoulli equation notation)

$$P_1 - P_2 = h_2\rho_2 g - h_1\rho_1 g$$

which, in the notation in Figure 5.5, becomes (noting that h_1 is negative downward)

$$\Delta P_1 = h_1(\rho_a - \rho_g)g \tag{5.6}$$

Similarly, the hydrodynamic pressure is written (using the Bernoulli equation notation)

$$P_2 - P_3 = \frac{1}{2}v_3^2\rho_3$$

which, in the notation in Figure 5.5, becomes

$$\Delta P_1 = \frac{1}{2}v_a^2\rho_a \tag{5.7}$$

The velocity into the enclosure through the lower vent is arrived at from Eq. (5.6) and (5.7) as

$$v_a = \sqrt{\frac{2h_1(\rho_a - \rho_g)g}{\rho_a}} \tag{5.8}$$

Note again that this is the velocity at a single point in the lower vent, at the height h_1, but since the pressure difference does not change much over the vent height, this can be said to be the mean velocity in the vent.

Relationship between temperature and density: The above equations have been expressed in terms of density of the gases inside and outside of the enclosure. In practice, we are more likely to express these properties of the gases in terms of temperature, since there is a direct relation between the two for ideal gases.

One form of the ideal gas law can be stated as

$$PM = \rho RT$$

where P is the pressure, M is the molecular weight of the gas in question, and R is the ideal gas constant, ≈ 8.314 J/(K mol). The atmospheric pressure changes very slightly with height, so we use the value of the standard atmospheric pressure, given as $101.3 \cdot 10^3$ N/m^2. The molecular weight of air is ≈ 0.0289 kg/mol; this value can also be used for fire gases, since to the greatest extent these consist of air. Noting that the units of R can also be written as [Nm/(K mol)], we find that

$$T = \frac{353}{\rho} \text{ and } \rho = \frac{353}{T} \tag{5.9}$$

where T is given in [K] and ρ in [kg/m^3]. As an alternative to using Eq. (5.9), one can simply look up the values of T and ρ in tables on the properties of air, supplied in many standard textbooks and handbooks.

5.2.3 MASS FLOW RATE THROUGH VENTS

Precise calculation of the mass flow through a vent, using the fundamental laws of nature, can be achieved by solving the Navier–Stokes equations in computers of considerable capacity. For all engineering purposes, however, the study of hydraulics provides the sufficiently accurate equations given above and the mass flow through the vent can be calculated using these, but a correction is required by using the flow coefficient, C_d. In most of the applications discussed in this text the appropriate value of the flow coefficient is between 0.6 and 0.7. We shall discuss the flow coefficient in greater detail later in this section.

In vents where the pressure difference is a constant value over the entire vent height, the velocity can also be said to be constant over the entire height (except for very close to the edges of the vent; the flow coefficient takes into account this edge effect). The mass flow can then be written as

$$\dot{m} = C_d A v \rho \tag{5.10}$$

where A is the vent area [m^2], v is a constant velocity over the area [m/s], and ρ is the density of the gases flowing through the vent. The mass flow rate, $\dot{m}$, is therefore given in [kg/s]. Note that the use of Eq. (5.10) requires the velocity to be constant over the vent height; when this is not the case we must integrate the velocity profile with respect to height to get the total mass flow rate.

In the example given in Figure 5.5 the mass flow through the top vent becomes, using Eq. (5.10) and (5.5),

$$\dot{m}_g = C_d A_u \rho_g \sqrt{\frac{2 h_u (\rho_a - \rho_g) g}{\rho_g}} \tag{5.11}$$

and the mass flow through the lower vent becomes

$$\dot{m}_a = C_d A_l \rho_a \sqrt{\frac{2 h_l (\rho_a - \rho_g) g}{\rho_a}} \tag{5.12}$$

Now, say that we are given the enclosure and vent geometry and the densities (or temperatures) involved. Since the position of the neutral plane is not given, the heights h_u and h_l are not known, and we will not be able to use Eq. (5.11) and (5.12) to calculate the mass flow rates. We must therefore find a way to express the height of the neutral plane.

Expression for the height of the neutral plane: We need two expressions in order to solve for the two unknowns, h_u and h_l. These expressions must link h_u and h_l to known variables. One such expression can be achieved by using the *conservation of mass*, saying that mass flow in must equal mass flow out.

Equating the mass flow rates in and out, using Eq. (5.11) and (5.12), results in

$$\frac{h_l}{h_u} = \left(\frac{A_u}{A_l}\right)^2 \frac{\rho_g}{\rho_a} \tag{5.13}$$

This gives the ratio of the heights we seek. The absolute values for these heights are obtained from a second expression, by observing that the total height between the centers of the two vents, H, must equal the sum of h_u and h_l (see Figure 5.5), so

$$H = h_u + h_l$$

Observe that by using Eq. (5.9) we can express the ratio of the heights in terms of temperature, and Eq. (5.13) becomes

$$\frac{h_l}{h_u} = \left(\frac{A_u}{A_l}\right)^2 \frac{T_a}{T_g} \tag{5.13a}$$

EXAMPLE 5.1

A building is 31 m high with 1 m high openings each at the top and the bottom; both openings are 2 m wide. The height between the centers of the openings is thus 30 m. The temperature outside is 0°C and the temperature inside is 20°C. Calculate

(a) the height of the neutral plane above some reference point

(b) the pressure difference across both vents

(c) the flow velocity at the center of both vents

(d) the mass flow rates in and out of the vents

SUGGESTED SOLUTION

Since most of the equations above use densities rather than temperatures, we convert temperatures to densities using Eq. (5.9) so that $\rho_a = 353/273 = 1.293$ kg/m^3 and $\rho_g = 353/293 = 1.205$ kg/m^3.

(a) Using Eq. (5.13) we find $\dfrac{h_l}{h_u} = \left(\dfrac{2}{2}\right)^2 \dfrac{1.205}{1.293} = 0.932$. We also know that $h_u + h_l = 30$ m, so

$\dfrac{30 - h_u}{h_u} = 0.932$, and therefore $h_u = \dfrac{30}{1 + 0.932} = 15.53$ m. This gives $h_l = 14.47$ m. If we take the center of the lower opening to be our reference point, the height of the neutral plane is

14.47 m above this point. If we take the floor of the 31 m high building to be our reference point, we must add 0.5 m to this height, since the center of the lower vent is 0.5 m above the floor. The height of the neutral plan above the floor is then 14.97 m.

(b) The pressure difference across the top vent is given by Eq. (5.3) as $\Delta P_u = 15.53$ $(1.293 - 1.205)9.81 = 13.4$ Pa. The pressure difference across the lower opening is given by Eq. (5.6) as $\Delta P_l = 14.47(1.293 - 1.205)9.81 = 12.49$ Pa.

(c) The flow velocity at the center of the top vent is given by Eq. (5.5) as

$$v_g = \sqrt{\frac{2 \cdot 15.53 \cdot (1.293 - 1.205)9.81}{1.205}} = 4.72 \text{ m/s. The flow velocity at the center of the lower}$$

vent is given as $v_a = \sqrt{\dfrac{2 \cdot 14.47 \cdot (1.293 - 1.205)9.81}{1.293}} = 4.40$ m/s.

(d) We assume the flow coefficient to be 0.7 in this case. The mass flow rate out of the top

opening is given by Eq. (5.11) as $\dot{m}_g = 0.7 \cdot 2 \cdot 1.205 \sqrt{\dfrac{2 \cdot 15.53 \cdot (1.293 - 1.205)9.81}{1.205}} = 7.96$ kg/s.

The mass flow out of the lower opening is given by Eq. (5.12) as $\dot{m}_a = 0.7 \cdot 2 \cdot 1.293$

$\sqrt{\dfrac{2 \cdot 14.47 \cdot (1.293 - 1.205)9.81}{1.293}} = 7.96$ kg/s. The conservation of mass requires the mass flow

rate in to equal the mass flow rate out. We have found this to be the case, which is a good check of our calculations.

The flow coefficient, C_d: The Bernoulli equation is only valid for idealized flow circumstances: stationary, incompressible, isothermal, friction-free, and without heat losses to the surrounding structure. In practice, deviation from these ideals is expressed through the use of the flow coefficient, C_d. In hydraulics, a distinction is made between two main types of openings: an orifice and a nozzle (see Figure 5.6).

Most fire vents are classified as orifices, and Figure 5.6 shows, schematically, that only a part of the vent area carries the flow efficiently. In fire applications the effective area is typically around 60–70% of the actual vent area, and this is expressed through the use of the flow coefficient.

The flow coefficient is a function of the Reynolds Number, expressed as $\text{Re} = \dfrac{vD\rho}{\mu}$, where v is velocity, ρ is the density, and μ the viscosity of the fluid approaching the vent. D is the hydraulic diameter, defined as $D = 4 A/P$, where A is area and P is perimeter of the vent. Figure 5.7 shows how the flow coefficient varies with the Reynolds Number for the two types of vents.

For ordinary openings, such as open doors and windows, where the velocity is typically in the range of 1–5 m/s, the Reynolds Number is in the vicinity of 10^6, and C_d is a little larger than 0.6. For very narrow openings, such as the gap under a door (say 0.01 m high and 1 m wide), and velocities in the range of 1 to 5 m/s, the Reynolds Number is in the vicinity of 10^3, and C_d gets closer to 0.7.

For conditions outside of those given above, or if the vent is of the nozzle type, special consideration must be given to the choice of flow coefficient. Since the value of $C_d \approx 0.6 - 0.7$ is valid for a very wide range of fire safety engineering applications, we shall assume C_d to be in this range for the remainder of this text.

5.2.4 SUMMARY

We have introduced some of the most important aspects of pressure-induced flows across vents that occur due to density differences. We have shown how classical hydraulics can be used to calculate pressure differences and velocities at certain heights above the neutral plane.

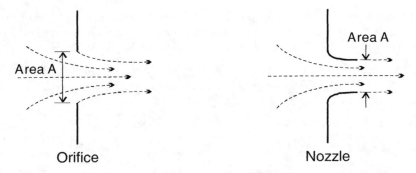

FIGURE 5.6 Two vent types: an orifice and a nozzle. Most fire vents are classified as orifices. (Adapted from Emmons[3]. With permission.)

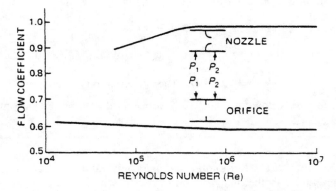

FIGURE 5.7 Flow coefficient, C_d, as a function of the Reynolds Number for the two vent types. (From Emmons[3]. With permission.)

Note that we have evaluated the pressure difference at the heights h_u and h_l and have assumed that the pressure difference is constant over the height of the vent. Since the vents are relatively narrow and their heights are small compared to h_u and h_l, we can safely make this assumption. In later sections we look at vents where the pressure changes considerably over the vent height, and in this case we must integrate some of the equations with respect to height.

5.3   EXAMPLES OF PRESSURE PROFILES IN A FIRE ROOM WITH A VENT

In the introduction to this chapter we mentioned five types of processes that cause pressure differences to occur in a building. Two of these were due to a fire in the enclosure and were caused by (1) thermal expansion of the gases and (2) density differences or buoyancy. A third type was due to natural causes: (3) the difference in temperature inside and outside of the building. Type 3 is controlled by the same physical phenomenon as type 2, but usually involving lower temperatures.

We examined a special case of types 2 and 3 in the previous section, where the enclosure openings were relatively narrow with regard to the height from the neutral plane. This allowed us to assume a constant pressure difference over the opening height, which simplified the mass flow calculations.

In this section we discuss qualitatively the more common enclosure fire scenario, where there is a large opening (i.e., door or window) involved. This means that in certain cases the pressure difference will vary considerably, from positive to negative, over the opening height, and that we must integrate to achieve analytical equations allowing calculation of the mass flow rates in and out of the enclosure.

We will describe the fire development in the enclosure in terms of pressure build-up, and thus identify four stages in the development, or four different categories of pressure profiles. Two of

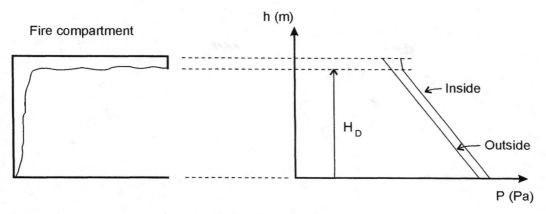

FIGURE 5.8   The pressure differences across the vent before the smoke starts flowing out the opening.

the initial cases are of type 1, where the pressure differences are due to expansion of the hot gases. These two cases usually last for short times and will be further addressed in Chapter 8. The later two stages will be termed "the stratified stage" and the "well-mixed stage." These are common in enclosure fires and are usually the cause of smoke movement in buildings. In Sections 5.4 and 5.5 we examine these in detail and develop equations in each of these two cases for calculating mass flow rates in and out of vents.

Figures 5.8 to 5.11 show how the pressure difference between the fire room and the surroundings develop as the fire in the compartment grows. As discussed in Chapter 2, a two-zone environment is created where the hot gases collect in the upper region and the cold, relatively uncontaminated air is in the lower region. The two layers are often stable at the height H_D from the floor. This height is often termed the height of the smoke layer or the height of the thermal discontinuity. Another height also characterizes the shape of the pressure profiles; this is the height of the neutral plane, which will be denoted H_N.

Stage A: The first stage is usually seen in the initial stages of the fire and is illustrated in Figure 5.8. Here, the hot layer has not reached the top of the opening. The energy release results in a pressure increase due to the thermal expansion of the gases, and air is pressed out through the opening. The whole room has higher pressure than the outside, and only cold air is pressed out through the opening. There is a small kink at the top of the inside pressure profile, since the hot gases have less weight than the cold gases.

Stage B: In the second stage, the smoke layer has just reached the top of the opening and the hot gases have started to flow out, as shown in Figure 5.9. The flow of cold air out through the lower part of the opening continues due to the thermal expansion of hot gases. There is still positive pressure in the whole compartment. However, this stage lasts only for a few seconds; the mass flow of hot gases out through the opening is comparatively large, and the mass balance in the room demands that this loss of mass be replaced by mass flow into the compartment.

Stage C: In the third stage, the flow has changed so that hot gases flow out through the top of the opening and fresh air enters through the lower part of the opening, as shown in Figure 5.10. At some height, which we will call H_N, there must be no flow at all. This is at the height where the pressure difference changes from being positive to negative. This height is called the *neutral plane height*, the height at which the pressure differences are 0 and the flow at that point is therefore 0. Below this height there is negative pressure in the room (compared to the outside pressure) so air flows in below H_N. Similarly, the hot gases flow out above H_N due to positive pressure in the room. In order to calculate the flow in and out of the room, one must have determined both H_N and H_D.

Stage D: In the fourth stage, the fire in the room is fully developed, and the hot gas layer has more or less reached the floor so that both pressure lines are linear, as shown in Figure 5.11. There

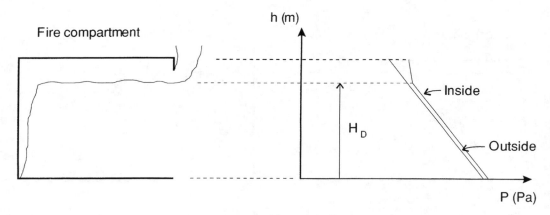

FIGURE 5.9 Pressure difference across the vent just after the smoke has started to flow out through the opening. This stage usually lasts only a few seconds.

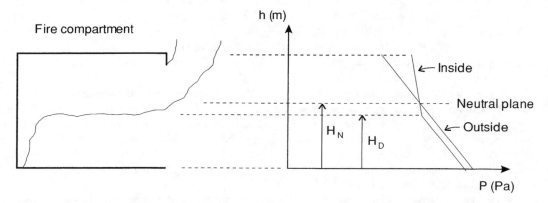

FIGURE 5.10 Pressure difference in a room with two layers, termed the stratified case.

is negative pressure in the lower part of the room and positive pressure in the upper part. The unknown height in this case is only H_N, so the pressures and mass flow rates are relatively easy to calculate for this type of a pressure profile. This stage is often termed the *fully developed fire*, or the *post-flashover fire*, indicating that flashover has occurred.

However, this type of pressure profile is also common in the case where the fire is weak in relation to the room volume. Much ambient air is entrained to the plume, the buoyancy force is weak and the smoke is not pumped up to the ceiling level but spreads evenly in the room. This situation is often referred to as *one-zone*, as opposed to the two-zone situation mentioned earlier.

We will refer to the post-flashover and the one-zone cases as the *well-mixed* case and treat the two identically, since both have the same type of pressure profile. The only difference between the two, with regard to flow calculations, is the magnitude of the temperatures, and therefore the mass flow rates.

Summary: We have given examples of how an enclosure fire may develop with respect to the pressure profiles across the vent. This development may, however, occur in a different order. A weak fire may not necessarily go through the three previous stages; it may simply cause the enclosure to be slowly filled with well-mixed homogeneous gases. We shall therefore refer to Stage D as "the well-mixed case" and the Stage C as "the stratified case." We shall derive equations for calculating the flow mass loss rate through the vents for these two cases in the next two sections.

The first two stages are typical for the initial enclosure fire development and are caused by the expansion of hot gases. These two stages are typically of a very short duration. When the vent is

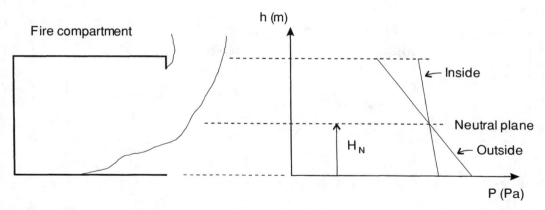

FIGURE 5.11 Pressure differences where the enclosure is filled with smoke, termed the well-mixed case.

relatively large (door, window), such pressure profiles will not contribute in any significant way to the smoke flow from the enclosure. However, for enclosures that are very tight, or with very small openings (for example, an engine room in a ship), the expansion of the hot gases can cause considerable pressures and mass flow rates. Equations for this case will be developed in Chapter 8.

5.4 THE WELL-MIXED CASE

In this section we consider the well-mixed case, where the enclosure is considered to have a uniform gas temperature over its entire volume. This is identical to the assumption made in Section 5.2, where equations for the pressure difference, velocities, and mass flow rates at a certain height were developed. The mixed case can be applied to the post-flashover fire, where the hot gases are assumed to fill the enclosure, or to the general case where the enclosure is assumed to have a uniform temperature that differs from the outside ambient temperature.

We therefore revisit the equations derived in Section 5.2 (applicable to relatively narrow openings) and reapply them to fit a relatively large, single opening. We then derive a very simple expression for the mass flow of gases into such an opening. We also consider a scenario where a ceiling vent has been added. These scenarios assume that the mass flow rate out of the enclosure equals the mass flow rate into the enclosure. No account is taken of mass that is produced in the enclosure. In fires, mass is always produced by the burning object, but in most cases it is insignificant compared to the mass flow rates in and out through the vent. Therefore, in the final section we consider this mass production in the equations and show the effect of the inclusion of this term.

5.4.1 MASS FLOW RATES AND THE HEIGHT OF THE NEUTRAL PLANE

Consider Figure 5.12, showing an enclosure with temperature and density T_g and ρ_g, a vent of height H_o, and a neutral plane situated at height H_N from the floor. The outside temperature and density are denoted T_a and ρ_a, and the mass flow rates in and out of the enclosure are denoted $\dot{m}_a$ and $\dot{m}_g$, respectively. The figure shows the velocity profile with maximum velocities at the bottom and the top of the opening, denoted $v_{a,max}$ and $v_{g,max}$. Similarly, the pressure profile shows the maximum pressure differences at the bottom and the top of the openings as $\Delta P_{l,max}$ and $\Delta P_{u,max}$. The height over which the flow into the vent occurs is denoted h_l and the corresponding height for the outflow of hot gases is denoted h_u, both measured from the neutral plane.

A note on sign convention and integration over the opening height: Since the flow velocity changes with height from the neutral plane, we must integrate with respect to height to arrive at the total mass flow through the vent. We therefore introduce the arbitrary height, z, originating from the neutral plane, where the velocity and pressure difference is zero. We will choose the height

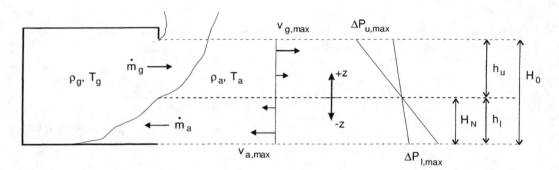

FIGURE 5.12 The well-mixed case: An enclosure with uniform temperature, T_g, higher than the outside temperature, T_a.

z to be positive upward from the neutral plane and mass flow rate, $\dot{m}_g$, to be positive out from the enclosure. As in Section 5.2, we can now choose between two sign conventions for the lower part of the vent. We can choose z to be *negative* downward from the neutral plane, resulting in negative mass flow rate, $\dot{m}_a$, out of the enclosure (meaning that flow is directed into the enclosure). Or, we can choose z to be *positive* downward from the neutral plane, resulting in positive mass flow rate into the enclosure. For our purposes, we choose the latter sign convention, as in Section 5.2.

Equation (5.10) gave the mass flow through a vent as $\dot{m} = C_d A v \rho$, where the velocity through the vent is constant over its area. Generally, when the velocity is not constant over the opening area, we can write

$$\dot{m} = C_d \int_A \rho v dA$$

We now set $dA = W \cdot dz$, where W is the width of the vent. In our case the velocity through the vent is a function of height, but is assumed to be constant across the width of the vent at any certain height.

The mass flow rate across a certain height z of the opening, from the neutral plane, is then achieved by integrating over the height z:

$$\dot{m} = C_d \int_0^z W \rho_g v(z) dz \tag{5.14}$$

Expressions for the pressure difference across the vent: From Eq. (5.3) we know that the maximum pressure difference at the top of the vent can be written as

$$\Delta P_{u,max} = h_u (\rho_a - \rho_g) g$$

and the maximum pressure difference at the bottom of the vent, from Eq. (5.6), as

$$\Delta P_{l,max} = h_l (\rho_a - \rho_g) g$$

where, according to our sign convention, h_u and h_l are both measured as positive entities from the neutral plane, and therefore $\Delta P_{u,max}$ is positive out of the vent and $\Delta P_{l,max}$ is positive into the vent.

Generally, the pressure difference as a function of height z can be written as

$$\Delta P(z) = z(\rho_a - \rho_g) g \tag{5.15}$$

with the sign convention discussed above.

Expressions for the velocity in the vent: Similarly, the maximum velocities at the top and the bottom of the vent can be written, using Eq. (5.5) and (5.8), as

$$v_{g,max} = \sqrt{\frac{2h_u(\rho_a - \rho_g)g}{\rho_g}}$$

and

$$v_{a,max} = \sqrt{\frac{2h_l(\rho_a - \rho_g)g}{\rho_a}}$$

The velocity as a function of height z above the neutral plane can be written as

$$v_g(z) = \sqrt{\frac{2z(\rho_a - \rho_g)g}{\rho_g}} \tag{5.16}$$

and below the neutral plane as

$$v_a(z) = \sqrt{\frac{2z(\rho_a - \rho_g)g}{\rho_a}} \tag{5.17}$$

Expression for the mass flow rates of hot gases out of the vent: Using Eq. (5.14) we can write the mass flow rate of hot gases out through the vent as

$$\dot{m}_g = C_d \int_0^{h_u} W\rho_g v_g(z)dz$$

where the limits of the integral range from the neutral plane ($z = 0$) to the top of the opening ($z = h_u$). Substituting in from Eq. (5.16) we find

$$\dot{m}_g = C_d \int_0^{h_u} W\rho_g \sqrt{\frac{2z(\rho_a - \rho_g)g}{\rho_g}} dz$$

The only entity within the integral sign that varies with height is z, so this can be rewritten

$$\dot{m}_g = C_d W\rho_g \sqrt{\frac{2(\rho_a - \rho_g)g}{\rho_g}} \int_0^{h_u} \sqrt{z}\,dz$$

The quantity inside the integral sign therefore simply becomes $\frac{2}{3}h_u^{3/2}$ and the mass flow rate of hot gases out through the vent becomes

$$\dot{m}_g = \frac{2}{3}C_d W\rho_g \sqrt{\frac{2(\rho_a - \rho_g)g}{\rho_g}} h_u^{3/2} \tag{5.18}$$

Expression for the mass flow rates of cold air in through the vent: In a similar fashion, we can express the mass flow rate of air through the vent as

$$\dot{m}_a = C_d \int_0^{h_l} W \rho_a v_a(z) dz$$

where the limits of the integration are from the neutral plane ($z = 0$) to the bottom of the opening ($z = h_l$). Substituting in from Eq. (5.17) and solving the integral we get

$$\dot{m}_a = \frac{2}{3} C_d W \rho_a \sqrt{\frac{2(\rho_a - \rho_g)g}{\rho_a}} h_l^{3/2} \tag{5.19}$$

Since the position of the neutral plane is not known, we do not know the two heights h_u and h_l. We must therefore determine the position of the neutral plane before we can use Eq. (5.18) and (5.19) to calculate the mass flows.

Expression for the height of the neutral plane: To solve for the two unknown variables, we need two expressions linking h_u and h_l to known variables. The height of the opening, H_o, is known and can be written $H_o = h_u + h_l$. We obtain the second expression by equating the mass flow rates into and out of the enclosure, so we write

$$\dot{m}_g = \dot{m}_a \tag{5.20}$$

Inserting Eq. (5.18) and (5.19) into Eq. (5.20), we find

$$\left(\frac{h_u}{h_l}\right)^{3/2} = \left(\frac{\rho_a}{\rho_g}\right)^{1/2} \tag{5.21}$$

Expressing h_u as $H_o - h_l$, we find

$$\frac{H_o - h_l}{h_l} = \left(\frac{\rho_a}{\rho_g}\right)^{1/3}$$

so the expression for h_l becomes

$$h_l = \frac{H_o}{1 + (\rho_a/\rho_g)^{1/3}} \tag{5.22}$$

The height of the neutral plane, H_N, from some reference point can then be decided. If this reference point is the bottom of the opening, then $H_N = h_l$.

EXAMPLE 5.2

A fire in an enclosure has reached flashover, and the gas temperature is assumed to be uniform at 1000°C. The enclosure has an opening 3 m wide and 2 m high. Calculate

(a) the position of the neutral layer, measured from the bottom of the opening

(b) the mass flow rates into and out of the opening.

SUGGESTED SOLUTION

We assume that the ambient temperature is 20°C. The equations above are given in terms of densities (they are often given in terms of temperatures), so we use Eq. (5.9) to translate the temperatures to densities and find $\rho_g = \dfrac{353}{1000+273} = 0.277$ kg/m³ and $\rho_a = \dfrac{353}{20+273} = 1.20$ kg/m³.

(a) Equation (5.22) gives $h_l = \dfrac{2}{1+\left(\dfrac{1.2}{0.277}\right)^{1/3}} = 0.76$ m. Since the reference point for the neutral plane is the bottom of the opening, the height of the neutral plane is $H_N = 0.76$ m. The height $h_u = 2 - 0.76 = 1.24$ m.

(b) Assuming the flow coefficient to be $= 0.7$, the mass flow into the opening is given by Eq. (5.18) as $\dot{m}_a = \dfrac{2}{3} \cdot 0.7 \cdot 3 \cdot 1.2 \sqrt{\dfrac{2(1.2-0.277)\cdot 9.81}{1.2}} \cdot 0.76^{3\,2} = 4.33$ kg/s. The mass flow rate of hot gases out through the opening is given by Eq. (5.19) as $\dot{m}_g = \dfrac{2}{3} \cdot 0.7 \cdot 3 \cdot 0.277$ $\sqrt{\dfrac{2(1.2-0.277)\cdot 9.81}{0.277}} \cdot 1.24^{3\,2} = 4.33$ kg/s.

5.4.2 A SIMPLIFIED EXPRESSION FOR THE MASS FLOW RATE IN THROUGH AN OPENING

Equation (5.19) gave an expression for the mass flow rate of air in through the opening in terms of h_l. Equation (5.22) gave an expression for the height h_l in terms of densities and the opening height, H_o. Combining the two equations, we get

$$\dot{m}_a = \frac{2}{3}C_d W \rho_a \sqrt{\frac{2(\rho_a-\rho_g)g}{\rho_a}} \left(\frac{H_o}{1+(\rho_a/\rho_g)^{1/3}}\right)^{3/2} [$$

Writing $A = W \cdot H_o$, where A is the opening area, this equation can be rewritten as

$$\dot{m}_a = \frac{2}{3}C_d A \sqrt{H_o} \sqrt{2g\rho_a} \sqrt{\frac{(\rho_a-\rho_g)\rho_a}{\left[1+(\rho_a/\rho_g)^{1/3}\right]^3}} \tag{5.23}$$

Figure 5.13 shows how the term $\sqrt{\dfrac{(\rho_a-\rho_g)/\rho_a}{\left[1+(\rho_a/\rho_g)^{1/3}\right]^3}}$ (which we will call the *density factor*) changes with the ratio of the gas temperature to the ambient temperature, T_g/T_a.

We see from the figure that the density factor changes very slightly once the gas temperature is twice the ambient temperature. The density factor reaches a maximum of 0.214 when the temperature ratio is 2.72. If the ambient temperature is taken to be 293 K, the corresponding gas temperature is ≈800 K. We can therefore assume that the density factor is roughly constant at the value 0.214 for gas temperatures that are twice the ambient temperature, or higher.

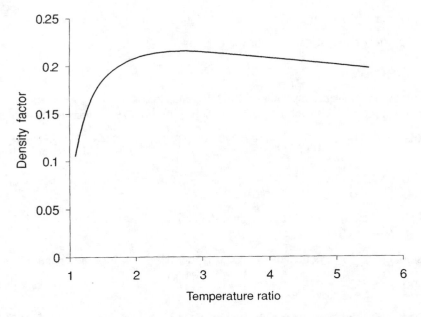

FIGURE 5.13 Density factor in Eq. (5.23) as a function of temperature ratio, T_g/T_a.

Equation (5.23) can then be rewritten as

$$\dot{m}_a = \frac{2}{3} C_d A \sqrt{H_o} \sqrt{2g\rho_a}\, 0.214$$

Taking the standard values of $C_d = 0.7$, $g = 9.81$ m/s², and $r_a = 1.2$ kg/m³, the constant in the above equation becomes ≈0.5. We therefore arrive at the very simple, useful, and well-known relationship for the mass flow into an opening:

$$\dot{m}_a = 0.5 \cdot A \sqrt{H_o} \qquad\qquad (5.24)$$

We can now estimate the mass flow rate out through an opening by knowing only the area and the height of the vent. This equation gives an estimate of the mass flow rate for fires where the gas temperature is at least twice the ambient temperature (measured in Kelvin) and where the enclosure temperature can be assumed to be uniformly distributed over the entire volume. In practice, this means that the gas temperature should be higher than 300°C (≈573 K, which is slightly less than twice the ambient temperature of 293 K). The two conditions are most often met in post-flashover fires, where temperatures are in excess of 800 K and the enclosure is more or less filled with smoke of roughly uniform temperature. Equation (5.24) has therefore been very useful in the analysis of post-flashover fires.

EXAMPLE 5.3

Use Eq. (5.24) to estimate the mass flow rate in through the vent in Example 5.2.

SUGGESTED SOLUTION

The mass flow rate becomes (irrespective of gas temperature) $\dot{m}_a = 0.5 \cdot 3 \cdot 2\sqrt{2} = 4.24$ kg/s, which compares well with the value of 4.33 kg/s arrived at in Example 5.2.

5.4.3 Taking into Account the Mass Produced in the Room
(the Burning Rate)

In the above sections we assumed that the conservation of mass could be expressed through Eq. (5.20), where the mass flow rate into the enclosure was equated to the mass flow rate out of the enclosure.

To be more correct, we should also take into account the mass produced inside the room, since the burning object will release some small amount of mass, equal to the mass loss rate or the burning rate $\dot{m}_b$. The conservation of mass therefore demands that the gases exiting the enclosure must equal the air entering the enclosure plus the mass produced in the enclosure, expressed as

$$\dot{m}_g = \dot{m}_a + \dot{m}_b \tag{5.25}$$

The burning rate is, for many fuels, typically in the range of 0.01–0.05 kg/s per square meter of fuel. The mass burning rate would typically be 1 to 10% of the mass flow rates through the vent. The production of mass inside the enclosure can therefore be ignored in most cases.

For completion, we will include the mass burning rate in the conservation equation for mass and derive an expression for the mass flow rate into the enclosure that accounts for the burning rate. This will result in a slight adjustment in the height of the neutral plane and in the mass flow rates into and out of the vent.

The mass flow of air in through the vent is still written as Eq. (5.19), reproduced here:

$$\dot{m}_a = \frac{2}{3}C_d W \rho_a \sqrt{\frac{2(\rho_a - \rho_g)g}{\rho_a}} h_l^{3/2} \tag{5.19}$$

The mass flow rate out of the vent would have to include the mass produced in the room, by using Eq. (5.18) and (5.25):

$$\dot{m}_g = \dot{m}_a + \dot{m}_b = \frac{2}{3}C_d W \rho_g \sqrt{\frac{2(\rho_a - \rho_g)g}{\rho_g}} h_u^{3/2} \tag{5.26}$$

Since h_u and h_l are unknowns we must use Eq. (5.19) and (5.26) as well as the relationship $H_o = h_u + h_l$ to arrive at an expression for h_l. This expression can then be substituted into Eq. (5.19) to arrive at an expression for the mass flow rate into the vent, in terms of H_o instead of h_l.

Dividing (5.26) by (5.19) gives

$$\frac{m_a + m_b}{m_a} = \left(\frac{h_u}{h_l}\right)^{3/2}\sqrt{\frac{\rho_g}{\rho_a}}$$

Setting $h_u = H_o - h_l$ and rearranging, we find

$$\frac{H_o - h_l}{h_l} = \left(\frac{1 + \dot{m}_b/\dot{m}_a}{\sqrt{\rho_g/\rho_a}}\right)^{2/3}$$

and therefore

$$\frac{H_o}{h_l} = \left(\frac{1 + \dot{m}_b/\dot{m}_a}{\sqrt{\rho_g/\rho_a}}\right)^{2/3} + 1$$

This leads to an expression for h_1 of the form

$$h_1 = \frac{H_o}{1 + \left(\dfrac{1 + \dot{m}_b/\dot{m}_a}{\sqrt{\rho_g/\rho_a}}\right)^{2/3}} \tag{5.27}$$

To get a general expression for $\dot{m}_a$ that is not in terms of the unknown h_1, we substitute the above expression into Eq. (5.19) to get

$$\dot{m}_a = \frac{\dfrac{2}{3} C_d W \rho_a \sqrt{\dfrac{2(\rho_a - \rho_g)g}{\rho_a}} \cdot H_o^{3\,2}}{\left[1 + \left(\dfrac{1 + \dot{m}_b/\dot{m}_a}{\sqrt{\rho_g/\rho_a}}\right)^{2/3}\right]^{3/2}} \tag{5.28}$$

Analyzing the density terms in Eq. (5.28), we find again that these do not vary much with temperature. In the range 600 to 1200°C, the density term in the denominator, $1/\left(\sqrt{\rho_a/\rho_g}\right)^{2/3}$, varies from 1.44 to 1.71, with an average value of ≈ 1.6. Similarly, for the same range in temperatures, the density term in the numerator, $\sqrt{(\rho_a - \rho_g) \cdot \rho_a}$, varies between 0.97 to 1.07, with an average value of ≈ 1.0. So, for a wide range of post-flashover temperatures, Eq. (5.28) can be written as

$$\dot{m}_a = \frac{2.1 \cdot A\sqrt{H_o}}{\left[1 + 1.6 \cdot (1 + \dot{m}_b/\dot{m}_a)^{2/3}\right]^{3/2}} \tag{5.29}$$

using $C_d = 0.7$, $r_a = 1.2$ kg/m³, and $g = 9.81$ m/s². Note that the term $\dot{m}_a$ is on both sides in the equation, so that it can only be solved by iteration. It is appropriate to use Eq. (5.24) as a first guess of $\dot{m}_a$ and to use this value in the denominator of Eq. (5.29). This will give a new value of $\dot{m}_a$, which again is used in the denominator, until a solution is arrived at by iteration.

EXAMPLE 5.4

A pool of transformer oil is on fire in the enclosure discussed in Example 5.2. In order to examine the influence of including the burning rate in the calculations, we will assume the very large burning rate of 0.4 kg/s. This is roughly the equivalent of 10 m² of transformer oil burning in the open (see Table 3.3). Give an evaluation of the mass flow rate of air into the opening, taking into account the mass produced in the enclosure.

SUGGESTED SOLUTION

Using Eq. (5.24), we found a rough value of $\dot{m}_a$ to be 4.24 kg/s. We now use this as an initial value in the denominator of Eq. (5.29) and calculate a new value of $\dot{m}_a$ to be $\dot{m}_a = \dfrac{2.1 \cdot 3 \cdot 2 \cdot \sqrt{2}}{\left[1 + 1.6 \cdot (1 + 0.4/4.24)^{2/3}\right]^{3/2}} = 4.01$ kg/s. By iterating and using this new value in the denominator and recalculating, we get $\dot{m}_a = 4.0$ kg/s, which is the answer to the example. We therefore see that even when the burning rate is very substantial, the influence on the mass flow rate is relatively small.

5.4.4 SUMMARY

The process we have followed to derive the mass flow rate expressions is as follows:

- Give an expression for the hydrostatic pressure differences.
- Equate this with an expression for the hydrodynamic pressure differences, arriving at an expression linking velocity, densities, and height.
- Give expressions for the mass flow rates in terms of two unknown heights.
- Equate the mass flow rate into and out of the enclosure, and use knowledge of the vent geometry to solve for the two unknown heights, thus arriving at an expression for the neutral plane height.
- Insert these heights into the expressions for mass flow rate to get numerical results.

We then saw that some of the density terms appearing in the equations varied little for a relatively wide range of temperatures. The above expressions for mass flow rate could therefore be simplified and expressed as a direct relation to the area of the opening times the root of its height. Finally we examined the influence of including the burning rate of the fuel in the mass balance calculations and found that the burning rate must be very high to make a significant impact on the numerical results.

5.5   THE STRATIFIED CASE

We now consider the case where the enclosure is only partly filled with hot gases, collected under the ceiling. This model of an enclosure fire is often called the *two-zone model* or the *stratified case*. The two-zone model represents the enclosure as consisting of two distinct gas zones: an upper volume with uniformly distributed hot gas, and a lower volume of ambient temperature, as is the temperature outside of the enclosure.

This approximation to reality has been found to fit experimental results quite reasonably for a wide range of enclosures and fire scenarios. Figure 5.14 shows one such experimental result, where the experimental temperature distribution and the experimental pressure profile is plotted. These compare favorably to the idealization of the two-zone model.

In this section we derive expressions for the mass flow rates in and out of the enclosure in terms of height from the neutral plane. We then use the conservation of mass to calculate the height of the neutral plane, and thus allow quantitative estimation of the mass flow rates. Finally, we consider the case where there is a ventilation opening in the ceiling.

5.5.1   MASS FLOW RATES INTO AND OUT OF THE VENT

Consider Figure 5.15. We assume the pressure profile discussed in Section 5.3 for the stratified case. All heights are measured from the bottom of the vent, which has the height H_o. What mathematically distinguishes this case from the well-mixed case is that both the height of the neutral plane, H_N, and the height of the smoke layer, H_D, are unknowns. We must therefore divide the pressure profile into three levels.

The first level in the pressure difference diagram is above the neutral plane, where mass flows out of the vent at the rate $\dot{m}_g$. The second level is between the neutral plane and the smoke layer height, where mass flows into the enclosure at the rate $\dot{m}_{a1}$. The third level is below the smoke layer height, where mass will flow into the enclosure at the rate $\dot{m}_{a2}$. The total mass flowing into the enclosure is therefore $\dot{m}_a = \dot{m}_{a1} + \dot{m}_{a2}$.

A note on sign convention and integration over the opening height: Since the pressure difference varies with height for the first two levels, we must integrate over the arbitrary height z, which again is measured positive upward from the neutral plane and negative downward from the

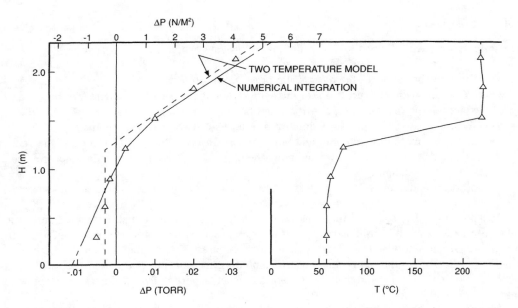

FIGURE 5.14 Experimentally measured temperature and pressure profile as a function of enclosure height, compared to the two-zone model idealization. (From McCaffrey and Rockett[4]. With permission.)

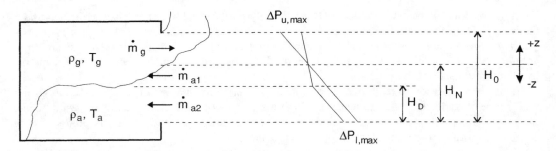

FIGURE 5.15 A schematic of the stratified case.

neutral plane. Consequently, when measuring the geometric quantities H_o, H_N, and H_D, our reference height and zero point is at the bottom of the opening. But when integrating with respect to z, our reference height and zero point is at the neutral plane.

The mass flow rate across a given opening height z from the neutral plane is then achieved by integrating over the height z, using Eq. (5.14)

$$\dot{m} = C_d \int_0^z W \rho_g v(z) dz \qquad (5.14)$$

We will develop mass flow rate equations over the following heights:

- $\dot{m}_g$ over the height H_N to H_o, which in terms of z is $0 < z < H_o - H_N$, measured upward from the neutral plane
- $\dot{m}_{a1}$ over the height H_D to H_N, which in terms of z is $0 < z < H_N - H_D$, measured downward from the neutral plane

- $\dot{m}_{a2}$ over the height 0 to H_D, which in terms of z is $H_N - H_D < z < H_N$, measured downward from the neutral plane

For height H_N to H_o: The velocity of gases leaving the vent, as a function of height z, can be expressed by rewriting Eq. (5.16) as

$$v(z) = \sqrt{\frac{2z(\rho_a - \rho_g)g}{\rho_g}} \tag{5.30}$$

The mass flow rate is achieved by integrating from the height $z = 0$ at the neutral plane up to the height $z = H_o - H_N$ (note that this is not the same as integrating from $z = H_N$ to $z = H_o$, since z is defined as 0 at the neutral plane). Using Eq. (5.14), the mass flow rate is then expressed as

$$\dot{m}_g = C_d \int_0^{H_o - H_N} W\rho_g v(z)dz$$

Inserting Eq. (5.30) for the velocity, we get

$$\dot{m}_g = C_d W\rho_g \sqrt{\frac{2(\rho_a - \rho_g)g}{\rho_g}} \int_0^{H_o - H_N} \sqrt{z}dz$$

which becomes

$$\dot{m}_g = \frac{2}{3} C_d W\rho_g \sqrt{\frac{2(\rho_a - \rho_g)g}{\rho_g}} (H_o - H_N)^{3\,2} \tag{5.31}$$

For height H_D to H_N: The velocity of gases entering the vent between H_D and H_N, as a function of height z, can be written by rewriting Eq. (5.17) as

$$v_{a1}(z) = \sqrt{\frac{2z(\rho_a - \rho_g)g}{\rho_a}} \tag{5.32}$$

The mass flow rate is achieved by integrating from the height $z = 0$ at the neutral plane down to the height $z = H_N - H_D$. The mass flow rate is then expressed as

$$\dot{m}_{a1} = C_d \int_0^{H_N - H_D} W\rho_a v(z)dz$$

Inserting Eq. (5.32) for velocity and integrating, this becomes

$$\dot{m}_{a1} = \frac{2}{3} C_d W\rho_a \sqrt{\frac{2(\rho_a - \rho_g)g}{\rho_a}} (H_N - H_D)^{3\,2} \tag{5.33}$$

For height 0 to H_D: The velocity of gases entering the vent below the hot layer, H_D, is constant, since the pressure difference over this height is constant. We will therefore not have to integrate over this height, but simply determine the magnitude of the constant velocity and calculate the mass flow rate from Eq. (5.10) (reproduced here for clarity)

$$\dot{m} = C_d A v \rho \tag{5.10}$$

where A is the area over which the flow occurs ($= W \cdot H_D$), v is the constant velocity, and ρ is the density of the gas entering the vent ($= \rho_a$). The velocity can be expressed using Eq. (5.32) at the height $z = H_N - H_D$. This results in the equation

$$v_{a2} = \sqrt{\frac{2 \cdot (H_N - H_D)(\rho_a - \rho_g)g}{\rho_a}} \tag{5.34}$$

which is the constant velocity from the bottom of the vent to the smoke layer, where the pressure difference is constant at $\Delta P = (H_N - H_D) \cdot (\rho_a - \rho_g) \cdot g$. The mass flow rate can then be written using Eq. (5.10) and (5.34) as

$$\dot{m}_{a2} = C_d W H_D \rho_a \cdot \sqrt{\frac{2 \cdot (H_N - H_D)(\rho_a - \rho_g)g}{\rho_a}} \tag{5.35}$$

Total mass flow rate in through the vent: The total mass flow rate in through the vent is given by the expression

$$\dot{m}_a = \dot{m}_{a1} + \dot{m}_{a2}$$

Adding Eq. (5.33) and (5.35) gives

$$\dot{m}_a = \frac{2}{3} C_d W \rho_a \cdot \sqrt{\frac{2 \cdot (\rho_a - \rho_g)g}{\rho_a}} \left[(H_N - H_D)^{3/2} + \frac{3}{2} H_D (H_N - H_D)^{1/2} \right]$$

Noting that the term in the brackets can be written as

$$(H_N - H_D)^{1/2} \left[(H_N - H_D) + \frac{3}{2} H_D \right] = (H_N - H_D)^{1/2} \left(H_N + \frac{1}{2} H_D \right)$$

results in the expression for the total mass flow rate in through the vent, written as

$$\dot{m}_a = \frac{2}{3} C_d W \rho_a \cdot \sqrt{\frac{2 \cdot (\rho_a - \rho_g)g}{\rho_a}} (H_N - H_D)^{1/2} \left(H_N + \frac{1}{2} H_D \right) \tag{5.36}$$

Position of the neutral plane and position of the smoke layer: In previous sections we have been able to give expressions for the height of the neutral plane by equating the mass flow rates into and out of the vent, and by using an equation relating the height of the vent, H_o, to the neutral plane height. In this case there are two unknowns, H_N and H_D, which are not internally related to H_o. We can therefore not give explicit solutions for the height of the neutral plane and the height of the smoke layer.

Solutions for these heights can, however, be obtained by iteration if the plume mass flow rate (see Chapter 4) is known at the smoke layer height, H_D. This plume mass flow rate defines the mass rate entering the hot layer and must equal the mass flow rate exiting the vent. This relationship is used in computer programs that solve for the neutral layer height and the smoke layer height by iteration or other numerical methods.

In the hand-calculation methods presented here, we must therefore be given one of these heights, usually the smoke layer height. This is often the case in design calculations, where a certain smoke layer height is desirable, and therefore given, and the system is solved for this given height. We shall give an example of such a case in the next section.

EXAMPLE 5.5

An enclosure on fire has a door opening of width 1 m and height 2 m. Smoke flows out through the opening, and the smoke layer is observed to be at the height 1.5 m from the floor. The gas temperature is 300°C. Calculate

(a) the position of the neutral plane

(b) the mass flow rate of gases into and out of the enclosure.

SUGGESTED SOLUTION

Equating the mass flow rates into and out of the vent, Eq. (5.31) and (5.36) gives

$\sqrt{\rho_a}(H_N - H_D)^{1/2}\left(H_N + \frac{1}{2}H_D\right) = \sqrt{\rho_g}(H_o - H_N)^{3\,2}$. We assume $\rho_a = 1.2$ kg/m^3, and using Eq. (5.9) we find $\rho_g = 353/(300 + 273) = 0.616$ kg/m^3. Inserting and solving for H_N gives

$H_N = \dfrac{(2 - H_N)^{3/2}\sqrt{0.616/1.2}}{(H_N - 1.5)^{1/2}} - \dfrac{1}{2}1.5$. Solving by iteration gives $H_N = 1.512$ m. Equation (5.31)

gives the mass flow rate out of the enclosure as $\dot{m}_g = \dfrac{2}{3}0.7 \cdot 1 \cdot 0.616\sqrt{\dfrac{2(1.2 - 0.616)9.81}{0.616}}$

$(2 - 1.512)^{3/2} = 0.423$ kg/s. The mass flow rate into the vent, using Eq. (5.36) becomes

$\dot{m}_a = \dfrac{2}{3}0.7 \cdot 1 \cdot 1.2\sqrt{\dfrac{2(1.2 - 0.616)9.81}{1.2}}(1.512 - 1.5)^{1\,2}\left(1.512 + \dfrac{1}{2}1.5\right) = 0.429$ kg/s. The very small difference in the mass flow rate in and out is due to the sensitivity of the term H_N.

5.5.2 SPECIAL CASE: MASS FLOW OUT THROUGH A CEILING VENT

Consider Figure 5.16, where an enclosure of height H is filled with smoke to the height H_D and the neutral plane is situated at height H_N. The only mass flow out of the enclosure occurs through a ceiling vent of area A_c and the inflow occurs through a vent near the floor with an area of A_l. The pressure difference across the ceiling vent is constant and denoted ΔP_c, the pressure difference across the lower vent is constant and given as ΔP_l. We assume that the mass flow in through the lower vent, $\dot{m}_l$, equals the mass flow rate out through the ceiling vent, $\dot{m}_c$. One could also make use of plume equations and equate these two flow rates to the plume flow rate at the height H_D, but we shall simply solve here for the condition $\dot{m}_l = \dot{m}_c$.

To give expressions for the mass flow rates out of and into the enclosure, we determine the pressure differences across the vents and the flow velocities.

Expressions for pressure differences: The pressure difference across the ceiling vent is given by

$$\Delta P_c = (H - H_N) \cdot (\rho_a - \rho_g) \cdot g \tag{5.37}$$

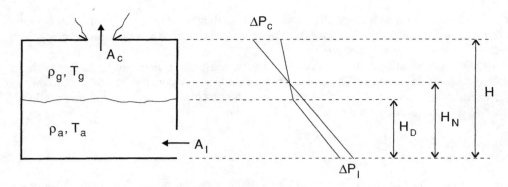

FIGURE 5.16 Enclosure with a ceiling vent for mass flow out and an opening near the floor for mass flow in.

and across the lower vent as

$$\Delta P_l = (H_N - H_D) \cdot (\rho_a - \rho_g) \cdot g \qquad (5.38)$$

Expressions for velocities: The velocities are given from the general expression

$$v = \sqrt{\frac{2\Delta P}{\rho}}$$

and using Eq. (5.37) and (5.38) we find the velocity out through the ceiling vent to be

$$v_c = \sqrt{\frac{2 \cdot (H - H_N) \cdot (\rho_a - \rho_g) \cdot g}{\rho_g}} \qquad (5.39)$$

and the velocity in through the lower vent to be

$$v_l = \sqrt{\frac{2 \cdot (H_N - H_D) \cdot (\rho_a - \rho_g) \cdot g}{\rho_a}} \qquad (5.40)$$

Expressions for mass flow rates: Since the pressure differences are constant across both openings, the mass flow rates through these can be given by using the general expression (also given in Section 5.2.3)

$$\dot{m} = C_d A v \rho \qquad (5.10)$$

The mass flow rate through the ceiling vent becomes

$$\dot{m}_c = C_d A_c \rho_g \cdot \sqrt{\frac{2 \cdot (H - H_N) \cdot (\rho_a - \rho_g) \cdot g}{\rho_g}} \qquad (5.41)$$

and the mass flow rate through the lower vent becomes

$$\dot{m}_l = C_d A_l \rho_a \cdot \sqrt{\frac{2 \cdot (H_N - H_D) \cdot (\rho_a - \rho_g) \cdot g}{\rho_a}} \qquad (5.42)$$

Expression for the height of the neutral plane: There are two unknowns in the above expressions, H_N and H_D. The system of equations therefore cannot be solved unless one of these heights are given. In practice, the layer depth is often a design criterion (for example, the smoke layer must be at least 2 m above the floor of the enclosure), and the engineer is asked to provide a ceiling vent large enough to ensure that the design criterion is met.

We shall therefore use the conservation of mass and derive an expression for H_N, assuming that all other variables are given. Equating the mass flow rates into and out of the enclosure, using Eq. (5.41) and (5.42), gives

$$A_c \sqrt{\rho_g} \sqrt{H - H_N} = A_I \sqrt{\rho_a} \sqrt{H_N - H_D}$$

By squaring the above expression and solving for H_N we get

$$H_N = \frac{A_I^2 \rho_a H_D + A_c^2 \rho_g H}{A_I^2 \rho_a + A_c^2 \rho_g} \tag{5.43}$$

This expression can be inserted into Eq. (5.41) to give the mass flow rate through the ceiling vent. Alternatively, if the area of the ceiling vent is sought, the system can be solved through an iteration process. For design purposes, the required mass flow rate through the vent is often equated to the plume mass flow at height H_D. When the required mass is known the required area of the ceiling vent can be sought.

Equivalent expressions, given in terms of temperature: A well-known expression, based on the same principles of hydraulics as the equations above, has been widely used for design purposes.[2] Through a similar derivation process the expression for mass flow rate through the ceiling vent is given as

$$\dot{m}_c = \frac{C_d A_c \rho_a \sqrt{2g(H - H_D)(T_g - T_a)T_a}}{\sqrt{T_g(T_g + A_c^2 T_a / A_I^2)}} \tag{5.44}$$

The expression is arrived at by writing the pressure differences in terms of velocities as $\Delta P_c = \frac{1}{2}\rho_g v_c^2$ and $\Delta P_I = \frac{1}{2}\rho_a v_I^2$ and realizing that the total pressure difference over both vents can be written $\Delta P_{tot} = (H - H_D)(\rho_a - \rho_g)g = \Delta P_c + \Delta P_I$. Combining these expressions results in a relationship between the two velocities and the two known heights, H and H_D, written as

$$\frac{1}{2}\rho_g v_c^2 + \frac{1}{2}\rho_a v_I^2 = (H - H_D)(\rho_a - \rho_g)g \tag{5.45}$$

Using the relationship between temperatures and densities given by Eq. (5.9) finally results in the well-known expression given by Eq. (5.44). The results from using this equation will be identical to those using Eq. (5.43) and (5.41).

If the inlet area, A_I, is much larger than the outlet area, A_c, Eq. (5.44) can be simplified, and becomes

$$\dot{m}_c = \frac{C_d A_c \rho_a \sqrt{2g(H - H_D)(T_g - T_a)T_a}}{T_g} \tag{5.46}$$

EXAMPLE 5.6

Example 4.3 discussed a 2 m² transformer oil fire and found that the plume mass flow rate at a height of 2.5 m above floor level was $\dot{m}_p$ = 5.6 kg/s. Assume this fire occurs in an enclosure of height 5 m, that there is an opening at the floor level of 5 m², and that the gas temperature is 300°C. What area of ceiling vent is needed to ensure that the smoke layer will stabilize at the height 2.5 m above the floor level?

SUGGESTED SOLUTION 1

To get an initial value for A_c we use Eq. (5.46), assuming $A_1 \gg A_c$. We assume that the mass entering the hot layer through the plume (= 5.6 kg/s) is equal to the mass exiting through the ceiling vent, and so we get

$$5.6 = \frac{0.7 \cdot A_c \cdot 1.2\sqrt{2 \cdot 9.81(5 - 2.5)(300 - 20) \cdot 293}}{(300 + 273)} \Rightarrow 5.6 = 2.94 \cdot A_c.$$

This gives A_c = 1.9 m². We now put this value of A_c into Eq. (5.44) and find $\dot{m}_c$ =

$$\frac{0.7 \cdot 1.9 \cdot 1.2\sqrt{2 \cdot 9.81(5 - 2.5)(573 - 293) \cdot 293}}{\sqrt{573(573 + 1.9^2 \cdot 293/5^2)}} = 5.39 \text{ kg/s}.$$ We must therefore increase A_c a

little and find after some iterations that a value of A_c = 2.0 m² will give a mass flow rate of 5.7 kg/s, which is close to our design criterion.

SUGGESTED SOLUTION 2

We know the mass flow rate into the smoke layer and we therefore set $\dot{m}_p = \dot{m}_c = \dot{m}_1 = 5.6$ kg/s. We use Eq. (5.42) to calculate a value for H_N. The gas density is calculated as 353/(300 + 273) = 0.616 kg/m³. Equation (5.42) becomes

$$5.6 = 0.7 \cdot 5 \cdot 1.2 \cdot \sqrt{\frac{2 \cdot (H_N - 2.5) \cdot (1.2 - 0.616) \cdot 9.81}{1.2}} \Rightarrow H_N = 2.69 \text{ m},$$

Inserting this into Eq. (5.41) gives

$$5.6 = 0.7 \cdot A_c \cdot 0.616 \cdot \sqrt{\frac{2 \cdot (5 - 2.69) \cdot (1.2 - 0.616) \cdot 9.81}{0.616}} \Rightarrow A_c = 2.0 \text{ m}^2.$$

This is the same answer arrived at by iteration in Solution 1.

Limitations: The main limitations of the above methodology for designing required ceiling vent areas are as follows:

- We assume steady-state conditions, i.e., a constant gas temperature. The conditions are, however, likely to be transient. In Chapter 8 we introduce iterative methods that partly take the temperature variation into account. Otherwise, computer models must be used for such calculations, but the vent flow equations still hold.
- The effect of wind on pressure differences between the inside and the outside of a building can be significant. Wind effects are not accounted for here.

- If the ceiling vent has a relatively large area, there is risk for two-way flow in the vent, where hot gases exit at the edges and cold air enters in the center. This effect is sometimes counteracted by splitting the vent into many smaller ones.
- The ceiling vent flow may become bidirectional or oscillatory if inlet vents in the lower part of the enclosure are limited; see Emmons.[3]

REFERENCES

1. Kandola, B.S., "Introduction to Mechanics of Fluids," *SFPE Handbook of Fire Protection Engineering*, 2nd ed., National Fire Protection Association, Quincy, MA, 1995.
2. Hinkley, H.P. and Gardiner, J.P., "Design Principles for Smoke Ventilation in Enclosed Shopping Centres," Building Research Establishment Report, Garston, BRE, 1990.
3. Emmons, H.W., "Vent Flows," *SFPE Handbook of Fire Protection Engineering*, 2nd ed., National Fire Protection Association, Quincy, MA, 1995.
4. McCaffrey, N.J. and Rockett, J.A., "Static Pressure Measurements of Enclosure Fires," *Journal of Research of the National Bureau of Standards*, June 1977.

PROBLEMS AND SUGGESTED ANSWERS

5.1 The stairwell of a tall building is 30 m high. The temperature inside is 20°C and the temperature outside is 0°C. The door at floor level is open with dimensions 1 m wide and 2 m high. Assume a flow coefficient $C_d = 0.7$.
(a) Estimate the pressure difference at the top of the building.
(b) At the ceiling an opening is provided with the same dimensions as the door. Estimate the mass flow rate of air through the building.

Suggested answer: (a) $\Delta P_{max} \approx 26$ Pa; (b) $\dot{m} \approx 7.8$ kg/s

5.2 An enclosure has openings at the bottom and at the top, as shown in the figure below. The height from the floor to the neutral plane is h_1 and from the neutral plane to the ceiling h_2. The temperatures inside and outside are T_g and T_a, respectively. Show that

$$\frac{h_1}{h_2} = K\left[\frac{A_2}{A_1}\right]^2$$ and give an expression for K.

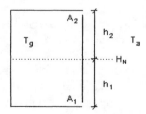

5.3 A large enclosure has geometric dimensions indicated by the figure below. The temperature in the room is 400 K above ambient temperature. $A_1 = A_2 = 6$ m^2.
(a) Set up equations for calculation of the height of the neutral plane, taking into account that the pressure difference varies over the height of the openings. Calculate the height of the neutral plane.
(b) Calculate the height of the neutral plane, but now assume that the pressure difference across each opening is constant over the opening height, evaluated at the center of each opening.

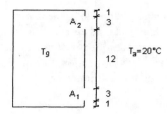

Suggested answer: (a) and (b) Height of neutral plane = 6.96 m from the floor

5.4 An enclosure has a facade and openings according to the figure below and an ongoing, fully developed fire with a gas temperature of T_g = 800°C. Set up the mass balance equations and calculate
(a) the position of the neutral plane
(b) the pressure difference across the openings
(c) the mass flow rates through the openings

The openings can be considered to be narrow with respect to height.

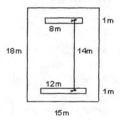

Suggested answer: (a) Neutral layer position is 1.51 m up from the center of the lower opening; (b) ΔP_{lower} = 12.9 Pa, ΔP_{upper} = 107 Pa; (c) $\dot{m}_a \approx$ 47 kg/s, $\dot{m}_g \approx$ 47 kg/s.

5.5 An atrium in a hotel building is 20 m high and has a floor area of 600 m². The atrium has relatively large openings at the floor level (≈10 m²). An elevator door, which is 15 m above the floor, has a leakage area of 0.06 m², and smoke can spread from the atrium to the rest of the building, mainly through this leak. If the gas temperature in the atrium is 250°C and the ambient temperature is 20°C, calculate the mass flow rate through the elevator door.

Suggested answer: Assuming the pressure difference across the lower opening to be ≈0, the pressure difference at 15 m height causes $\dot{m}_g \approx$ 0.43 kg/s.

5.6 An apartment building has 16 levels that are each 3 m high, connected by a stairwell. The apartments have relatively large leakage areas to the outside and the temperature in the apartments is 20°C, as is the ambient outside temperature. The pressure in the apartments is therefore assumed to be equal to the pressure outside. A relatively weak fire in some rubbish fills the stairwell with smoke at a temperature of 60°C. The leakage area between the apartments and the stairwell is ≈0.03 m² per level. Assume $C_d \approx$ 0.7 and $\rho_a \approx$ 1.2 kg/m³. Calculate the mass flow rate of smoke into the top most apartment if
(a) the stairwell has no other openings
(b) a door of 2 m² at the bottom level is open

Suggested answer: Assuming the top opening to be situated 1.5 m from the top of the building, we get (a) $\dot{m}_g \approx$ 0.17 kg/s; (b) $\dot{m}_g \approx$ 0.26 kg/s

5.7 A fire in a 12-m high building with a 2-m high door at the floor level develops successively and goes through the four stages A, B, C and D; smoke filling without outflow of hot gases (A), outflow of hot and cold gases (B), and outflow of hot gases and inflow of cold gases (C and D). Draw the pressure difference profile across the building height for each stage, and explain how these occur. Indicate the position of the neutral plane, smoke layer height, and flow directions through the opening.

5.8 An enclosure is shown in the figure below. The enclosure is filled with hot smoke with density $\rho_g = 0.6$ kg/m³. Both openings have a height of 0.5 m and can be considered to be narrow with respect to the total height of the building. Opening A is 2 m². Draw a schematic (no numbers necessary) of the pressure profile across the building facade for the three cases
(a) opening B = 0 m²
(b) opening B = 2 m²
(c) opening B = 4 m²

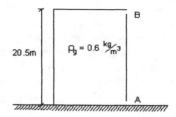

5.9 Equations (5.23) and (5.24) assume that the fuel burning rate can be ignored when calculating the mass flow rate in through an opening. Derive Eq. (5.29)

$$\dot{m}_a = \frac{2.1 \cdot A \sqrt{H_o}}{\left[1 + 1.6 \cdot (1 + \dot{m}_b / \dot{m}_a)^{2/3} \right]^{3/2}}$$

showing how the fuel burning rate can be taken into account when calculating mass flow rate in through an opening.

5.10 The fuel load in an office can result in an energy release rate of 5 MW (mostly wood-type fuel). The room has an opening 2 m wide and 2 m high.
(a) Give a rough estimate of the mass flow into the enclosure, ignoring the burning rate.
(b) Calculate the mass flow into the enclosure, taking into account the burning rate.
(c) 30 m² of the walls are now lined with wood panel with $Q'' = 200$ kW/m² and $\Delta H_{eff} = 16$ MJ/kg, and this burns in addition to the 5 MW. How will this added burning rate influence the mass flow rate into the room?
(d) Compare the mass flow rate of unburnt gases out of the opening for cases (b) and (c).

Suggested answer: (a) 2.8 kg/s; (b) burning rate $\dot{m}_b = 0.31$ kg/s, mass flow rate $\dot{m}_a = 2.64$ kg/s; (c) burning rate $\dot{m}_b = 0.678$ kg/s, mass flow rate $\dot{m}_a = 2.42$ kg/s; (d) the maximum possible energy release rate is given by assuming that all in-flowing oxygen is used for combustion, which gives for (b) $Q_{max} = 2.64 \cdot 0.23 \cdot 13.1 = 7.95$ MW. Since the fire develops 5 MW we assume only small amounts of unburnt gases flow out. For case (c) $Q_{max} = 2.42 \cdot 0.23 \cdot 13.1 = 7.3$ MW. The fire can potentially develop ≈ 11 MW. If all the oxygen is used for combustion inside the compartment (unlikely), the amount of unburnt gases flowing out is (11 − 7.3)/16 = 0.23 kg/s.

5.11 An enclosure has two openings, A_1 and A_2, as shown in the figure below. A_1 has a height $h_1 = X_1$ and width b_1. A_2 has a height $h_2 = X_3 - X_2$ and width b_2. For a fully developed fire with $T_g = 900°C$, the neutral plane lies exactly at the height X_1. This is only possible for a certain relation between X_1, X_2, X_3, b_1 and b_2. Give this relationship.

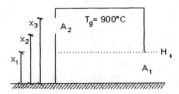

Suggested answer: $\quad 2.0 \dfrac{b_1}{b_2} = \dfrac{\left(X_3 - X_1\right)^{3/2} - \left(X_2 - X_1\right)^{3/2}}{X_1^{3/2}}$

5.12 The figure below illustrates the mass flow rates of gases for a fully developed fire. If the area of the ceiling vent increases, a situation will arise where the neutral plane is exactly in line with the top of the window opening and all hot gases exit through the ceiling vent. The window height is H and its area is A, the height from the window soffit to the ceiling vent is H_c and the area of the ceiling vent is A_c Show that the relation $\dfrac{A\sqrt{H}}{A_c\sqrt{H_c}} = 1.5\sqrt{\dfrac{\rho_g}{\rho_a}}$

applies in this situation.

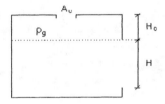

5.13 A fire in an enclosure, with a door at floor level 2 m wide and 2 m high, causes the upper layer to stabilize at a height of 1.7 m from the floor with an upper layer gas temperature of 200°C. Calculate the position of the neutral plane and estimate the flow rate of gases out of the enclosure.

Suggested answer: Due to the sensitivity in the equations used, we give numbers with four decimal places: Using $\rho_g = 0.7463$ kg/m³, we get $H_N = 1.7025$ m, $\dot{m}_g = 0.3904$ kg/s, $\dot{m}_a = 0.3893$ kg/s.

5.14 An enclosure is 6 m high and has an opening at the floor level 2 m wide and 2 m high. In case of fire the smoke gases are not to descend further than 2 m from the floor level. Assuming a fire that causes a plume mass flow rate of 7 kg/s at the height of 2 m and an upper layer temperature of 300°C, determine the needed area of ceiling ventilation.

Suggested answer: A first guess can be arrived at by assuming $A_1 >> A_c$, which gives a rough value of $A_c \approx 1.9$ m². Using a value of $A_c = 2$ m² in Eq. (5.43) gives $H_N = 2.455$ m. Using this in Eq. (5.41) and (5.42) gives mass flow rate in equal to mass flow rate out equal to mass flow rate in plume at height 2 m.

6 Gas Temperatures in Ventilated Enclosure Fires

It is of considerable importance to the fire protection engineer to be able to roughly predict the hot gas temperature in a fire compartment. This knowledge can be used to assess when hazardous conditions for humans will occur, when flashover may occur, when structural elements are in danger of collapsing, and the thermal feedback to fuel sources or other objects. Any prediction of the hot gas temperature in a compartment fire must be based on the conservation of energy and mass; this usually leads to a number of differential equations that can be solved numerically to predict the temperature. By simplifying the energy and mass balances, one can arrive at fairly simple analytical equations that can predict the hot gas temperature for a number of compartment fire scenarios. These scenarios usually require that the fire is well ventilated, i.e., that there is an opening to the outside. This chapter reviews a number of methods that have been developed for predicting temperatures in both pre- and post-flashover phases of well-ventilated compartment fires.

CONTENTS

6.1 Terminology 115
6.2 Introduction 116
6.3 The Pre-Flashover Fire 117
    6.3.1 A Simplified Energy Balance 117
    6.3.2 Experiments and Statistical Correlation 120
    6.3.3 Calculation of the Effective Heat Transfer Coefficient.........................120
    6.3.4 Calculational Procedure 123
    6.3.5 Limits of Applicability...124
    6.3.6 Predicting Time to Flashover...124
    6.3.7 Some Related Expressions for Special Cases 125
6.4 The Post-Flashover Fire...127
    6.4.1 Definitions of Some Terms 128
    6.4.2 The Energy and Mass Balance 129
    6.4.3 Method of Magnusson and Thelandersson...132
    6.4.4 Other Related Methods 135
References 138
Problems and Suggested Answers 139

6.1 TERMINOLOGY

Fire load — The fire load for an enclosure is a measure of the total energy released by combustion of all combustible materials in the enclosure. It is assigned the symbol Q and is given in [MJ].

Fire load density — The fire load density is the fire load per unit area. The fire load density is assigned the symbol Q'' and is given in [MJ/m²]. In some countries, the fire load is given per unit floor area of the enclosure; in other countries this is given in terms of the total enclosure surface area, A_t.

The MQH method — A hand-calculation method for calculating gas temperatures in the pre-flashover stage.

Nominal temperature–time curves — These are standard temperature–time curves to which structural components can assumed to be subjected during the post-flashover fire. The best known such curve is termed the "ISO 834" curve. The standard curves, defined by different standards organizations, do not take into account different compartment geometries, openings, fuel content, and thermal properties.

The opening factor — The opening factor is given as the ventilation factor divided by the total enclosure surface area A_t. The opening factor is therefore $A_o\sqrt{H_o}/A_t$. This factor has been found to be very useful when systemizing calculated temperature–time curves for the post-flashover case.

The post-flashover stage — When the objective of fire safety engineering design is to ensure structural stability and safety of firefighters, the post-flashover stage is of greatest concern. The design load in this case is characterized by the temperature–time curve assumed for the fully developed fire stage.

The pre-flashover stage — The growth stage of a fire, where the emphasis in fire safety engineering design is on the safety of humans. The design load in this case is characterized by an energy release rate curve, where the growth phase of the fire is of most importance.

Simulated natural fire exposure — Refers to methods for calculating or approximating temperature–time curves for the post-flashover stage, where the calculations are based on a solution of the enclosure mass and energy balances. These methods take into account different compartment geometries, openings, fuel content, and thermal properties.

Total enclosure surface area — The total surface area bounding the enclosure, including openings, is given the symbol A_t. Note that in the previous sections we have used the term A_T to represent the enclosure surface area, not including openings.

The ventilation factor — The factor $A_o\sqrt{H_o}$ (where A_o is opening area and H_o is opening height) has been found to be directly proportional to the mass flow rate of air in through an opening during the post-flashover stage. This factor is termed the ventilation factor.

6.2 INTRODUCTION

When the environmental conditions due to a fire in an enclosure are to be evaluated, estimation of the temperature of the gases is of central importance. Not only will the temperature have a direct impact on human safety and structural safety, but an estimate of temperature also is necessary for predicting mass flow rates in and out through openings, thermal feedback to the fuel and other combustible objects, and thermal influence on detection and sprinkler systems, to name a few important processes.

Heat poses a significant physical danger to humans, both when the skin or lungs are in direct contact with heated air and when heat is radiated from a distance. Building regulations therefore typically specify a certain maximum temperature to which humans may be exposed, as well as some maximum radiative heat flux. Calculations of the load-bearing and separating capacity of structural elements are also based on the temperature to which the elements are exposed in case of fire. Building fire regulations thus commonly require that two main objectives be met: life safety of the occupants and structural stability of the building. Two distinctly different design procedures are applied in each case, the former to do with the pre-flashover stage of the fire, the latter with the post-flashover stage.

Pre-flashover stage: In the case where the objective is to facilitate escape for the occupants the time frame is usually relatively short (most often less than 30 minutes) and the design fire is specified as energy release rate vs. time. When the energy release rate is known, the conservation equations of mass and energy will allow the gas temperature to be calculated. A number of engineering assumptions must be made along the way to allow solution of these equations.

Enclosure gas temperatures resulting from fire can vary greatly depending on position in the enclosure. Temperatures within the flame zone will be very high; ambient air entrained through openings will be heated only slightly as it flows toward the flame. Once entrained to the plume, the gases will flow upward and cool due to mixing with colder gases and heat transfer to surfaces.

Despite this extreme range of spatially varying gas temperatures in an enclosure fire, a simplified description can be achieved by using the **two-zone model** concept: a uniform upper gas region within the enclosure, a lower region at ambient temperature, and a plume region represented as a localized heat source. This is the most commonly used assumption when calculating gas temperatures in enclosure fires.

In all cases, calculation of the temperature must be based on the energy and mass balance in the fire compartment. This will often lead to a number of coupled differential equations that must be solved by computer, especially if effects on rooms other than the fire compartment are to be studied. For the case where the temperature is to be calculated in the fire compartment, some simple methods can be used for calculations, and we discuss these in Section 6.3.

Post-flashover stage: In the case of structural stability the objective is to protect property and ensure that firefighters can gain entry to the building without the risk of a structural collapse. Here, the time frame is relatively long (often 0.5 to 3 hours), the fire is assumed to have caused flashover at a very early stage, and the design fire is usually given as a temperature–time curve.

Here, the most commonly used assumption is the **one-zone model,** where the entire compartment is assumed to be filled with fire gases of uniform temperature. Again, the calculations have a basis in the energy and mass balance of the compartment, and the objective is to arrive at a temperature–time curve covering the whole process of fire development.

We discuss the energy balance usually applied to post-flashover fires and review some methods for assessing post-flashover temperatures in Section 6.4.

6.3   THE PRE-FLASHOVER FIRE

Calculation of the enclosure gas temperatures must be based on the enclosure energy balance. The mass balance must also be considered, since mass flow into and out of the enclosure has an effect on the energy balance. McCaffrey, Quintiere, and Harkleroad[1] developed a simple method for estimating the gas temperature in a ventilated enclosure that we review in the next sections.

They set up a simplified energy balance for the room and thus derived a set of easily computed dimensionless variables upon which the temperature depends. Using experimental results, where the temperature was a measured quantity and all the dependent dimensionless variables were known, they determined a number of constants through regression analysis and thus arrived at an expression where the gas temperature can be computed directly by hand.

An excellent review of this method, with calculated examples, is given by Walton and Thomas in the *SFPE Handbook,*[2] and Drysdale also discusses the method in some detail.[3] Our description will be similar to their descriptions.

6.3.1   A SIMPLIFIED ENERGY BALANCE

A schematic of the problem under consideration is given in Figure 6.1. A room with an opening of height H_o and area A_o contains a fire of energy release rate $\dot{Q}$. The mass flow rate out through the opening is $\dot{m}_g$. The compartment walls have a thickness δ and material properties k (conductivity), ρ (density), and c (specific heat).

An energy balance for a fire compartment can be set up in many different forms. An energy balance may be reasonably complete, accounting for many minor processes, or it may consider

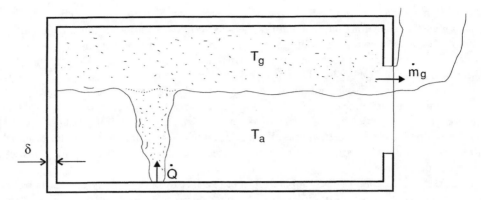

FIGURE 6.1   Schematic of a fire compartment with upper-layer temperature T_g.

only the most dominant processes. A simple energy balance for the enclosure depicted in Figure 6.1 may be stated as follows:

$$
\begin{bmatrix} \text{Rate of energy} \\ \text{released in the} \\ \text{compartment} \end{bmatrix} = \begin{bmatrix} \text{Rate of energy} \\ \text{lost due to fluid} \\ \text{flow out through} \\ \text{opening} \end{bmatrix} + \begin{bmatrix} \text{Rate of heat loss} \\ \text{by the hot gases} \\ \text{to compartment} \\ \text{boundaries} \end{bmatrix} \tag{6.1}
$$

A number of other terms can be taken into account, for example the rate of heat lost due to flame radiation out through the opening and the rate at which energy is stored in the gas, but these can be said to be minor and the terms given in Eq. (6.1) may be considered the dominant terms. We discuss the energy balance further in Chapter 8 and present it in a different form. For now, we are satisfied with the simple energy balance given above, which can be written as

$$
\dot{Q} = \dot{m}_g c_p \left(T_g - T_a \right) + \dot{q}_{loss} \tag{6.2}
$$

where $\dot{Q}$ is energy release rate (kW), $\dot{m}_g$ is the mass flow rate out through the opening (kg/s), c_p is the specific heat of the gases (kJ/kgK), T_g and T_a are the upper layer and ambient gas temperatures (°C or K), and $\dot{q}_{loss}$ is the heat lost by the gases to the compartment boundaries (kW).

Energy release rate: For our purposes the energy release rate is assumed to be known. In a design situation the energy release rate is specified and arrived at using methods described in Chapter 3. In a situation where experiments have been carried out, the energy release rate is either measured or calculated from the mass loss rate of fuel, also, discussed in Chapter 3.

Heat lost to the compartment boundaries: The heat lost to the boundaries involves many heat transfer processes. Radiation and convective heat transfer occurs at the solid boundaries, followed by conduction of the heat into the solid. Radiative heat loss also occurs at openings. The dominant term is heat lost by conduction to the solid. We shall therefore define the term h_k as an effective heat conduction term for the solid boundaries and A_T as the boundary surface area to be used for heat transfer considerations and write this heat loss term as

$$
\dot{q}_{loss} = h_k A_T \left(T_g - T_a \right) \tag{6.3}
$$

We discuss the term, the assumptions involved, and how it is to be computed in Section 6.3.3.

Energy lost due to fluid flow through openings: In Chapter 5 we investigated flow rates through openings for the stratified case: the case where the upper part of the compartment is filled with hot gases and the lower with cold air. We found that the flow rate of hot gases out through the opening could be written as Eq. (5.31)

$$\dot{m}_g = \frac{2}{3} C_d W \rho_g \sqrt{\frac{2(\rho_a - \rho_g)g}{\rho_g}} (H_o - H_N)^{3/2}$$

where H_0 is the height of the opening, H_N is the height of the neutral layer, and W is the width of the opening (see Figure 5.15). The densities can be transformed to temperatures using the ideal gas law relation $r_a T_a = r_g T_g$ and Eq. (5.31) becomes

$$\dot{m}_g = \frac{2}{3} C_d W \rho_a \sqrt{2g \frac{T_a}{T_g} \left(1 - \frac{T_a}{T_g}\right)} (H_o - H_N)^{3/2} \tag{6.4}$$

Once this mass flow is known, the energy content of the gases exiting the opening (or the enthalpy) can simply be assessed by multiplying by the specific heat of the gases and the temperature increase.

Since H_N in Eq. (6.4) is not known, we must write $\dot{m}_g$ as some function of the known variables. We note that $W \cdot H_o^{3/2}$ can be written as $A_o \sqrt{H_o}$ (often termed the ventilation factor) where A_o is the area of the opening, and write

$$\dot{m}_g = \rho_a \sqrt{g} A_o \sqrt{H_o} \cdot f\left(T_g, \dot{Q}, A_o, H_o\right) \tag{6.5}$$

where f stands for "a function of."

Expression for gas temperatures: We can write Eq. (6.2) in terms of temperature increase ($DT = T_g - T_a$) as

$$\Delta T = \frac{\dot{Q}}{\dot{m}_g c_p + h_k A_T} \tag{6.6}$$

Equation (6.6) cannot be solved directly, since the height of the neutral layer in Eq. (6.4) is not known. This could be amended by equating $\dot{m}_g$ with $\dot{m}_a$, the expression for the in-flowing air (Eq. (5.36)) and then solving by iteration on a computer, but we are interested in a simpler procedure.

Dimensionless variables: Due to the unknowns, McCaffrey and colleagues defined a number of dimensionless variables, rewrote Eq. (6.6) expressing the temperature rise in terms of these, and used experimentally measured values to determine a number of constants, allowing the gas temperature to be calculated directly.[1]

The temperature increase can be made dimensionless by dividing Eq. (6.6) through by T_a, giving

$$\frac{\Delta T}{T_a} = \frac{\dot{Q}}{\dot{m}_g c_p T_a + h_k A_T T_a} = \frac{\dot{Q}/\dot{m}_g c_p T_a}{1 + \frac{h_k A_T}{\dot{m}_g c_p}} \tag{6.7}$$

Substituting the known dimensions from Eq. (6.5) into this expression and rearranging, DT/T_a can be expressed as a function of two dimensionless groups:

$$\frac{\Delta T}{T_a} = f\left(\frac{\dot{Q}}{\sqrt{g}\rho_a c_p T_a A_o \sqrt{H_o}}, \frac{h_k A_T}{\sqrt{g}\rho_a c_p A_o \sqrt{H_o}}\right) \tag{6.8}$$

6.3.2 Experiments and Statistical Correlation

McCaffrey and colleagues termed the two dimensionless groups in Eq. (6.8) X_1 and X_2 and assumed a relationship for the dimensionless temperature rise as

$$\frac{\Delta T}{T_a} = CX_1^N X_2^M \tag{6.9}$$

To determine the appropriate numerical values for the coefficients C, N, and M, they analyzed over 100 sets of experimental data, where all the variables in Eq. (6.8) were known.[1] The fuel used in the experiments was gas, wood, or plastics; the scale ranged from conventional room sizes down to nearly 1/8 of that; both door and window openings were used; and the construction material had a wide range of thermal properties.

Through regression analysis of the experimental data, the constants C, N, and M were found, so that Eq. (6.8) could be rewritten as

$$\frac{\Delta T}{T_a} = 1.63 \left(\frac{\dot{Q}}{\sqrt{g}\rho_a c_p T_a A_o \sqrt{H_o}} \right)^{2\,3} \cdot \left(\frac{h_k A_T}{\sqrt{g}\rho_a c_p A_o \sqrt{H_o}} \right)^{-1/3} \tag{6.10}$$

Figure 6.2 shows the results of the correlation. The line in the figure represents the least squares fit to the experimental data, which results in the coefficients C, N, and M given in Eq. (6.9) and (6.10).

A more convenient form of Eq. (6.10) is achieved by using conventional values for some constant quantities ($g = 9.81$ m/s^2, $r_a = 1.2$ kg/m^3, $T_a = 293$K, and $c_p = 1.05$ kJ/(kg K)). This results in the expression

$$\Delta T = 6.85 \left(\frac{\dot{Q}^2}{A_o \sqrt{H_o} h_k A_T} \right)^{1/3} \tag{6.11}$$

Here specific units must be used, $\dot{Q}$ in [kW], h_k in [kW/(m^2 K)], areas in [m^2] and the opening height in [m].

When calculating heat conduction into a solid, only the area in contact with the hot gas should be used, since this is the affected area for heat transport. This area changes through time due to the descent of the smoke layer, and it is unknown since the smoke layer height is unknown. McCaffrey and colleagues chose to represent this area by using the total internal surface area in the compartment, A_T, which is in strong relation to the affected area for heat transport.[1] However, the effects of this approximation are taken up by the experimentally determined coefficients in Eq. (6.10).

6.3.3 Calculation of the Effective Heat Transfer Coefficient

The simplest way of taking account of the heat conducted into the solid is to neglect any cooling from the boundaries and to assume that the boundary surfaces have a temperature $T_s = T_g$. Solving the general heat conduction equation for this case, assuming that the boundaries are semi-infinite in thickness, results in an expression for the heat flux per unit area conducted into the wall given by

$$\dot{q}'' = \frac{1}{\sqrt{\pi}} \cdot \sqrt{\frac{k\rho c}{t}} \cdot (T_g - T_a) \tag{6.12}$$

where k, r, c are the conductivity, density, and specific heat of the compartment surface material, respectively. The total heat transfer coefficient for this case would therefore be $h = \sqrt{\frac{k\rho c}{\pi t}}$.

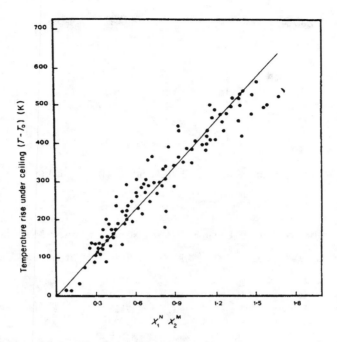

FIGURE 6.2 Correlation of the upper-layer gas temperatures with the two dimensionless variables X_1 and X_2. (From Drysdale[3]. With permission.)

For very thin solids, or for conduction through solid that goes on for a long time, the process of conduction becomes stationary, and the rate of heat conducted through the solid becomes

$$\dot{q}'' = \frac{k}{\delta}(T_g - T_a) \tag{6.13}$$

where δ is the thickness of the solid, resulting in $h = k/\delta$.

The time at which the conduction can be considered to be approaching stationary heat conduction is termed the *thermal penetration time*, t_p. This time can be given as

$$t_p = \frac{\delta^2}{4\alpha} \tag{6.14}$$

and indicates the time at which $\approx 15\%$ of the temperature increase on the fire-exposed side has reached the outer side of the solid. Here α is the thermal diffusivity, also given by the relation $\alpha = k/\rho c$, and given in [m^2/s] (see Table 6.1). Figure 6.3 shows a schematic of the temperature distribution in the solid for the two cases.

h_k **as defined by McCaffrey and colleagues:** McCaffrey and colleagues analyzed the surface materials used in the experiments, which Eq. (6.10) is based on, and defined h_k in the following manner:[1]

For $t < t_p$ $\qquad h_k = \sqrt{\dfrac{k\rho c}{t}}$ $\tag{6.15}$

and for $t \geq t_p$ $\qquad h_k = \dfrac{k}{\delta}$ $\tag{6.16}$

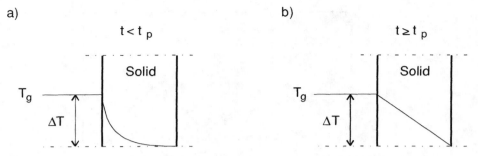

FIGURE 6.3 Assumed temperature distribution in the solid for (a) $t < t_p$ and (b) $t \geq t_p$.

TABLE 6.1
Typical Values of Thermal Properties for Some Common Materials

Material	k (W/m·K)	c (J/kg·K)	r (kg/m³)	krc (W²s/m⁴K²)	a (m²/s)
Aluminium	218	890	2700	$5.2 \cdot 10^8$	$9.1 \cdot 10^{-5}$
Copper	395	385	8920	$1.4 \cdot 10^9$	$1.2 \cdot 10^{-4}$
Steel (mild)	45	460	7820	$1.6 \cdot 10^8$	$1.3 \cdot 10^{-5}$
Brick (common)	0.69	840	1600	$9.3 \cdot 10^5$	$5.2 \cdot 10^{-7}$
Concrete	0.8–1.4	880	1900–2300	$2 \cdot 10^6$	$5.7 \cdot 10^{-7}$
Lightweight concrete	0.15	1000	500	$7.5 \cdot 10^4$	$3.0 \cdot 10^{-7}$
Glass (plate)	0.8	840	2600	$1.8 \cdot 10^6$	$3.7 \cdot 10^{-7}$
Cork plates	0.08	1000	500	$4.0 \cdot 10^4$	$1.6 \cdot 10^{-7}$
Fiber insulating board	0.041	2090	229	$2.0 \cdot 10^4$	$8.6 \cdot 10^{-8}$
Gypsum plaster	0.48	840	1440	$5.8 \cdot 10^5$	$4.1 \cdot 10^{-7}$
Mineral wool, plates	0.041	800	100	$3.3 \cdot 10^3$	$5.1 \cdot 10^{-7}$

Note that the use of the constant $\dfrac{1}{\sqrt{\pi}}$ has been dropped from Eq. (6.15), but the experimentally determined coefficients in Eq. (6.10) make up for that.

Combinations of different materials: For an enclosure bounded by different building materials the overall value of h_k must be weighted with respect to areas. For example, if the walls and ceiling (W, C) are of a different material to the floor (F), the value of h_k is calculated, for $t < t_p$,

$$h_k = \frac{A_{W,C}}{A_T} \sqrt{\frac{(k\rho c)_{W,C}}{t}} + \frac{A_F}{A_T} \sqrt{\frac{(k\rho c)_F}{t}} \qquad (6.17)$$

and for $t \geq t_p$,

$$h_k = \frac{A_{W,C}}{A_T} \frac{k_{W,C}}{\delta_{W,C}} + \frac{A_F}{A_T} \frac{k_F}{\delta_F} \qquad (6.18)$$

For a composite of many layers, h_k should be computed as appropriate for a series of thermal conductors. For a composite consisting of n layers,

$$h_k = \frac{1}{\displaystyle\sum_{i=1}^{n} \frac{1}{h_{k,i}}} \qquad (6.19)$$

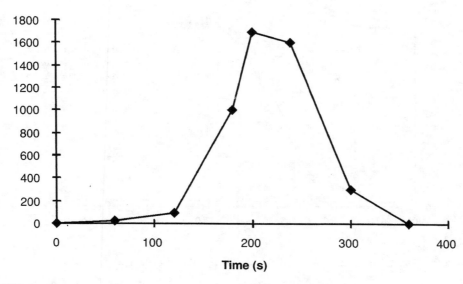

FIGURE 6.4   Energy release rate for the experiment discussed in Example 6.1.

Thermal properties for materials: Table 6.1 gives some thermal properties for common materials.

6.3.4 CALCULATIONAL PROCEDURE

1. Calculate t_p from Eq. (6.14). Use SI units.
2. Calculate h_k from Eq. (6.15) or (6.16) using values of $k\rho c$ from Table 6.1, using SI units. Divide by 1000 to get units of [kW/m²K].
3. Calculate $A_T = A_{walls} + A_{ceiling} + A_{floor} - A_{opening}$ in [m²].
4. Use Eq. (6.11) to calculate ΔT, where $\dot{Q}$ is given in [kW]. Add T_a to get T_g.

EXAMPLE 6.1

Experiments were conducted at Lund University where furniture was burnt in a compartment of 3.6 m by 2.4 m, 2.4 m high. The opening was 0.8 m wide and 2.0 m high. The compartment boundaries were made of 0.15 m thick lightweight concrete. The energy release rate was measured by oxygen calorimetry during the experiments and is given in Figure 6.4. Calculate the upper layer gas temperature at times $t = 3$ minutes and $t = 5$ minutes.

SUGGESTED SOLUTION

The time of penetration is calculated from Eq. (6.14) as $t_p = \dfrac{0.15^2}{4 \cdot 3 \cdot 10^{-7}} = 18750$ s, which is

over 5 hours, so the conduction will be transient for a long time. For $t = 180$ s, $h_k = \sqrt{\dfrac{7.5 \cdot 10^4}{180}}$

$= 20.4$ W/m²K $= 0.0204$ kW/m²K. The total interior surface area is $A_T = 4 \cdot (3.6 \cdot 2.4) +$

$2 \cdot (2.4 \cdot 2.4) - 0.8 \cdot 2 = 44.48$ m². At $t = 180$, $\dot{Q} = 1000$ kW. The gas temperature increase

is therefore $\Delta T = 6.85 \left(\dfrac{1000^2}{1.6\sqrt{2} \cdot 0.0204 \cdot 44.48} \right)^{1/3} = 539°C$. Assuming $T_a = 20°C$, we get

$T_g = 539 + 20 = 559°C$. For time $t = 300$ s, $h_k = 0.0158$ kW/m²K, $\dot{Q} = 300$ kW, and $T_g = 283°C$.

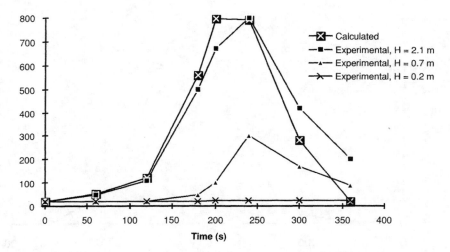

FIGURE 6.5 Comparison of measured and calculated upper-layer gas temperatures.

Figure 6.5 shows the results of the calculations compared to the temperatures measured in the experiments mentioned in Example 6.1. The temperatures were measured at three heights: 2.1 m, 0.7 m, and 0.2 m from the floor. W can see from the figure that Eq. (6.11) gives a good estimate of the temperature near the ceiling for this experiment. The temperatures shown go somewhat beyond the limits of applicability of the calculation method, which we shall discuss below.

6.3.5 LIMITS OF APPLICABILITY

The data on which the method is based were taken from experiments in conventional-sized rooms. The temperature differences varied from $\Delta T = 20°C$ to $600°C$. The fire source was away from walls. We can summarize the limitations as follows:

1. The rise in temperature must be at least 20°C and at most 600°C. Figure 6.5, however, indicates that the limits may be extended upward somewhat.
2. The method applies to both transient and steady-state fire growths; the energy release rate must be known.
3. The method assumes that there is heat loss due to mass flowing out through openings. It is therefore not applicable to situations where significant time passes before hot gases start leaving the compartment through openings. This may occur in large enclosures where it may take considerable time for the smoke layer to reach the opening height.
4. The fire is assumed to be fuel-controlled. Once it becomes ventilation-controlled and the gas temperature exceeds 600°C, some of the energy will be released outside of the compartment openings and will not be used to heat the gases in the enclosure. This can be amended by estimating the energy release rate occurring inside the compartment and using that value in the calculations.
5. If the burning fuel is flush with a wall or in a corner of the enclosure, the values of the coefficients in Eq. (6.10) must be changed. This will be discussed in Section 6.3.7.

6.3.6 PREDICTING TIME TO FLASHOVER

The method of McCaffrey, Quintiere, and Harkleroad can also be used to predict whether a fire in an enclosure may cause flashover.[1] Flashover is usually considered to occur when the upper layer temperature is in the range 500°C to 600°C, although in reality flashover may occur when the gas temperature is outside these limits. By choosing a temperature rise of $DT = 500°C$ as our flashover criteria and substituting this into Eq. (6.11), we get an expression for the energy release rate necessary to cause this temperature rise, or cause flashover. We assign this energy release rate the symbol $\dot{Q}_{FO}$ and rewrite Eq. (6.11) as

$$\dot{Q}_{FO} = 610(h_k A_T A_o \sqrt{H_o})^{1/2} \qquad (6.20)$$

The result is given in [kW] and the effective heat conduction coefficient, h_k, is given in [kW/(m²K)].

This expression also requires that some time is known, or that the fire has been ongoing for a long time and the heat conduction has become stationary. For compartments with thick concrete walls, the thermal penetration time is many hours, and it is unlikely that a fire slowly and gradually grows up to $\dot{Q}_{FO}$ in a number of hours. A reasonable time frame for estimating the likelihood of flashover is in the range of a few minutes up to around 30 minutes. We note that firefighter reaction time is usually also within this range.

EXAMPLE 6.2

Calculate the energy release rate necessary for flashover in a compartment 3.6 m by 2.4 m, 2.4 m high, with an opening 0.8 m wide and 2 m high. The boundary material can be take to be light-weight concrete.

SUGGESTED SOLUTION

The geometry of the compartment is the same as given in Example 6.1. We found that the thermal penetration time was many hours, and it is unreasonable to assume that a fire will grow slowly for many hours to finally reach flashover. We therefore choose some characteristic time. Here, we choose 2 minutes (to see which energy release rate will cause rapid flashover) and 10 minutes (which is a typical time at which the firefighters will arrive) and we compare results for the two times. After 2 minutes, $h_k = \sqrt{\dfrac{7.5 \cdot 10^4}{120}} = 25$ kW/m²K $= 0.025$ kW/m²K, and Eq. (6.20) gives $\dot{Q}_{FO} = 610(0.025 \cdot 44.48 \cdot 1.6\sqrt{2})^{1/2} = 967$ kW.

After 10 minutes and $h_k = \sqrt{\dfrac{7.5 \cdot 10^4}{600}} = 11.2$ W/m²K $= 0.0112$ kW/m²K and $\dot{Q}_{FO} = 610$ $(0.012 \cdot 44.48 \cdot 1.6\sqrt{2})^{1/2} = 647$ kW. We see that for a rapidly growing fire, the energy release rate must be ≈ 1 MW to cause flashover in this compartment. For a more slowly growing fire, the boundaries will warm up, and after 10 minutes, when the energy release rate is ≈ 650 kW, the upper layer temperature will be ≈ 500 °C.

Equation (6.20) will give somewhat conservative estimates, since flashover often does not occur until the gases have reached $\approx 600°C$. In the experiments discussed in Example 6.1, flames came out through the opening after roughly 3 minutes, indicating that flashover had occurred. The upper layer gas temperature was $\approx 600°C$ at that time and the energy release rate was ≈ 1 MW.

6.3.7 SOME RELATED EXPRESSIONS FOR SPECIAL CASES

Mechanically ventilated compartments: Foote, Pagni, and Alvares followed the methodology of McCaffrey, Quintiere, and Harkleroad and collected data from experiments carried out in a mechanically ventilated enclosure.[4] Only a single enclosure was considered, 4 m wide, 6 m long, and 4.5 m high. Fresh air was introduced at floor level and mechanically pulled out at ceiling level by a fan. Experimental data was collected for a range of energy release rates and fan flow velocities. However, the fuel source was the same for all tests (methane gas) as well as the properties of the boundary materials.

Foot and colleagues found that the following expression best fitted the data:[4]

$$\frac{\Delta T}{T_a} = 0.63 \left(\frac{\dot{Q}}{\dot{m} c_p T_a} \right)^{0.72} \left(\frac{h_k A_T}{\dot{m} c_p} \right)^{-0.36} \tag{6.21}$$

where $\dot{m}$ is the ventilation mass flow rate (in kg/s), c_p is the specific heat of air, often taken to be 1.0 kJ/(kg K), and h_k is calculated as discussed in Section 6.3.3. Using the units [kJ/(kgK)] for specific heat requires h_k is given in [kW/(m²K)] and $\dot{Q}$ in [kW]. When using the expression, one must make certain that the flow velocity allows enough air so that combustion of all the fuel is possible.

Fires by walls and in corners: When the entrainment of cold air into the plume is restricted by a wall or a corner, one can expect higher temperatures in the upper layer. Mowrer and Williamsson[5] discussed how the work of McCaffrey et al.[1] could be applied to fires that were flush by walls or were positioned in a corner. They found that the upper layer temperature could be calculated using Eq. (6.11), multiplied by a factor.

For fires flush to walls they recommended that the results from Eq. (6.11) be multiplied by 1.3, and for fires in corners, the equation should be multiplied by 1.7. The resulting expressions follow. For fires flush with walls

$$\Delta T = 1.3 \cdot 6.85 \left(\frac{\dot{Q}^2}{A_o \sqrt{H_o} h_k A_T} \right)^{1/3} \tag{6.22}$$

and for fires in corners

$$\Delta T = 1.7 \cdot 6.85 \left(\frac{\dot{Q}^2}{A_o \sqrt{H_o} h_k A_T} \right)^{1/3} \tag{6.23}$$

Combustible wall lining materials: Karlsson[6] showed that the methodology of McCaffrey et al.[1] could also be used for the case where a room was lined with combustible lining material and ignited in a corner. He analyzed temperature and energy release data from 24 experiments with different lining materials. A regression analysis of the data revealed that the coefficients N and M were roughly the same as those found by McCaffrey et al.,[1] but that the coefficient C was somewhat higher, as was found by Mowrer and Williamsson.[5] Karlsson[6] found that for the case of combustible lining materials, Eq. (6.11) should be multiplied by a factor of 2 (as opposed to the factor of 1.7 found by Mowrer and Williamsson[5]).

Karlsson also found that the correlation could be used for post-flashover calculations, but energy released outside of the compartment should not be taken into account in the calculations.[6] Equation (6.20) can be used to calculate a maximum energy release rate that can be combusted within the enclosure, which is used as a constant value once the experimentally measured values exceed this maximum.

Results from this approach are shown by Karlsson[6]; experimentally measured and calculated values for 24 sets of experiments were comparable.

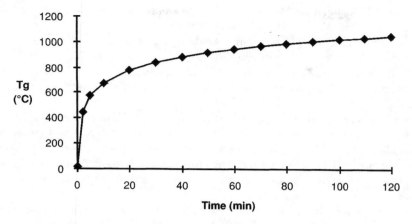

FIGURE 6.6 The standard temperature–time curve as defined by ISO 834.[9]

6.4 THE POST-FLASHOVER FIRE

Calculations of the load-bearing and separating capacity in the design of structural components exposed to fire must be based on knowledge of the thermal exposure to which the structural component is subjected; this is usually quantified by temperature–time curve. Once this is known, the load-bearing and separating capacity can be determined either by testing in a furnace heated according to the temperature–time curve, or by various calculation methods.

In this section we discuss the temperature–time curves structural components are subjected to when they are designed against fire. The two most commonly used approaches are based on either the "nominal temperature–time curves" (also "standard temperature–time curves") or "simulated natural fire exposure" (where the temperature–time curve is calculated or approximated, based on a solution of the enclosure mass and energy balances).

We only briefly discuss the nominal temperature–time curves. The remainder of this chapter deals with simulated natural fire exposure, which allows a more rational design of structural elements exposed to fire. In this case the temperature–time curve can basically be achieved by using three approaches: using a method developed by Magnusson and Thelandersson,[7] using the so-called "parametric fire exposure" (which is basically a compact formulation of [6.9]), or solving the energy and mass balances for each practical case (on which [6.9] is based). Most of this section will therefore deal with the work of Magnusson and Thelandersson.[7]

Nominal temperature–time curves: The *Standard temperature–time curve* is defined by different national standards organizations. The European Committee for Standardization[8] and the International Standards Organization[9] give the "standard" curve as

$$T_g = 20 + 345 \log_{10}(8t + 1) \tag{6.24}$$

where t is time in minutes. Figure 6.6 shows how the temperature to which the structural element is exposed is assumed to change with time. This is also sometimes called the "ISO 834" curve, after the number of the international standards document that gives the curve.

A manufacturer of structural elements will have the element tested in a furnace heated according to the standard curve, and the structural element will then be classified, depending on the time to structural collapse or failure of separating function. Alternatively, determination of the time to failure can be done by various calculation methods, using the standard temperature–time curve as input. A designer or architect will then consult an approved list of structural elements and choose an element that meets with demands set forth in the appropriate building regulations.

The European Committee for Standardization specifies two additional such curves: the "external fire curve" and the "hydrocarbon curve," which are used in specified cases.

Simulated natural fire exposure: The "standard" curve does not take into account the fact that the fire will at some stage decline. Also, it does not account for the fact that different compartment geometries, different sizes of ventilation openings, different fuels, and differences in thermal properties of the compartment boundaries will result in a wide variety of thermal exposure to the structural elements.

In the late 1960s and in the 1970s considerable experimental and theoretical efforts were made to account for the above-mentioned factors, to allow a more rational design of structural elements than the "standard" curve allows. One of these methods was developed by Magnusson and Thelandersson.[7] This method has since become the most frequently used design basis, given by a set of gas temperature–time curves as a function of the fire load density, the opening factor of the compartment, and the thermal properties of the bounding surfaces.

Other methods developed either are not used as frequently or are simplifications or derivations of the method proposed by Magnusson and Thelandersson.[9] Therefore, in the next sections we give an overview of the method developed by Magnusson and Thelandersson and briefly discuss some other methods. In Section 6.4.1 we define some necessary terms; in Section 6.4.2 we discuss the energy balance that is the basis for all such methods; in Section 6.4.3 we give an overview of the method developed by Magnusson and Thelandersson[7]; and in Section 6.4.4 we discuss other related methods for calculating temperature–time curves for the fully developed enclosure fire.

6.4.1 DEFINITIONS OF SOME TERMS

In this section we define a number of terms that are commonly used when determining temperature–time curves for the fully developed enclosure fire, as well as some terms Magnusson and Thelandersson used in their treatment.

Total enclosure surface area — The total surface area bounding the enclosure, including openings, is given the symbol A_t. Note that in the previous sections we used the term A_T to represent the enclosure surface area, not including openings.

Fire load — The fire load for an enclosure is a measure of the total energy released by combustion of all combustible materials in the enclosure. It is assigned the symbol Q and is given in [MJ].

Fire load density — The fire load density is the fire load per unit area. The fire load density is assigned the symbol $Q \leq$ and is given in [MJ/m²]. In some countries, the fire load is given per unit floor area of the enclosure; we call this $Q_f \leq$. In other countries this is given in terms of the total enclosure surface area, A_t; we call this $Q_t \leq$. Since the method of Magnusson and Thelandersson uses the latter,[7] in the following we use $Q_t \leq$.

The fire load density can be arrived at by multiplying the mass of each combustible object with its effective heat of combustion, summing these up, and dividing by A_t, according to the following equation:

$$Q_t'' = \frac{\sum_{i}^{n} M_i \Delta H_{eff,i}}{A_t}$$
(6.25)

where M_i is the mass of object i and $DH_{eff,i}$ the effective heat of combustion of object i.

In a design situation the contents of the enclosure are often not known. Statistical studies have provided tables of fire load densities for different types of buildings and occupancies. A comprehensive list of such data is given by CIB W14.[10] Table 6.2 gives a few values, taken from a more detailed description by Pettersson et al.[11]

TABLE 6.2
Typical Fire Load Density, $Q_t\leq$, for Different Building Occupancies

Type of Occupancy	Fire Load Density, Q_t'' (MJ/m²)
Dwelling, 2 rooms and kitchen	168
Dwelling, 3 rooms and kitchen	149
Offices	709[a]
Schools	96.3
Hospitals	147
Hotels	81.6

Note: The values given are those that were obtained in 80% of observed cases.

Source: Adapted from Petersson et al.[11]

[a] Per floor area, not total enclosure surface area.

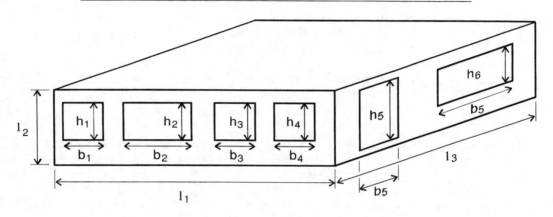

$$A_o = A_1 + A_2 + \ldots + A_6 = b_1h_1 + b_2h_2 + \ldots + b_6h_6$$

$$H_o = (A_1h_1 + A_2h_2 + \ldots + A_6h_6)/A_o$$

$$A_t = 2(l_1l_2 + l_1l_3 + l_2l_3)$$

FIGURE 6.7 Determination of a weighted value of $A_o\sqrt{H_o}$ for enclosures with more than one opening. (Adapted from Pettersson et al.[11])

The opening factor — The ventilation factor was earlier given as $A_o\sqrt{H_o}$. We now define the opening factor, which is the ventilation factor divided by the total enclosure surface area A_t. The opening factor is therefore $\dfrac{A_o\sqrt{H_o}}{A_t}$. For rooms with more than one opening, a weighted value of $A_o\sqrt{H_o}$ is determined according to Figure 6.7. Horizontal openings, such as ceiling vents, can be taken into account in a similar fashion; the details are given in the literature (for example, Refs. 7, 11, and 19).

6.4.2 THE ENERGY AND MASS BALANCE

Figure 6.8 shows a schematic exemplifying the energy balance for an enclosure fire. The energy balance is given by the relation

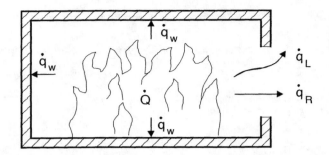

FIGURE 6.8 Energy balance for a fully developed compartment fire.

$$\dot{Q} = \dot{q}_L + \dot{q}_W + \dot{q}_R + \dot{q}_B \tag{6.26}$$

where $\dot{Q}$ = energy release rate due to combustion
$\dot{q}_L$ = rate of heat lost due to replacement of hot gases by cold
$\dot{q}_W$ = rate of heat lost to compartment boundaries
$\dot{q}_R$ = rate of heat lost by radiation through openings
$\dot{q}_B$ = rate of heat storage in the gas volume (most often neglected)

The mass balance is simply given by the relation

$$\dot{m}_g = \dot{m}_a + \dot{m}_b$$

where $\dot{m}_g$ is the mass rate of outflow of hot gases, $\dot{m}_a$ is the mass rate of inflow of air and $\dot{m}_b$ is the fuel burning rate. The mass balance is not discussed further here, but we consider the energy balance of Eq. (6.26) in greater detail below.

Energy release rate: For the fully developed fire, the energy released in the compartment will be ventilation-limited. In Chapter 5 we showed that the flow rate of air in through an opening, for the fully developed fire, could be estimated by Eq. (5.24):

$$\dot{m}_a = 0.5 \cdot A_o \sqrt{H_o} \tag{5.24}$$

Knowing that each kilogram of oxygen used for combustion produces $\approx$13.2 MJ (Huggett[12]) and that 23% of the air mass entering the opening is oxygen, we find that a maximum energy release rate of $1.518 \cdot A_o \sqrt{H_o}$ MW is possible in the compartment, assuming that all the oxygen is used for combustion. Dividing this by an effective heat of combustion for wood of $\approx$17 MJ/kg, we conclude that the maximum energy release rate can be written as

$$\dot{Q} = 0.09 A_o \sqrt{H_o} \cdot \Delta H_{eff,wood} \tag{6.27}$$

where the result is in [MW]. The constant 0.09 is roughly equivalent to the constant in the expression for maximum burning rate for wood given by Kawagoe[13] as $\dot{m}_{f,wood} = 5.5\, A_o \sqrt{H_o}$, expressed in [kg/min] (multiplying 0.09 by 60 seconds gives 5.4, which is roughly equivalent with 5.5).

However, Eq. (6.27) only gives an absolute maximum value for the energy release rate. We are interested in including a fire growth period, a fully developed period, and a decay period. All this depends on a number of factors to do with fuel type, amount, spacing, orientation, and position. Magnusson and Thelandersson provided a method by which the phases of growth, ventilation control, and decay could be taken into account,[7] and we review this in the next section.

Heat lost due to flow of hot gases out of opening: Again we make use of Eq. (5.24) expressing the mass flow rate of gases entering the compartment. By assuming that this equals the mass flow rate exiting the compartment (ignoring the mass produced by the burning fuel), we conclude that the heat lost is the energy content of these gases, written as

$$\dot{q}_L = \dot{m}_g c_p (T_g - T_a) = 0.5 A_o \sqrt{H_o} \, c_p (T_g - T_a) \tag{6.28}$$

where the results are in [kW] if the specific heat of air is used as $c_p = 1.0$ kJ/(kgK).

Heat lost to the compartment boundaries: This term was briefly discussed in Section 6.3.3. For a rough estimate of this term, Eq. (6.12), (6.13), and (6.14) can be used. As an example, for $t < t_p$, Eq. (6.12) can be multiplied by the area that is in contact with the boundaries (which is the total internal area minus the opening areas). The term can then be written as

$$\dot{q}_w = (A_t - A_o) \frac{1}{\sqrt{\pi}} \cdot \sqrt{\frac{k \rho c}{t}} \cdot (T_g - T_a) \tag{6.29}$$

However, the limitations of Eq. (6.12) and (6.13) are such that they cannot be used for the purpose of deriving temperature–time curves for the complete duration of the enclosure fire. These equations were derived from the general heat conduction equation with the assumption that T_g is constant for all times, but in our case the gas temperature changes dramatically with time. Also, the boundary condition assumed no convective and radiative cooling from either surface.

Magnusson and Thelandersson therefore solved the general heat conduction equation numerically by computer, allowing much more realistic boundary conditions and initial conditions to be applied.[7] The numerical solution of this equation is a standard application in many engineering disciplines and will not be discussed further here.

Heat lost by radiation through opening: The radiation losses through the opening can be calculated in a conventional way as

$$\dot{q}_R = A_o \varepsilon_f \sigma (T_g^4 - T_a^4) \tag{6.30}$$

where e_f is an average emissivity for the flames and gases radiating out through the opening, s is the Stefan–Boltzmann constant, and A_o the area of the opening.

Heat lost due to storage in the gas volume: This term arises since the total mass in the compartment changes as the gases heat up. The change of mass causes a change in the energy balance. Hägglund gives this term explicitly when describing a computer program for simulating fires in enclosures.[14] Running the computer program allows the term to be evaluated, and for typical fires in enclosures the term is less than 1% of the total energy balance. The term is therefore most often ignored.

EXAMPLE 6.3

A room has a window, 3 m wide and 2 m high. The bounding surfaces have an area of 400 m²
($= A_t - A_o$) and consist of 20-cm-thick concrete, $krc = 2 \cdot 10^6$ (W²s/m⁴K²) from Table 6.1. The gas temperature during a fully developed fire $T_g = 1000°C$ and is assumed constant throughout the fire duration. Calculate the terms in the energy balance at 330 seconds and at 120 minutes.

SUGGESTED SOLUTION

Equation (6.28) $\dot{q}_L = 0.5 A_o \sqrt{H_o} \, c_p (T_g - T_a)$ gives, using $c_p = 1.2$ kJ/(kg K), the heat capacity of air at a temperature of $\approx 1000°C$:

$$\dot{q}_L = 0.5 \cdot 6\sqrt{2} \cdot 1.2(1000 - 20) = 5000 \text{ kW} = 5 \text{ MW}.$$

Equation (6.29) $\dot{q}_W = (A_t - A_o)\dfrac{1}{\sqrt{\pi}} \cdot \sqrt{\dfrac{k\rho c}{t}} \cdot (T_g - T_a)$ gives, at 330 seconds, $\dot{q}_W =$

$400\dfrac{1}{\sqrt{\pi}} \cdot \sqrt{\dfrac{2 \cdot 10^6}{330}} \cdot (1000 - 20) = 17.2 \cdot 10^6 \text{ W} = 17.2 \text{ MW}$. At 120 minutes $\dot{q}_W =$

$400\dfrac{1}{\sqrt{\pi}} \cdot \sqrt{\dfrac{2 \cdot 10^6}{120 \cdot 60}} \cdot (1000 - 20) = 3.7 \cdot 10^6 \text{ W} = 3.7 \text{ MW}$. Note that the thermal penetration time for the wall is ≈ 5 hours.

Equation (6.30) $\dot{q}_R = A_o e_f \sigma(T_g^4 - T_a^4)$ gives

$$\dot{q}_R = 6 \cdot 1.0 \cdot 5.67 \cdot 10^{-8}(1273^4 - 293^4) = 0.89 \cdot 10^6 \text{ W} = 0.89 \text{ MW}.$$

The total energy balance gives 23 MW at 330 seconds and 9.6 MW at 120 minutes. Equation (6.27) shows that the maximum energy release rate is $\dot{Q} = 0.096 \sqrt{2} \cdot 17.5 = 13.4$ MW. This imbalance is serious and to some extent due to the very rough approximations made in Eq. (6.29).

The energy balance discussed above shows that the temperature T_g can be solved from the equations, since the other terms can be estimated. But Example 6.3 shows that far more detailed calculations are needed to arrive at expressions for temperature–time curves involving the entire fire duration, especially with regard to the heat lost to the boundaries. We shall therefore, in the next section, review the method developed by Magnusson and Thelandersson,[7] which is the most referred to method of its kind.

6.4.3 METHOD OF MAGNUSSON AND THELANDERSSON

Magnusson and Thelandersson made the following assumptions when setting up a computer model for solving the energy balance given by Eq. (6.26):[7]

- The energy release rate is ventilation-controlled during the fully developed stage, but based on data from full-scale experiments during the growth and decay stages.
- Combustion is complete and takes place entirely within the confines of the enclosure.
- The temperature is uniform within the enclosure at all times.
- A single surface heat transfer coefficient is used for the entire inner surface of the enclosure.
- The heat flow to and through the enclosure boundaries is one-dimensional (i.e., effects of corners and edges are ignored) and the boundaries are assumed to be "infinite slabs."

Other models developed to solve the energy balance equation (for example, Babrauskas and Williamsson[15]) also make the above assumptions. There are two main advantages of the method derived by Magnusson and Thelandersson:

- They expressed the energy release rate as a function of time, when the only known input parameters are the fuel load density, the opening factor, and the thermal properties of the bounding materials.
- They managed to present the results in a simple and systematic way through graphs and tables, eliminating the need for direct computer simulations.

The energy release rate: The release rate of energy as a function of time determines to a very large extent the shape of the temperature–time curve. The intensity and duration of a fully developed fire depends mainly on the amount, type, geometry, position, and orientation of the fuel; the size and geometry of the ventilation openings; and the thermal properties of the boundary surface materials.

If the combustible components have a relatively large exposed surface area, a short but intensive fire will result, while the opposite conditions will result in a longer but less intensive fire. An increase in ventilation will, if flashover occurs, result in a shorter but more intensive fire.

Information on the fire load alone will therefore not be sufficient to determine the shape of the energy release rate curve, but will determine the total area under the curve, i.e. the total amount of energy released. Information on the ventilation will give the maximum possible energy release rate.

Magnusson and Thelandersson found that the only way to determine the shape of the energy release rate curve would be to use available experimental data in conjunction with the above conditions. They therefore collected data from four test series consisting of approximately 30 experimental data sets. They postulated that the energy release rate would have an ignition phase, a flame phase (ventilation-controlled), and a decay phase, and that the energy release rate would

- be polygonal under the ignition phase, increasing from zero to maximum (determined from the opening factor)
- be constant during the flame phase (corresponds to a ventilation-controlled fire)
- decrease as a polygonal function from maximum to zero under the decay phase

A computer program was then used to solve the energy balance and calculate the temperature–time curves, assuming different shapes of the energy release rate curves. The calculated temperatures were compared to the measured values and the shape of the energy release rate curve was varied until a satisfactory result was obtained.

The results in Figure 6.9 shows one of the experimental data sets used to prescribe the shape of the energy release rate curve. The smaller diagram shows the energy release rate normalized by the maximum energy release rate (at ventilation control). The larger diagram shows the agreement between the calculated (dashed line) and measured (solid lines) gas temperatures.

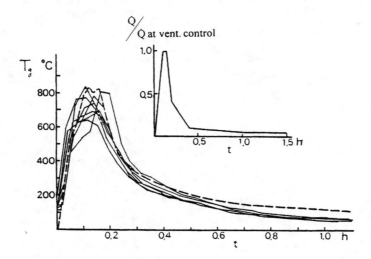

FIGURE 6.9 Calculated (dashed line) and experimentally measured temperature–time curves for a fire load density of 96 MJ/m^2 and opening factor of 0.068 m$^{1/2}$. Smaller diagram shows the derived energy release rate normalized by the maximum energy release rate at ventilation control. (Adapted from Pettersson et al.[11])

Having thus derived the energy release rate curve for experiments with a wide range of values for fuel load, ventilation factor, total enclosure surface area, and boundary material properties, they arrived at a method for generally postulating the shape of the energy release rate curve, depending on these variables.

The energy release rate could therefore be determined when $Q_t\leq$, $A_o\sqrt{H_o}$, A_t, and boundary material properties were known. Magnusson and Thelandersson could therefore systematically use their computer program for deriving temperature–time curves for a wide variation of these factors.

Systematic presentation of the temperature–time curves: A presentation of the calculated temperature–time curves for a wide range of the above-mentioned variables would require an enormous amount of diagrams or tables. Magnusson and Thelandersson therefore had to diminish the number of input factors to allow a simple and systematic presentation. This was achieved by

- Dividing the fire load and the ventilation factor by the total enclosure surface area. These three factors were therefore reduced to two: the fire load density, $Q_t\leq$, and the opening factor $\dfrac{A_o\sqrt{H_o}}{A_t}$.
- Defining a set of eight fire compartment types with respect to boundary material properties: types A, B, C, D, E, F, G, and H. The temperature–time curves were presented only for type A compartments, and a multiplying factor, K_f, was defined, allowing temperature–time curves for the other compartment types to be determined from the type A curves.

Figure 6.10 shows typical temperature–time curves for a range of opening factors and fuel load densities for type A fire compartment.

The factor K_f allows an equivalent fuel load density and equivalent opening factor to be calculated for types of fire compartments other than type A. The fuel load density for the actual compartment is determined and multiplied by K_f to give an equivalent fuel load density.

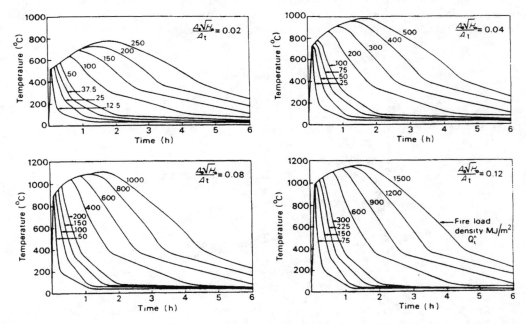

FIGURE 6.10 Calculated temperature–time curves for a range of fuel load densities and opening factors, fire compartment type A. (Adapted from Pettersson et al.[11])

TABLE 6.3
Conversion to Equivalent Fire Load Density and Equivalent Opening Factor

		Factor K_f					
Fire Compartment		**Actual Opening Factor (m$^{1/2}$)**					
Type	**Description of Enclosing Construction**	**0.02**	**0.04**	**0.06**	**0.08**	**0.1**	**0.12**
A	Thermal properties taken as average values for concrete, brick, and lightweight concrete (standard fire compartment)	1.0	1.0	1.0	1.0	1.0	1.0
B	Concrete	0.85	0.85	0.85	0.85	0.85	0.85
C	Lightweight concrete	3.0	3.0	3.0	3.0	3.0	2.5
D	50% lightweight concrete, 50% concrete	1.35	1.35	1.35	1.5	1.55	1.65
E	50% lightweight concrete, 33% concrete and 17% (13 mm gypsum plasterboard, 100 mm mineral wool and brickwork [from the inside outward])	1.65	1.5	1.35	1.5	1.75	2.0
F	80% uninsulated steel sheeting, 20% concrete (typically a warehouse with uninsulated ceiling and walls of steel sheeting and a concrete floor)	1.0–0.5	1.0–0.5	0.8–0.5	0.7–0.5	0.7–0.5	0.7–0.5
G	20% concrete, 80% (2 ¥ 13 mm gypsum plasterboard, 100 mm air gap and 2 ¥ 13 mm gypsum plasterboard)	1.5	1.45	1.35	1.25	1.15	1.05
H	100% (steel sheeting, 100 mm mineral wool, steel sheeting)	3.0	3.0	3.0	3.0	3.0	2.5

Note: Equivalent fire load density = $K_f \cdot$ actual fire load density; equivalent opening factor = $K_f \cdot$ actual opening factor.

Similarly, the actual opening factor is multiplied by the same K_f to give an equivalent opening factor. These can then be used in figures such as Figure 6.10 to determine the actual temperature–time curve. Table 6.3 gives the values for K_f to be used and defines the compartment types.

EXAMPLE 6.4

A compartment is 10 m by 5 m, 2.5 m high. It has a single ventilation opening 2.5 m wide and 2 m high. The fire load density $Q_t\leq$ is 75 MJ/m^2. Determine the temperature–time curve to which structural elements in the compartment will be exposed (a) if the materials in the bounding structure correspond with fire compartment type A; (b) if they correspond with fire compartment type C.

SUGGESTED SOLUTION

$A_t = 2 \cdot (10 \cdot 2.5 + 5 \cdot 2.5 + 5 \cdot 10) = 175$ m^2. $\dfrac{A_o\sqrt{H_o}}{A_t} = 2.5 \cdot 2 \cdot \sqrt{2}/175 = 0.04$ m$^{1/2}$.

(a) Figure 6.10 gives the temperature–time relationship for opening factor 0.04 m$^{1/2}$ and fire load density 75 MJ/m^2.

(b) Table 6.3 gives $K_f = 3.0$ for this case. The equivalent opening factor is $3 \cdot 0.04 = 0.12$ m$^{1/2}$, and the equivalent fire load density $Q_t\leq$ is $3 \cdot 75 = 225$ MJ/m^2. Figure 6.10 gives the temperature–time relationship.

6.4.4 OTHER RELATED METHODS

Several other methods exist for estimating temperature–time curves or thermal loads to which structural elements are subjected in case of fire. Here, we briefly mention only three of these.

Computer program by Babrauskas and Williamsson: Babrauskas and Williamsson also solved the energy balance for the fully developed fire by computer.[15] Their model is possibly somewhat more sophisticated than the Magnusson and Thelandersson model, but it gives very similar results for the same input variables.

Hand-calculation method by Babrauskas: Babrauskas developed a relatively simple method for calculating the post-flashover gas temperature at some time t.[17] He expressed the temperature as

$$T_g - T_a = (T^* - T_a)\theta_1 \theta_2 \theta_3 \theta_4 \theta_5 \tag{6.31}$$

where $T^* = 1725K$ and the factors q_1, q_2, q_3, q_4, and q_5 are given by separate expressions to do with energy release rate, losses to boundaries, opening factor, and combustion efficiency. As input, the energy release rate or the burning rate of the fire must be known. An excellent description of the method, with calculated examples, is given by Walton and Thomas in the *SFPE Handbook*,[2] and the interested reader is referred to their treatment.

EUROCODE method (or the parametric fire exposure): The method for calculating temperature–time curves according to the European Committee for Standardization[8] is to a considerable degree based on comparison with the temperature–time curves presented by Magnusson and Thelandersson.[7] The EUROCODE method divides the fire development into two phases: the heating phase and the cooling phase.

The heating phase follows the ISO 834 "standard curve," but instead of using Eq. (6.24) directly, the "standard curve" is expressed in terms of natural logarithms. Pettersson found such an expression for the "standard curve" to be very useful, since natural logarithms lend themselves easily to differentiation.[18]

The EUROCODE temperature–time curve in the heating phase is given by

$$T_g = 1325(1 - 0.324e^{-0.2t^*} - 0.204e^{-1.7t^*} - 0.472e^{-19t^*}) \tag{6.32}$$

where t^* is a modified time given by

$$t^* = t \cdot \left(\frac{A_0 \sqrt{H_0/A_t}}{\sqrt{k\rho c}}\right)^2 \left(\frac{1160}{0.04}\right)^2 \tag{6.33}$$

so that $t^* = t$ when $A_o \sqrt{H_o}/A_t = 0.04$ m$^{1/2}$ and $= \sqrt{k\rho c} = 1160$ Ws$^{1/2}$/m^{2}K.

The EUROCODE temperature–time curve in the cooling phase is given by

$$
\begin{aligned}
T_g &= T_{g,max} - 625(t^* - t_d^*) & \text{for } t_d^* \leq 0.5 \\
T_g &= T_{g,max} - 250(3 - t_d^*)(t^* - t_d^*) & \text{for } 0.5 < t_d^* < 2 \\
T_g &= T_{g,max} - 250(t^* - t_d^*) & \text{for } t_d^* \geq 2
\end{aligned}
\tag{6.34}
$$

where $T_{g,max}$ is the maximum temperature in the heating phase for $t^* = t_d^*$ and

$$t_d^* = \frac{0.13 \cdot 10^{-3} \cdot Q_t''}{A_0 \sqrt{H_0}/A_t} \cdot \left(\frac{A_0 \sqrt{H_0} A_t}{\sqrt{k\rho c}}\right)^2 \left(\frac{1160}{0.04}\right)^2 \tag{6.35}$$

In terms of real time, the duration of the heating phase is given by

$$t_d = \frac{0.13 \cdot 10^{-3} Q_t''}{A_0 \sqrt{H_0}/A_t} \tag{6.36}$$

so the modified duration time can be written as (also, see Eq. (6.33))

$$t_d^* = t_d \left(\frac{A_0 \sqrt{H_0}/A_t}{\sqrt{k\rho c}} \right)^2 \left(\frac{1160}{0.04} \right)^2 \tag{6.35a}$$

Pettersson has compared the results from Eq. (6.32)–(6.35) with the temperature–time curves from Magnusson and Thelandersson for an opening factor of 0.04 $m^{1/2}$ and found that better agreement for the cooling phase would be desirable.[18] He also compared this with equations for the cooling phase suggested by Anderberg and Pettersson[19] and found that their equations showed a much better agreement. His comparisons for opening factor = 0.04 $m^{1/2}$ and fuel load density 400 MJ/m^2 are shown in Figure 6.11.

Observe that Eq. (6.32)–(6.35) give the temperature at a modified time and not at real time. Instead of using these equations and transforming the result to real times, it may in some cases be more convenient to simply use the Magnusson and Thelandersson curves directly. The advantage of the EUROCODE method, on the other hand, is that the equations can be used directly on hand-calculators or computers.

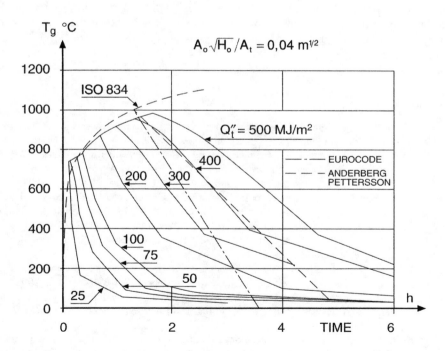

FIGURE 6.11 Temperatures in the decay phase of the compartment fire for two different approaches, compared with the temperature–time curves according to Magnusson and Thelandersson for the opening factor 0.04 $m^{1/2}$ and fuel load density 400 MJ/m^2. (Adapted from Pettersson[18].)

EXAMPLE 6.5

A compartment has an opening factor $A_o\sqrt{H_o}/A_t = 0.08$ m$^{1/2}$, fuel load density $Q_{\leq} = 400$ MJ/m^2 and the internal surface area can be assumed to be concrete. Use the EUROCODE method to calculate the maximum gas temperature during a fully developed fire.

SUGGESTED SOLUTION

Table 6.1 gives $krc = 2 \cdot 10^6$ W^{2}s/m^{4}K^2. Equation (6.36) gives the duration of the heating phase as $t_d = \dfrac{0.13 \cdot 10^{-3} Q_t''}{A_0\sqrt{H_0}/A_t} = \dfrac{0.13 \cdot 10^{-3} 400}{0.08} = 0.65$ hours. This is the time at which the temperature will be at maximum. The modified duration time can be calculated from Eq. (6.35a) (or Eq. (6.33))

$$t_d^* = t_d\left(\frac{A_0\sqrt{H_0}/A_t}{\sqrt{k\rho c}}\right)^2\left(\frac{1160}{0.04}\right)^2 = 0.65\left(\frac{0.08}{\sqrt{2\cdot 10^6}}\right)^2\left(\frac{1160}{0.04}\right)^2 = 1.75 \text{ hours.}$$

The temperature at this modified time is calculated from Eq. (6.32)

$T_g = 1325(1 - 0.324e^{-0.2t^*} - 0.204e^{-1.7t^*} - 0.472e^{-19t^*})$ which at $t^* = 1.75$ gives $T_g = 1325$ $(1 - 0.324e^{-0.2 \cdot 1.75} - 0.204e^{-1.7 \cdot 1.75} - 0.472e^{-19 \cdot 1.75}) = 1008.7°C$.

A rough comparison with the Magnusson and Thelandersson method can be made by taking the appropriate maximum temperature from Figure 6.10 (opening factor 0.08 m$^{1/2}$, fuel load density 400 MJ/m^2). The maximum temperature is slightly higher and occurs a little later than the above calculations show. However, Figure 6.10 is valid for fire compartment type A (where krc is an average value for concrete, brick, and lightweight concrete), our example is valid for a concrete structure.

REFERENCES

1. McCaffrey, B.J., Quintiere, J.G., and Harkleroad, M.F., "Estimating Room Fire Temperatures and the Likelihood of Flashover Using Fire Test Data Correlations," *Fire Technology*, Vol. 17, No. 2, pp. 98-119, 1981.
2. Walton, W.D. and Thomas, P.H., "Estimating Temperatures in Compartment Fires," *SFPE Handbook of Fire Protection Engineering*, 2nd ed., National Fire Protection Association, Quincy, MA, 1995.
3. Drysdale, D., *An Introduction to Fire Dynamics*, Wiley-Interscience, New York, 1992.
4. Foot, K.L., Pagni, P.J., and Alvares, N.J., "Temperature Correlations for Forced-Ventilated Compartment Fires," *Proceedings of the First International Symposium, International Association of Fire Safety Science*, Hemisphere Publishing, pp. 139–148, Washington, D.C., 1986.
5. Mowrer, F.W. and Williamsson, R.B., "Estimating Room Temperatures from Fires along Walls and in Corners," *Fire Technology*, Vol. 23, No. 2, 1987.
6. Karlsson, B., "Modeling Fire Growth on Combustible Lining Materials in Enclosures," Department of Fire Safety Engineering, Lund University, Sweden, Report TVBB-1009, Lund 1992.
7. Magnusson, S.E. and Thelandersson, S., "Temperature-Time Curves for the Complete Process of Fire Development—A Theoretical Study of Wood Fuels in Enclosed Spaces," *Acta Polytechnica Scandinavica*, Ci 65, Stockholm, 1970.
8. EUROCODE 1: Basis of Design and Actions on Structures. Part 2-2: Actions on Structures Exposed to Fire. ENV 1991-2-2: 1995 E, European Committee for Standardization, Brussels, 1995.
9. International Standards Organization, "Fire Resistance Tests. Elements of Building Construction," ISO 834, International Organisation for Standardisation, Geneva, 1975.
10. CIB W14, "Design Guide for Structural Fire Safety," *Fire Safety Journal*, Vol. 10, No. 2, 1986.

11. Pettersson, O., Magnusson, S.E., and Thor, J., "Fire Engineering Design of Steel Structures," Swedish Institute of Steel Construction, Publication No. 50, Stockholm, 1976.
12. Huggett, C., "Estimation of Rate of Heat Release by Means of Oxygen Consumption Measurements," *Fire and Materials*, Vol. 4, pp. 61–65, 1980.
13. Kawagoe, K., "Fire Behaviour in Rooms", Report No. 27, Building Research Institute, Tokyo, 1958.
14. Hägglund, B., "Simulating Fires in Natural and Forced Ventilated Enclosures," FOA Rapport C 20637-2.4, National Defense Research Institute, Stockholm, 1986.
15. Babrauskas, V. and Williamsson, R.B., "Post Flashover Compartment Fires: Bases of a Theoretical Model," *Fire and Materials*, Vol. 3, pp. 1–7, 1978.
16. Shields, T.J. and Silcock, G.W.H., *Buildings and Fire*, Longman Scientific and Technical, London, 1987.
17. Babrauskas, V., "A Closed-Form Approximation for Post-Flashover Compartment Fire Temperatures," *Fire Safety Journal*, Vol. 4, pp. 63–73, 1981.
18. Pettersson, O., "The Parametric Temperature-Time Curves According to ENV 1991-2-2:1995 E—A Summary Evaluation," International Standards Organization for Standardization, ISO/TC92/SC2/WG2:N261, Department of Fire Safety Engineering, Lund University, Sweden,1996.
19. Anderberg, Y. and Pettersson, O., "Fire Engineering Design of Concrete Structures" (in Swedish), Swedish Council for Building Research, T13:1992, Stockholm, 1992.

PROBLEMS AND SUGGESTED ANSWERS

6.1 A steady fire with an energy release rate of 500 kW occurs in a room 3 m wide, 3 m long, and 2.4 m high, with a door opening which is 1.8 m high and 0.6 m wide. The ambient temperature is 20°C. Calculate the average gas temperature in the upper layer of the room after 60 seconds and 600 seconds if

(a) the construction is made of lightweight concrete with wall thickness 0.15 m.

(b) the construction is made of double gypsum plaster board, each plate 0.012 m thick.

(c) as in (b), but insulated with mineral wool of thickness 0.05 m, with $k = 0.041 W/(mK)$, $r = 100$ kg/m^3, and $c = 800$ J/(kgK).

Suggested answer: At 60 s: (a) 345°C, (b) 250°C, (c) 580°C. At 600 s: (a) 500°C; (b) 415°C; (c) above 600°C (840°C according to Eq. (6.11)).

6.2 An industrial building has floor area of 5 m by 8 m and is 3 m high with a door 1 m wide and 2 m high. Leakage of transformer oil from a pump occurs and is ignited. What is the maximum area of spill in [m^2] if flashover is to be avoided. Assume the construction material to be lightweight concrete and perform the calculations at a characteristic time of 10 minutes after ignition.

Suggested answer: Roughly 1.7 m^2 (70% combustion efficiency, maximum allowable energy release rate $\approx$1350 kW, and burning rate $\approx$0.025 kg/s).

6.3 A room of concrete has dimensions 5 m by 3 m by 2.5 m high with two windows which are 0.9 m by 0.9 m each. A fire in the room causes an energy release rate which increases linearly from zero to 1 MW in 10 minutes and is constant after this.

(a) Will the fire become ventilation controlled?

(b) Is there a risk for flashover in 10 minutes?

(c) Calculate the gas temperature at time 5 minutes from ignition.

Suggested answer: (a) No; (b) no; (c) 230°C

6.4 A building is used for storing commodities at floor level, which are ignitable if a hot upper layer of fire gases develops, with a temperature in excess of 300°C. A design fire is assumed to develop a maximum of 5 MW. The floor area is 20 m by 10 m and the ceiling height is 6 m, the building is constructed of lightweight concrete. Assume that one or more large doors, with a height of 2 m, are open when the fire starts. Calculate the required area of openings if flashover is to be avoided, using a characteristic time of 5 minutes.

Suggested answer: 22 m^2

6.5 A room has the dimensions 6 m by 8 m by 2.5 m high and is constructed of lightweight concrete. The room is used for fire experiments. In one experiment the fire load consists of four wooden pallets of weight 450 kg. The total burning rate after 5 minutes is 0.15 kg/s and the effective heat of combustion is 12 MJ/kg. The room has an opening 3 m wide and 2 m high.
(a) Will flashover occur within 5 minutes?
(b) Calculate the gas temperature after 5 minutes.

Suggested answer: (a) No; (b) 385°C

6.6 A fire compartment has a given geometry and an opening. The room is to be constructed of lightweight concrete.
(a) How many times higher is the energy release rate required for flashover, if the room is instead constructed of ordinary concrete?
(b) How many times lower will the temperature be if the room is constructed of ordinary concrete?

Suggested answer: (a) $\dot{Q}_{FO,concrete} = 2.3 \cdot \dot{Q}_{FO,lwc}$; (b) $DT_{concrete} = 0.58 \, DT_{lwc}$

6.7 A concrete building has floor area 10 m by 20 m and is 4 m high. The building has four windows with a height 1.6 m and a width 2 m. After 10 minutes the gases in the room have an average temperature of 700°C. Estimate the energy release rate by
(a) using the regression Eq. (6.11)
(b) calculating the terms in the energy balance for the fully developed fire

Suggested answer: (a) 24 MW; (b) 20.2 MW, from $\dot{q}_L = 5.5$ MW, $\dot{q}_W = 14$ MW, $\dot{q}_R = 0.65$ MW

6.8 A hotel room has a floor area of 10 m by 4 m and is 2.5 m high and has a window which is 1.6 m high and 3 m wide. The room contains some furniture and other commodities containing 50 kg polyurethane (with $DH_{eff} = 25$ MJ/kg) and 350 kg wood (with $DH_{eff} = 18$ MJ/kg). Calculate the fuel load density and the opening factor and give a temperature–time curve for the design of structural stability
(a) if the enclosing construction can be assumed to be of fire compartment type A
(b) if the enclosing construction is of lightweight concrete

Suggested answer: (a) Opening factor = 0.04 m$^{1/2}$, fuel load density = 50 MJ/m^2, see design curve in Fig. 6.10; (b) opening factor = 0.12 m$^{1/2}$, fuel load density = 150 MJ/m^2, see design curve in Figure 6.10

7 Heat Transfer in Compartment Fires

The ability to numerically estimate how much heat is transferred between a fire, the fire gases, the fuel bed, and the surfaces in an enclosure is essential for any fire hazard assessment. The energy balance is very much affected by heat being transferred from the flames and the hot gases to the enclosure surfaces and out through the enclosure openings. The heat transferred from these sources toward a fuel package will to a considerable extent control the rate at which fuel evaporates and heat is released. This chapter briefly discusses convection heat transfer as applied to enclosure fires, but most of the text is dedicated to radiative heat transfer in enclosures. It is presumed that the student already has a background in the fundamentals of heat transfer.

CONTENTS

7.1 Terminology 141
7.2 Introduction 142
    7.2.1 Background 142
    7.2.2 Modes of Heat Transfer 142
    7.2.3 Measurements 145
7.3 Convective Heat Transfer in Fire 148
    7.3.1 Specific Convective Studies: Fire Plume Heat Transfer to Ceilings 149
7.4 Radiative Heat Transfer in Fire 154
    7.4.1 Two Approximate Methods for Calculating Radiation from Flame to Target 155
    7.4.2 Basic Principles of Radiative Transfer 159
7.5 Enclosure Applications 170
    7.5.1 Electrical Circuit Analogy 172
    7.5.2 First Example: Heat Flux to a Sensor at Ceiling Level 172
    7.5.3 Second Example: Heat Flux to a Sensor at Floor Level 176
7.6 Summary 178
References 179
Problems and Suggested Answers 180

7.1 TERMINOLOGY

Absorptivity — The fraction of the radiation energy incident on a surface that is absorbed by the surface, given the symbol α.

Configuration factor, view factor, shape factor — The configuration factor is a purely geometric quantity. It gives the fraction of the radiation leaving one surface that strikes another surface directly. In other words, it gives the fraction of hemispherical surface area seen by a differential element when looking at another differential element on the hemisphere.

Emissivity — The fraction of the radiation energy emitted in relation to the maximum possible emission from a surface, given the symbol ε.

Extinction coefficient and absorption coefficient — The extinction coefficient is a measure of how the intensity of a beam of radiation passing through a media changes with length;

it is given the symbol κ. The extinction coefficient is generally a sum of two parts: the absorption coefficient and the scattering coefficient. In our applications scattering is negligible, and the extinction coefficient represents only absorption. Here, the extinction coefficient and the absorption coefficient are synonymous.

Gray gas and gray body assumption — Assuming that the radiative properties of a gas or a surface are independent of wavelength. This simplifies the radiation transfer equation enormously. For enclosure fire radiative exchanges gray gas and gray body assumptions are generally satisfactory.

Mean beam length — A measure of the length that a beam of radiation travels, through a media, given the symbol *L*. This measure, together with the absorption coefficient, can be used to estimate the radiation intensity emitted by a body of gas or flames.

7.2   INTRODUCTION

This section provides a brief background, discusses the different modes of heat transfer, and introduces the techniques used to measure temperatures and heat transfer.

7.2.1   BACKGROUND

It is presumed that the student already has a background in the fundamentals of heat transfer. Hence, in this chapter we do not include the basic development, which would necessitate a book in itself. Instead, we include some basics more relevant to compartment fires. This should reinforce principles and show their importance. Standard texts on heat transfer should be consulted to expand on our presentation.

In this chapter we also present a short discussion about the two different modes of heat transfer, conduction and radiation. Conduction of heat between a moving fluid and a solid boundary is sometimes taken to be the third mode and is termed *convection*. This chapter does not discuss conduction within a solid; there are excellent textbooks available in that field with methods directly applicable to the problem of fires in enclosures.

The chapter briefly discusses convective heat transfer, since this has some special applications in enclosure fires. Convective heat transfer can be the dominating mechanism for the heating up of small devices such as detectors and temperature measuring devices. This mode of heat transfer is also important for heat transfer from a flame or hot gases impinging on a surface.

Due to the high temperatures of flames and fire gases, radiative heat transfer is very often the dominating mechanism for transferring heat in enclosure fires. The radiative transfer equation is in its most complete form an integro-differential equation, and its solution even for a one-dimensional, planar, gray medium is far too complex for engineering calculations. In fires the multidimensional combustion system consists of highly nonisothermal and nonhomogeneous medium where spectral variation of the radiative properties of the medium should be accounted for. It is therefore necessary to introduce some simplifying assumptions and strike a compromise between accuracy and computational effort.

The latter part of this chapter therefore concentrates on simplified engineering methods for estimating the radiative properties of the medium, allowing the radiating intensity emitted to be calculated. Once this is done, methods are available to calculate configuration factors so that the energy received by a distant object can be estimated.

7.2.2   MODES OF HEAT TRANSFER

Fundamentally, there are only two mechanisms of heat transfer. Heat is energy transmitted due to a temperature difference. It can occur due to **conduction** and **radiation** where

- conduction is a molecular energy transport
- radiation is an electromagnetic energy transfer arising from a body possessing thermal energy indicated by (absolute) temperature.

The former depends on a matter for transfer, while the latter can be propagated through vacuum.

Conduction is expressed by Fourier's Law, developed in the early 1800s. It states that the heat flux–energy flow rate per unit area, $\bar{q}''$- is directly proportional to the temperature gradient, ∇T. Conduction can also be induced by concentration gradient (Dufour effect) and electropotential gradient (Seebeck effect), but these are usually negligible. However, the latter is the basis of the operation of a thermocouple, the tool most often used to measure temperatures in fire experiments. The proportionality constant with ∇T is the thermal conductivity, k, and the heat flux is written as

$$\bar{q}'' = -k\nabla T \tag{7.1}$$

A knowledge of the temperature field is required to compute $\bar{q}''$. Note the heat flux is a vector quantity having magnitude and direction. In the following, we drop the vector notation for convenience and write $\bar{q}''$ as $\dot{q}''$.

Convection heat transfer is merely conduction heat transfer in a moving fluid to a solid boundary. The motion of the fluid effects the temperature field, and therefore the heat flux depends on the fluid velocity and properties. It is convenient to represent convective heat transfer through a coefficient, h, that contains all of the specific fluid effects, i.e.,

$$\dot{q}'' = h(T - T_s) \tag{7.2}$$

where h is the convective heat transfer coefficient,

        T is the fluid stream temperature,

        T_s is the solid boundary temperature.

Thermal radiation was first adequately described in the early 1900s by Planck's Law, which gives the energy ideally emitted per unit area and wavelength of the electromagnetic spectrum. This is called the *monochromatic emissive power*,

$$E_{b,\lambda} = \frac{C_1 \lambda^{-5}}{e^{C_2/\lambda T} - 1} \tag{7.3}$$

where $C_1 = 3.743 \cdot 10^8$ W-$\mu m^4/m^2$,

        $C_2 = 1.4387 \cdot 10^4 \cdot \mu m \cdot K$,

        T is the body's temperature (K),

        λ is wavelength of the energy (μm).

Adding (integrating Eq. (7.3) from 0 to ∞) all of the energy in the spectrum gives the familiar blackbody (ideal) emissive power,

$$E_b = \sigma T^4 \tag{7.4}$$

where σ is the Stefan–Boltzmann constant, $5.67 \cdot 10^{-8}$ W/m²K⁴.

Radiation can be absorbed, reflected, or transmitted through matter. This is illustrated in Figure 7.1 for a surface and a medium (i.e., gas, fog, or window). The properties of *reflectivity* (ρ), *absorptivity* (α), and *transmissivity* (τ) represent the fractions of incident energy reflected, absorbed, and transmitted, respectively. All of these properties depend on the material, the wavelength of the incident energy, and the materials temperature. Some surfaces can be opaque, $\tau = 0$, and gases and other media can absorb selectively over discrete wavelength bands, while aerosols such as soot or oily tars in combustion products are continuous spectral attenuators.

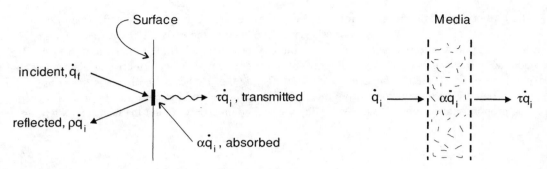

FIGURE 7.1 Destiny of incident thermal radiation.

Surface or absorbing media are not ideal radiators (blackbodies), necessarily, so the Planck emission must be reduced by the property called *emissivity*, ε. This, too depends on the medium and its temperature, and further varies with wavelength. Hence, real emissive power, in terms of the blackbody emission power, is given as

$$E = \varepsilon E_b \tag{7.5}$$

Kirchhoff's Law states that the spectral emissivity and absorptivity are equal for a medium at the same temperature, i.e.,

$$\alpha_\lambda = (\lambda, T) = \varepsilon_\lambda(\lambda, T) \tag{7.6}$$

Nevertheless, the total (spectrally integrated) properties are not necessarily equal, since the emission depends on the source temperature while absorption depends on the temperature and characteristics of the *received* radiation. However, because of the complex nature of radiative heat transfer, and the general lack of spectral properties, the assumption

$$\alpha(T) = \varepsilon(T) \tag{7.7}$$

for a material is a useful expedient. This is called the *gray body assumption*. But in analysis spectral considerations must always be kept in mind.

As an example of these spectral effects, consider the radiant heat absorbed by a fabric due to a particular source. This fabric might be employed in protective firefighter coats, and the source of energy could be the sun (Figure 7.2a), a large fire (Figure 7.2b), or hot surfaces in a fire approximately black (Figure 7.2c).

The energy flux absorbed and transmitted through the fabric would be given as

$$\dot{q}'' = \int_0^\infty (1-\rho_\lambda)\dot{q}''_{i,\lambda} d\lambda \tag{7.8}$$

where $q''_{i,\lambda}$ is the incident radiant heat flux representative of the source. Examples of reflectivities for black, white, worn, and wet cotton duck fabrics are shown in Figure 7.3. The spectral range of the solar energy would be most discriminative for these various fabrics, and hence affect the solar heat load. White would be preferred over black in substantially reducing the solar load, yet radiation from fire sources would have less of an effect. But note that the wet fabric is the worst reflector over nearly the entire spectrum. One should further consider that the wet fabric may not be the worst under firefighting conditions, since the evaporative heating of the moisture would add protection. Thus one sees that spectral effects can be important.

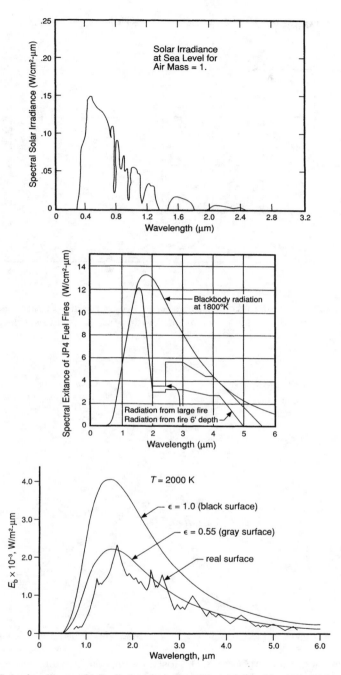

FIGURE 7.2 (a) Solar irradiance. (b) Radiation from fuel fires. (c) Blackbody radiation.

7.2.3 MEASUREMENTS

The measurements of temperature and heat flux are important aspects of heat transfer. The accuracy is crucial to ensure the correct interpretation to data and the confidence in the many empirical correlations that are used.

Although temperature can be measured in different ways, both intrusive and nonintrusive, the thermocouple is the predominant device used. Its bimetallic element produces a voltage relative to

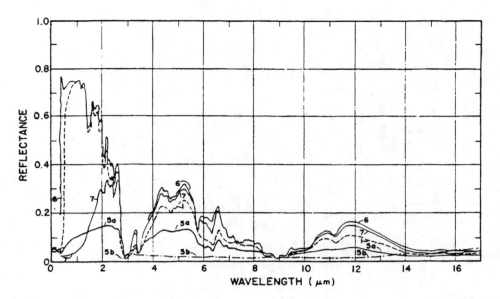

FIGURE 7.3 Spectral reflectance for cotton fabrics. The wavelength range of data was $\rho_\lambda = 0.35 - 22$ μm. Notation: 1 = Yellow cotton; 5a = Yellow cotton, worn, dry; 5b = Yellow cotton, worn, wet; 6 = White cotton, army duck; 7 = Black cotton, army duck. (From Quintiere[22].)

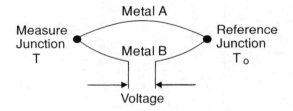

FIGURE 7.4 Thermocouple.

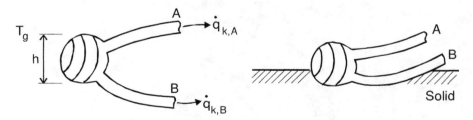

FIGURE 7.5 Gas measurement (a) and solid surface measurement (b) by a thermocouple.

its reference junction temperature (Figure 7.4). Most high-impedance voltmeters will cause negligible current flow in the circuit, and thus the voltage measured will be directly related to the temperature difference $T - T_o$. Thermocouples are usually spherical or cylindrical if a perfectly built weld is made for the wires A and B. The purpose of the thermocouple is to make it measure the temperature of the medium in which it is placed. This is not necessarily the case, since heat transfer plays a role in its response. Such effects are illustrated in Figure 7.5 and for gas and surface measurements.

It should be obvious that the thermocouple junction need not measure the point in the medium where it is placed. For example, a heat balance on the gas-measuring thermocouple depends on its mass (m), convection, and radiation transfers, e.g.,

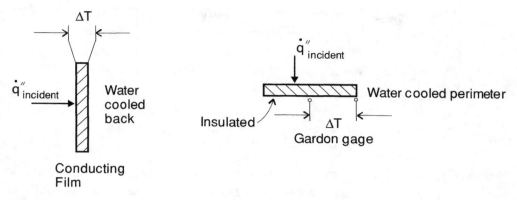

FIGURE 7.6 Heat flux gage principles.

$$mc = \frac{dT}{dt} = hA(T_g - T) + \dot{q}(\lambda, \text{absorbed}) - \dot{q}(\lambda, \text{emitted}) - \dot{q}_{k, \text{ A and B}} \qquad (7.9)$$

The thermocouple will have a time response, and its equilibrium temperature will depend on the various convection, conduction, and radiative interactions.

Heat flux can also be measured by various devices. Separating radiative and convective contributions is always a challenge, and various windows and coatings complicate the interpretation of the measurement. Two robust devices used to measure both radiative and convective incident heat fluxes in fire environments are thermocouple-based devices measuring a temperature difference. They are illustrated in Figure 7.6.

The film gage works from the temperature difference through a film of known conductivity. The back is usually water-cooled, but the film can be bonded to a surface to measure the absorbed heat flux into the solid. Of course the properties of the film affect the heat flow. The Gardon gage uses a thin metal circular disk in which the center to edge temperature is proportional to a *uniform* heat flux. Note that convection would not be uniform, since the gage temperature varies over the surface. However, the gage can be designed to have both convective and radiative heat flux property measured. Radiation will depend on the surface properties of the gage, and convective effects are sometimes screened by transmitting windows. The spectral characteristics of these components are very important for the general use of the gage in different heating environments. One should be aware of these gages and their limitations, and one must always assess the measurement techniques in interpreting or designing experiments. A heat flux gage is not always applicable to all environments.

EXAMPLE 7.1

A thermocouple is placed in a large heated duct to measure the temperature of the gas flowing through the duct. The duct walls are at 450°C and the thermocouple indicates a temperature of 170°C. The convective heat transfer coefficient from the gas to the thermocouple was roughly calculated to be 150 W/m² °C. The emissivity of the thermocouple material is 0.5. What is the temperature of the gas? Assume the gas is not participating in the radiative transfer.

SUGGESTED SOLUTION

Since conditions are steady, $dT/dt = 0$ and Eq. (7.9) can be reduced to a simple energy balance, where the heat gain to the thermocouple is assumed to be due to convection by gas and radiation from walls. The heat loss is due to radiation. The balance can be written

$$\left[\begin{array}{l}\text{Convective}\\\text{heat transfer}\\\text{to the}\\\text{thermocouple}\end{array}\right]+\left[\begin{array}{l}\text{Radiation to the}\\\text{thermocouple from}\\\text{wall (assumed as}\\\text{black body)}\end{array}\right]=\left[\begin{array}{l}\text{Radiation}\\\text{emitted}\\\text{from the}\\\text{thermocouple}\end{array}\right]+\left[\begin{array}{l}\text{Radiation}\\\text{reflected}\\\text{from the}\\\text{thermocouple}\end{array}\right]$$

or

$$h\left(T_g - T_t\right) + \tau T_w^4 = \varepsilon \sigma T_t^4 + (1-\varepsilon)\sigma T_w^4$$

Assuming that the areas over which the cooling occurs are equal for both terms leads to a simple energy balance: $0 = h\left(T_g - T_t\right) + \sigma\varepsilon(T_w^4 - T_t^4)$. Inserting our values, we get $0 = 150$ $(T_g - 443) + 5.67 \cdot 10^{-8} \cdot 0.5(723^4 - 443^4)$. This gives $T_g = 399\text{K} = 126°\text{C}$. This is significantly less than the measured value, due to the radiation from the duct to the thermocouple.

Note that we have assumed that the gas does not participate in the radiative transfer. This means that none of the radiation from the wall is absorbed in the gas, and the gas itself does not radiate toward the thermocouple. This may be a reasonable assumption for clear air, but for thick black smoke the assumption will not hold. Later we discuss the emissivity and absorptivity of gases containing combustion products.

In the following we present heat transfer results and theory that are important in compartment fires. This will be done with some rigor, but not in sufficient depth and breadth to cover the field of heat transfer. Other texts should be used for background, and the heat transfer chapters of the *SFPE Handbook of Fire Protection Engineering* should be consulted.[1,2,3]

7.3   CONVECTIVE HEAT TRANSFER IN FIRE

Most of natural fire and flows associated with fire are in the domain of natural convection. Although wind can be an important factor for fires such as forest fires, forced flow convection is not generally relevant. In natural convection we can represent a characteristic velocity

$$u \approx \sqrt{(1 - T_\infty/T_s)gl}$$

So for a flow length of 1 m, $g = 9.81$ m/s^2, and $(1 - T_\infty/T_s)$ of 0.5 to 0.7, we have u ranging from 1 to 3 m/s. Such flows do not produce very large convective coefficients. Moreover, turbulent natural convection on a heated vertical plate is

$$\text{Nu} = \frac{hl}{k} = 0.1\,\text{Ra}^{1/3},\ \text{Ra} > 10^9 \qquad (7.10)$$

where   Ra = Rayleigh Number, $\dfrac{g\beta\left(T_s - T_\infty\right)l^3}{\nu^2}\text{Pr},\ \beta = \dfrac{1}{T_\infty}$

Pr = Prantl Number, $\nu/\alpha,\ \alpha = \dfrac{k}{\rho c}$

and where k, α, ν are evaluated at an average film temperature, $T_f = (T_s + T_\infty)/2$. This allows an approximate estimate of typical convective heat transfer coefficients in two conditions: the heated wall being cooled by air and the flame against a cold wall. The two situations can often be treated similarly, since the flow field can be considered approximately similar.

EXAMPLE 7.2

A vertical flat plate of length 1 m and temperature $T_s = 800°C$ is exposed to air at $T_\infty = 20°C$. Calculate the convection coefficient.

SUGGESTED SOLUTION

The Prantl number has a range of 0.71 to 0.68 in the temperature range 300K to 1000K. It is common to assume Pr = 0.7. The film temperature is 410°C and the properties of air at that temperature are $k = 57.8 \cdot 10^{-3}$ W/mK and $v = 82.3 \cdot 10^{-6}$ m²/s. Hence Ra = $\dfrac{(9.81 \text{ m/s}^2)(780 \text{ K})(1 \text{ m})^3}{293 \text{ K}(82.3 \times 10^{-6} \text{ m}^2/\text{s})^2} \cdot (0.7) = 3.94 \cdot 10^9$. Using Eq. (7.10) and solving for h we get

$$h = \left(\frac{57.8 \cdot 10^{-3} \text{ W/mK}}{1 \text{ m}} \right) 0.1 \left(3.94 \cdot 10^9 \right)^{1/3} \text{ or } h = 9.12 \text{ W/m}^2\text{K}.$$

This is a typical convective coefficient value, and a range of 5 to 15 W/m²K is to be expected for these gas velocities and associated natural convection conditions.

It is interesting to assess the level of convective heat flux likely under fire conditions. Within flames, and with direct continuous flame impingement we would expect maximum turbulent temperatures of roughly 800°C. In very large fire plumes (exceeding several meters in diameter) we might achieve temperatures of up to 1200°C. In laminar flames, we can expect temperatures on the thin flame of roughly 2000°C at most. Also, 800° to 1000°C is typical of fully involved fire conditions in a compartment. For a cold target surface at 20°C, we would then expect convective heat fluxes of 10 W/m²K · 1000 to 2000 K, or 10 to 20 kW/m². Moreover, at a turbulent flame tip the maximum temperature is only 350°C approximately. Hence, the associated convective heat flux is only approximately 3 kW/m²; not sufficient to ignite anything. However, direct flame contact could convectively ignite objects. But we will see that the radiative component in turbulent fires and in fully developed compartment fires can be much more, and can drown the convective contribution. For that reason, convective heat transfer is not a principal factor in fire heat transfer, but its presence cannot be overlooked. Also, there are not many studies of convective heat transfer in fire. Its relative unimportance and its ease in making estimates compensates for its lack of depth of study.

However, some specific applications are of importance, such as convective heat transfer to ceilings or detection devices at ceiling level, and we discuss these briefly below.

7.3.1 SPECIFIC CONVECTIVE STUDIES: FIRE PLUME HEAT TRANSFER TO CEILINGS

Several investigators, for example Cooper,[4] Alpert,[5] and Kokkala,[6] have studied heat transfer to a ceiling from an axisymmetric turbulent fire plume. All of these studies have been a relatively small-scale, involving burning to ceiling heights (H) less than 1 m, and energy release rate of less than 10 kW. Fuel sources ranged from a solid, liquids, and gases. Investigators either measured the total heat flux to the ceiling or the convection coefficient and ceiling temperature, and tried to separate convection from radiation in some cases. Cases of no impingement and with flame impingement were studied. For flame impingement, Kokkala reports the radiative contribution to the total incident

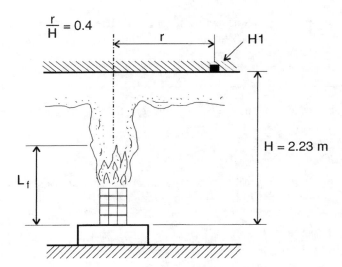

FIGURE 7.7 A description of the measurement of total incident ceiling heat flux to a water cooled sensor (H1) during compartment fire experiments with crib fuel configurations. (Adapted from Quintiere and McCaffrey[7].)

ceiling heat flux ($\dot{q}''$) to be approximately 50%.[6] The total heat flux in this range was roughly constant at 60 kW/m² up to the point where the free (no ceiling) flame height (L_f) is three times the ceiling height. This again shows the significance of radiative heat flux, especially under flame contact.

An example illustrating heat transfer to the ceiling in a room fire is taken from Quintiere and McCaffrey,[7] as shown in Figure 7.7, which depicts a water-cooled heat flux sensor (H1) heated by a ceiling jet due to a crib fire in a room. The formula given by Veldman et al. was selected,[8] which gives the convective flux as

$$\dot{q}_c'' = h_c (T_{ad} - T_w) \tag{7.11a}$$

$$h_c = C\rho_\infty c_p \sqrt{(gH)} Q^{*1/3} \tag{7.11b}$$

$$Q^* = \frac{\dot{Q}_c}{\rho_\infty c_p T_\infty \sqrt{(gH)H^2}} \tag{7.11c}$$

where $\dot{Q}_c$ is the convective energy release rate of the fire. It was assumed that the cold sensor does not significantly affect the ceiling boundary layer, the adiabatic wall temperature (T_{ad}) was taken as the average ceiling jet temperature, and T_w is the temperature of the cooling water. The constant C depends on r/H, and is 0.025 for the dimensions shown in Figure 7.7 ($r/H = 0.4$).

EXAMPLE 7.3

In Example 4.2 a 100 kW wastebasket fire was found to result in a plume temperature increase of 70°C. With an ambient temperature of 20°C and assuming that the average ceiling jet temperature is ≈90°C at the ceiling height of 2.23 m, what is the convective heat flux to the ceiling at the point $r/H = 0.4$?

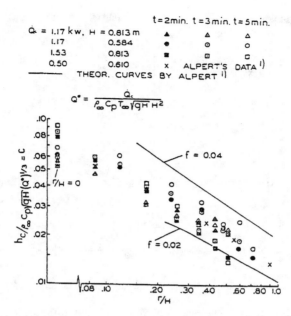

FIGURE 7.8 Dimensionless ceiling heat transfer coefficient, C, as a function of r/H. The factor f is an assumed friction factor from theoretical work by Alpert[10]. (From Veldman et al.[8] With permission.)

SUGGESTED SOLUTION

Assuming that $\dot{Q}_c = 70$ kW, we find that

$$Q^* = \frac{70}{1.2 \cdot 1.0 \cdot 293 \cdot \sqrt{9.81 \cdot 2.23} \cdot 2.23^2} = 0.0086$$

$$h_c = 0.025 \cdot 1.2 \cdot 1.0 \cdot \sqrt{9.81 \cdot 2.23} \cdot 0.0086^{1/3} = 0.029 \, \text{kW/m}^2\text{K}$$

Assuming that the ceiling surface temperature equals the average ceiling jet temperature and that the cooling water temperature equals the ambient temperature, we find that the convective heat flux to the ceiling at the point is

$$\dot{q}_c'' = 0.029(90 - 20) = 2.03 \, \text{kW/m}^2$$

For Example 7.3 the constant C is 0.025 for the dimensions shown in Figure 7.7. Veldman et al. give further data for the value of C when r/H ranges from 0.1 to 1.0, and this is shown in Figure 7.8.[8] Also, Figure 7.9 gives the adiabatic wall temperature in terms of Q^* and r/H, which should be used for a ceiling of unlimited extent.

In general, T_{ad} is the adiabatic surface temperature that would occur for a perfectly insulated ceiling. For significant cooling to the ceiling T_{ad} must be determined. Such correlations exist (e.g., Cooper[4]). Later we return to the specific example of Figure 7.7 estimating both the convective and radiative contributions to the $H1$ measurement.

Others have developed correlations for incident heat flux to a ceiling from a fire plume. These results should be taken as alternatives to Eq. (7.11). Such results are characterized by the works

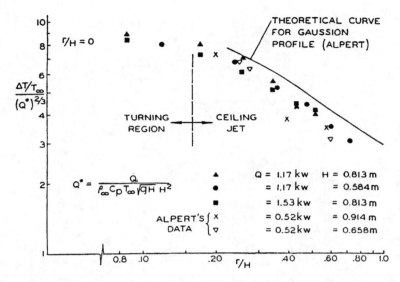

FIGURE 7.9   Dimensionless adiabatic ceiling jet temperature as a function of Q^* and r/H. (From Veldman[8].)

of You and Faeth[9] and by Alpert.[10] The results of You and Faeth in Figure 7.10 are for the stagnation point heat flux at the ceiling along the plume centerline, i.e., $r = 0$, where r is the radius measured from the centerline. The Rayleigh Number (Ra) used in Figure 7.10 is defined as

$$Ra = g\beta\dot{Q}H^2/\rho_\infty c_p v^3 \tag{7.12}$$

where β is the coefficient of volume expansion ($= 1/T_\infty$) and v is the kinematic viscosity.

It should be noted that this is not the usual Rayleigh Number, since it is defined in terms of combustion energy release rate $\dot{Q}$. It is related to Q^* (combustion to advective energy) by

$$Q^* = Ra \cdot \left(\frac{v^2}{g\beta T_\infty H^3} \right)^{3/2} \tag{7.13}$$

The term in the parentheses is an inverse Grashof Number (viscous to buoyant force).

A formula developed for the turbulent case is given below, where the gas properties are to be evaluated at ambient air conditions:

$$\dot{q}''H^2/\dot{Q} = 31.2 \, Pr^{-3/5} \, Ra^{-1/6} \tag{7.14}$$

For $L_f/H > 1$, $\dot{q}''H^2Ra^{1/6}/\dot{Q}$ decreases significantly from the results of Eq. (7.14) (Figure 7.11). When $L_f/H > 1$, the flame is impinging on the ceiling. This decrease in stagnation point heat flux can be explained due to the decrease in velocity as one descends in the flame.

EXAMPLE 7.4

Use Eq. (7.14) to estimate the convective heat flux at the stagnation point ($r = 0$) for the case discussed in Example 7.3.

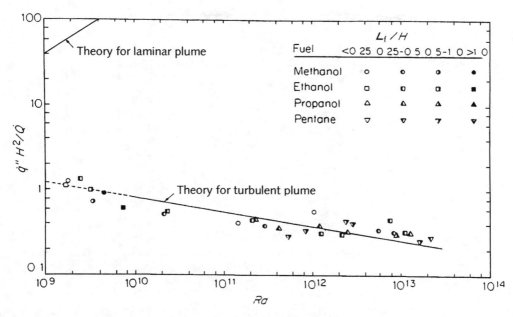

FIGURE 7.10 Turbulent (and theoretical laminar) stagnation heat transfer to a ceiling by a fire plume. (From You and Faeth[9].)

SUGGESTED SOLUTION

Taking the average gas temperature to be 90°C, we find that the kinematic viscosity $\nu = 21.8 \cdot 10^{-6}$ m²/s. The Rayleigh Number then becomes $Ra = 9.81 \cdot \dfrac{1}{293} \cdot \dfrac{100 \cdot 2.23^2}{1.2 \cdot 1.0 \cdot (21.8 \cdot 10^{-6})^3}$ $= 1.34 \cdot 10^{15}$. Assuming, as before, that $Pr = 0.7$, we find from Eq. (7.14) that the convective heat flux to the stagnation point at the ceiling is $\dot{q}'' = 31.2(0.7)^{-3/5} (1.34 \cdot 10^{15})^{-1/6} \cdot \dfrac{100}{2.23^2}$ $= 2.34$ kW/m². This is somewhat higher than the estimate from Eq. (7.11), but the flux should be higher at the stagnation point.

The heat flux decreases also with radial distance (r) along the ceiling from the stagnation point. The results with fire plume impingement on an unconfined ceiling are shown in Figure 7.12, where the work of Alpert is also used.[10] The results for $\dot{q}''H^2/\dot{Q}$ depend on Ra in the stagnation zone $0 \le r/H \le 0.2$, as given by Figure 7.11, and for $r/H > 0.2$ approximately follow

$$\dot{q}''H^2/\dot{Q} = 0.04(r/H)^{-1/3} \tag{7.15}$$

Results for $\dot{q}''H^2/\dot{Q}$ tend to be slightly higher without flame impingement, and with a confined ceiling for the same r/H.

At larger physical scales (>1 m), we would expect the case of flame impingement to have considerably higher heat fluxes due to increasing radiation effects. No study has been made under these conditions. For hazard analysis purposes it would be recommended to consider that the above equations represent solely convective heating. Then, for the large scale flames, a radiative heat flux component should be estimated and added.

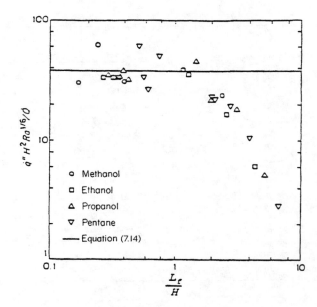

FIGURE 7.11 Effect of flame impingement on stagnation point heat flux for an unconfined ceiling. (From You and Faeth[9].)

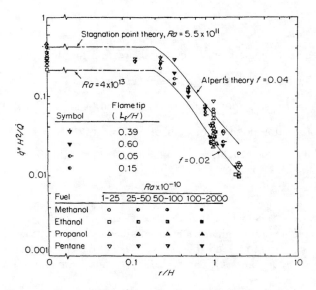

FIGURE 7.12 Heat flux as a function of position for fire plume impingement on an unconfined ceiling. Data from You and Faeth[9] and Alpert,[10] where f is an assumed friction factor.

7.4 RADIATIVE HEAT TRANSFER IN FIRE

We cannot present all of the detail required to understand radiative heat transfer and all of the fire-related work, albeit small in relation to the field. For such detail one should consult standard heat transfer tests, the excellent review chapter by Tien and colleagues in the *SFPE Handbook of Fire Protection Engineering*,[3] and the general literature. Here we will present some basic details and some relevant results for fire in compartments.

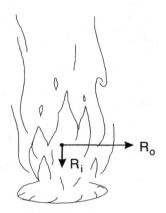
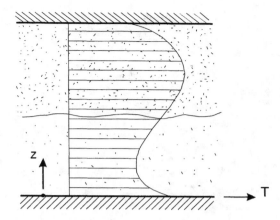

FIGURE 7.13 The significant radiative transport processes in room fires: internal (R_i) and external (R_o) flame radiation and radiation from hot combustion products.

Figure 7.13 displays two aspects of radiation in fires. One is flame radiation, in which radiative transfer can be divided into internal (R_i) and external (R_o) energy transfers. Of the internal transfers, radiation feedback to the fuel surface is most significant.

For pool fire orientations in which the base diameter is of the order of 1 m, radiative transfer from the flame primarily controls the rate of evolution of fuel gases at the surface (Burgess et al.[11]). For wall fires of any practical size, radiative feedback begins to compete with, then dominate the surface convective heat flux as the pyrolysis height increases (Orloff et al.[12]). Flame shape and thickness, respectively, play a key role in these burning orientations.

Flames and gases can participate in radiation heat transfer in several ways. Gases emit and absorb over discrete, specific wavelength (λ) bands of the spectrum. This applies to the combustion products, typically CO_2 and H_2O as the principal components, and to fuel gases. The former, due to high temperatures in the flame, can be a significant source of radiation, while the latter at relatively cooler temperatures near the decomposed surface of liquid or solid fuel can absorb (block) radiation from the flame back to the fuel. This limits the decomposition and therefore the burning rate. In contrast, particulates such as soot (mainly carbon), condensed tars (high molecular weight hydro-carbon), and water droplets (from condensed moisture in solids such as wood fuels) emit and absorb over the entire spectrum. The extinction (absorption and scattering) coefficient for soot depends roughly on λ^{-1} and liquid aerosols depend roughly on λ^{-4}. Both the gas and particulate radiation depend on their concentration as well. Hence, temperature, wavelength, and concentration of a complex mixture must be resolved to predict radiation heat transfer in such a medium. Simplifications must be made to make such a problem treatable. However, large-scale modern computers are making such problems more solvable using first principles.

7.4.1 TWO APPROXIMATE METHODS FOR CALCULATING RADIATION FROM FLAME TO TARGET

In this section we give as examples two approximate methods that are often used to calculate flame heat flux to a target. These two methods do not take into account radiation through an intervening media, such as smoke, where radiation can be absorbed and emitted and thus influence the radiation received at the target. In later sections we expand on this and discuss methods for taking into account any intervening media, such as smoke.

Modak's simple method: Except near the base of pool fires, radiation to the surroundings can be approximated as isotropic, or as emanating from a point source (Modak[13]). Modak's study allows

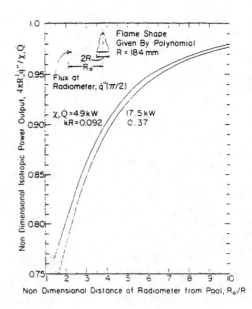

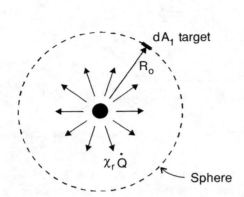

FIGURE 7.14   (a) Isotropic radiative source, (b) Modak's analysis showing the basis for Eq. (7.16).[13] The quantity $4\pi R_0^2 \dot{q}''/\chi_r \dot{Q}$ is plotted vs. the nondimensional distance, R_0/R, of an external sensor from the center of a pool of radius $R = 184$ mm. The plot is parametric in nondimensional flame absorption coefficient κR. The symbol $\dot{q}''$ is the radiative heat flux received by the external sensor located at a distance R_0 from the pool, in the horizontal plane of the base of the pool.

a very simple formula to be applied for computing radiant heat to a remote target (normal to the source), as shown in Figure 7.14.

The formula is given as

$$\dot{q}'' = \frac{\chi_r \dot{Q}}{4\pi R_0^2} \tag{7.16}$$

where R_0 is the distance of the target from the "center" of the flame, and χ_r is the fraction of total energy radiated. In general, χ_r depends on fuel and flame size and configuration. It can vary from approximately 0.15 for low sooting fuels, such as most alcohols, to 0.60 for high sooting fuels. For very large fires (greater than several meters in diameter), cold soot enveloping the luminous flame can reduce χ_r considerably.

Radiation from a cylindrical flame to a target: A more specific analysis of radiant heat flux to a target remote from a flame was developed by Dayan and Tien.[14] They considered the flame to be approximated as a homogenous cylinder of uniform temperature and other properties. Their model is depicted in Figure 7.15. The analysis in Dayan and Tien yields the incident radiant flame flux to a target element dA with a unit normal vector $n = u\vec{i} + v\vec{j} + w\vec{k}$ located a distance L from the center of the cylinder for $L/r \geq 3$. The heat flux to the target is written as

$$\dot{q}'' = \sigma T_f^4 \varepsilon (F_1 + F_2 + F_3) \tag{7.17a}$$

where

$$\varepsilon \cong 1 - \exp(-0.7\mu) \tag{7.17b}$$

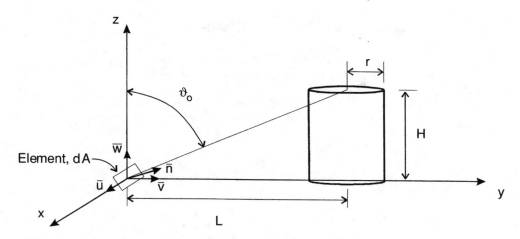

FIGURE 7.15 Schematic of radiation exchange between a target element, dA, and a homogeneous cylindrical flame. (Adapted from Dayan and Tien[14].)

for both $\mu > 0.15$ and $\varepsilon\, F_3$ small,

$$\mu = \frac{2r\kappa}{\sin\beta} \tag{7.17c}$$

where $\beta = \dfrac{\theta_0 + \pi/2}{2}$

$$F_1 = \frac{u}{4\pi}\left(\frac{r}{L}\right)^2 (\pi - 2\theta_0 + \sin 2\theta_0) \tag{7.17d}$$

$$F_2 = \frac{v}{2\pi}\left(\frac{r}{L}\right)(\pi - 2\theta_0 + \sin 2\theta_0) \tag{7.17e}$$

$$F_3 = \frac{w}{\pi}\left(\frac{r}{L}\right)\cos^2\theta_0 \tag{7.17f}$$

where κ is the effective flame absorption coefficient, T_f the flame temperature, and r the radius of the flame cylinder. All angles are in radians.

The quantity ε represents a mean emissivity and $2r/\sin\beta$ is an effective path length through the flame as seen by the target. The factors F_1, F_2, and F_3 are effective geometric factors indicative of the field of view of the flame from the target. We will develop the theory for these concepts later, but it should be noted that the heat flux involves an emissivity based on the flame or gas and a geometric factor.

Estimated properties for wood and plastic (polyurethane) crib fires are shown in Figure 7.16, in which experimental data, from the arrangement of sensors R1, R2, H1, H2 shown in Figure 7.17, are compared to calculated radiative heat flux to these targets using Eq. (7.17). An effective crib flame cylindrical radius was taken as the $\sqrt{\text{top area}/\pi}$. The cribs ranged in peak energy release rate of 60 to 345 kW for wood and 115 to 690 kW for the polyurethane cribs. For both, their fraction of energy radiated χ_r is approximately 0.35. Note that the heat flux for the R1 and R2 targets are as high as 25 kW/m² for the 690 kW crib.

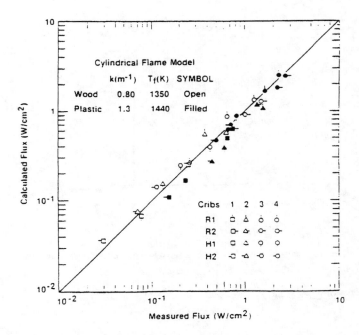

FIGURE 7.16 A comparison of measured and calculated radiative heat flux to targets for crib fires of wood and polyurethane construction. (From Quintiere and McCaffrey[7].)

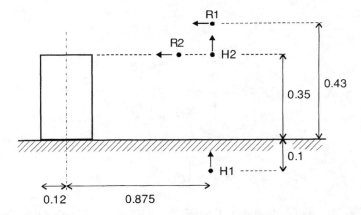

FIGURE 7.17 A sketch of the arrangement of heat flux gages (R1, R2, H1, H2) to measure flame radiation from free-burning crib fires. R1 and R2 are directed toward the fire, H1 and H2 are directed upward. Exact locations are given in Quintiere and McCaffrey[7].

EXAMPLE 7.5

Calculate, using Eq. (7.16), the heat flux to the R1 and R2 targets given in Figure 7.17 and compare this to the measured values given in Figure 7.16.

SUGGESTED SOLUTION

For $R_0 \approx 0.9$ m from the center of the crib to R1 and R2, we find $\dot{q}'' = \dfrac{(0.35)(690)}{4\pi(0.9)^2} = 23.7 \text{ kW/m}^2$, which is consistent with the peak values measured in Figure 7.16.

EXAMPLE 7.6

A plastic crib with sides 0.245 m (as depicted in Figure 7.17) burns with an effect of 690 kW. The flame height, denoted H here, is found to be roughly 2.9 m (from Eq. (4.3)). Use Eq. (7.17) to calculate the incident radiant heat flux to a small differential element lying flat on the floor at a distance of 0.875 m from the center of the crib fire.

SUGGESTED SOLUTION

The radius of the cylinder is taken to be $r = \sqrt{0.245^2/\pi} = 0.138$. We find that the condition $L/r = 0.875/0.138 > 3$ is satisfied. The angle $\theta_0 = \tan^{-1}(L/H) = \tan^{-1}(0.875/2.9) = 16.8° = \pi 16.8/180 = 0.293$ radians. Since the normal vector of the element is parallel to the cylinder axis, we know that $u = v = 0$ and $w = 1$, so $F_1 = F_2 = 0$ and $F_3 = \frac{1}{\pi}\left(\frac{0.138}{0.875}\right)\cos^2(0.293) = 0.046$. The angle $\beta = (0.293 + \pi/2)/2 = 0.932$ radians, so $\sin\beta = 0.803$. The absorption coefficient is taken from Figure 7.16 for plastics as $\kappa = 1.3$. Therefore $\mu = \frac{2 \cdot 0.138 \cdot 1.3}{0.803} = 0.45$ and $\varepsilon = 1 - \exp(-0.7 \cdot 0.45) = 0.27$. The flame temperature from Figure 7.16 is $T_f = 1440$, and Eq. (7.17) gives the heat flux as $\dot{q}'' = 5.67 \cdot 10^{-8} \cdot 1440^4 \cdot 0.27 \cdot 0.046 = 3027$ W/m². This corresponds roughly with the highest calculated value for sensor H1 in Figure 7.16.

Note that if the normal vector of the sensor had instead been perpendicular to the cylinder axis (pointing toward the axis of the flame cylinder), then the normal vector would have a component in the v direction only and $F_1 = F_3 = 0$.

7.4.2 BASIC PRINCIPLES OF RADIATIVE TRANSFER

In this section we consider some basic principles of radiative transfer and briefly discuss configuration factors, spectral absorptivity, spectral emissivity, gray gases, real gas properties, and radiation through an intervening medium.

Configuration factors: Radiation transfer is determined by using a quantity called radiative intensity, I. This depends on energy directed at a target element, and pertains to energy at a given wavelength (spectral intensity). Consider rate of energy $d\dot{q}_o$ leaving dA_1 in the $\hat{r}$ direction intercepted by dA_2. This is shown in Figure 7.18 where $\bar{n}_1$ and $\bar{n}_2$ are the unit outward normals to dA_1 and dA_2 accordingly. Intensity I is defined as

$$I \equiv \frac{d\dot{q}_o}{(\cos\theta_1 dA_1)d\omega} \tag{7.18}$$

where $d\omega$ is the solid angle subtended by dA_2 from dA_1.

The differential solid angle is defined as

$$d\omega = \frac{dA_2 \cos\theta_2}{r^2} \tag{7.19}$$

Note the area segments $dA_2 \cos\theta_2$ and $dA_1 \cos\theta_1$ are the projections of the areas on planes normal to the r direction. Also note $\bar{n}_1 \cdot \bar{r}_1 = \cos\theta_1$. Therefore, the rate of energy leaving dA_1 intercepted by dA_2 can be written as

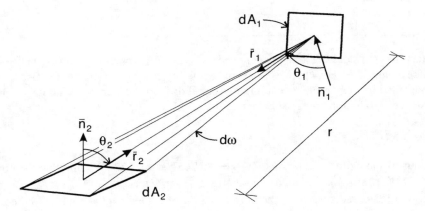

FIGURE 7.18 Configuration of radiation exchange between dA_1 and dA_2.

$$d\dot{q}_{1\to2} = \frac{I_1\left(\bar{r}_1 \cdot \bar{n}_1 dA_1\right)\left(\bar{r}_2 \cdot \bar{n}_2 dA_2\right)}{r^2} \tag{7.20}$$

and the rate of energy leaving dA_2 intercepted by dA_1 is

$$d\dot{q}_{2\to1} = \frac{I_2\left(\bar{r}_2 \cdot \bar{n}_2 dA_2\right)\left(\bar{r}_1 \cdot \bar{n}_1 dA_1\right)}{r^2} \tag{7.21}$$

Intensity from a point (dA) can depend on direction, especially for mirror-like (specular) reflection produced intensities. Alternatively, a surface (not mirror-like, or optically rough) may have intensities independent of direction. Such surfaces are said to be *diffuse*. For a diffuse surface, it can be shown that the rate of energy per unit area leaving a surface is

$$\left(\frac{d\dot{q}}{dA}\right)_{\text{Diffuse}} = \pi I \tag{7.22}$$

This can be shown by intercepting all of the energy leaving dA_1 in Figure 7.18 with a hemisphere of fixed radius. This result, along with Eq. (7.20) and (7.21), allows for computing radiation exchange between "black" ($\varepsilon = 1$) or "gray" ($\varepsilon = \alpha$) diffuse surfaces. For example, for black surfaces separated by a vacuum where nothing effects the I transported between the surfaces,

$$\left(\frac{d\dot{q}}{dA}\right)_{\text{Black}} = \sigma T^4 = \pi I_{\text{b}} \tag{7.23}$$

where I_b is the blackbody intensity, and the *net* rate of energy leaving dA_1

$$d\dot{q}_{1,\text{net}} = d\dot{q}_{1\to2} - d\dot{q}_{2\to1}$$

or from Eq. (7.20) and (7.21)

$$d\dot{q}_{1,\text{net}} = \sigma\left(T_1^4 - T_2^4\right)\frac{\left(\bar{r}_1 \cdot \bar{n}_1 dA_1\right)\left(\bar{r}_2 \cdot \bar{n}_2 dA_2\right)}{\pi r^2} \tag{7.24}$$

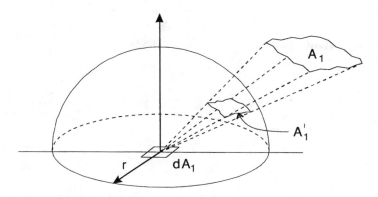

FIGURE 7.19 Physical representation of F_{12}.

The term involving the geometry is called the *configuration factor* (shape factor or geometric factor) and is designated as $F_{12} dA_1$ or $F_{21} dA_2$. The physical significance of F_{12} is the fraction of hemispherical surface area seen by dA_1 looking at dA_2. The case where A_2 is finite is illustrated in Figure 7.19. This suggests that configuration factors can be estimated by "eye-balling." Formally, they require complex integration analyses over finite areas for A_1 and A_2. Table 7.1, taken from the *SFPE Handbook for Fire Protection Engineering*,[3] gives some results. For radiative exchange where fluxes from surfaces are assumed to be uniformly and diffusely distributed, two commonly used rules for configuration factors can be obtained from the equations in Table 7.1.

The reciprocity relation states that for any given pair in a group of exchanging surfaces,

$$A_i F_{i-j} = A_j F_{j-i}$$

The summation rule can be used for surfaces $1, 2, \ldots, N$ that subtend a closed system so that $\displaystyle\sum_{j=1}^{N} F_{i-j} = 1$.

EXAMPLE 7.7a

Use Table 7.1 to calculate the configuration factor from a cylinder of height 1 m and radius 0.25 m to a differential element 2 m away from the cylinder axis. The differential element has a normal vector directed toward the cylinder axis and is perpendicular to it.

SUGGESTED SOLUTION

This is the third case shown in Table 7.1. We calculate the factors $L = 1/0.25 = 4$, $H = 2/0.25 = 8$, $X = (1 + H)^2 + L^2 = 97$, and $Y = (1 - H)^2 + L^2 = 65$. The equation given in Table 7.1 has several arguments that are best calculated separately; we shall call these arguments A1, A2, A3, and A4, where A1 $= \dfrac{L}{\sqrt{H^2 - 1}} = 0.504$, A2 $= \dfrac{X - 2H}{H\sqrt{XY}} = 0.1275$, A3 $= \sqrt{\dfrac{X(H-1)}{Y(H+1)}} = 1.1607$, and A4 $= \sqrt{\dfrac{H-1}{H+1}} = 0.882$. Substituting this into the equation in Table 7.1, remembering that angles are given in radians, gives

$$F_{c-dA} = \frac{1}{8\pi} \tan^{-1}(0.504) + \frac{4}{\pi} \left[0.1275 \tan^{-1}(1.1607) - \frac{1}{8} \tan^{-1}(0.882) \right] = 0.043.$$

TABLE 7.1
Common Configuration Factors

Configuration	Equation

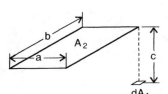

$$F_{d1-2} = \frac{1}{2\pi}\left[\frac{a}{\sqrt{a^2+c^2}} \tan^{-1}\left(\frac{b}{\sqrt{a^2+c^2}} \right) + \frac{b}{\sqrt{b^2+c^2}} \tan^{-1}\left(\frac{a}{\sqrt{b^2+c^2}} \right) \right]$$

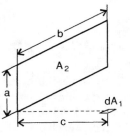

$$F_{d1-2} = \frac{1}{2\pi}\left[\tan^{-1}\left(\frac{b}{c} \right) + \frac{c}{\sqrt{a^2+c^2}} \tan^{-1}\left(\frac{b}{\sqrt{a^2+c^2}} \right) \right]$$

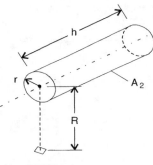

$$F_{d1-2} = \frac{1}{H\pi} \tan^{-1}\frac{L}{\sqrt{H^2-1}} + \frac{L}{\pi}\left[\frac{X-2H}{H\sqrt{XY}} \tan^{-1}\sqrt{\frac{X(H-1)}{Y(H+1)}} - \frac{1}{H}\tan^{-1}\sqrt{\frac{H-1}{H+1}} \right]$$

where $L = 1$ r, $H = h/r$, $X = (1+H)^2 + L^2$, $Y = (1-H)^2 + L^2$

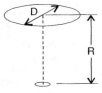

$$F_{d1-2} = \frac{D^2}{4R^2 + D^2}$$

EXAMPLE 7.7b

Recalculate the example above, but now use the configuration factors given by Eq. (7.17).

SUGGESTED SOLUTION

Since the normal vector is pointed toward the cylinder axis and is perpendicular to it, we see from Figure 7.15 that v is the only vector component, and therefore that $F_1 = F_3 = 0$ and

$$F_2 = \frac{1}{2\pi} \cdot \frac{0.25}{2} (\pi - 2 \cdot 1.107 + \sin(2 \cdot 1.107)) = 0.0344.$$ We see that the approximating Eq. (7.17) gives, in this case, a lesser value than the exact solution given in Table 7.1 and calculated in Example 7.7a.

EXAMPLE 7.7c

Now let us investigate the error introduced if the cylindrical flame is replaced by an equivalent rectangular sheet, placed at the cylinder axis. Assume that there are two rectangles side by side of height 1 m and each of width 0.25 m, 2 m from the differential element.

SUGGESTED SOLUTION

This is the first case given in Table 7.1. $X = 0.25/2 = 0.125$ and $Y = 1/2 = 0.5$. There are four arguments in the equation which we call A1, A2, A3, and A4. These get the values A1 = $\dfrac{X}{\sqrt{1+X^2}} = 0.1240$, A2 = $\dfrac{Y}{\sqrt{1+X^2}} = 0.4961$, A3 = $\dfrac{Y}{\sqrt{1+Y^2}} = 0.1240$, A4 = $\dfrac{X}{\sqrt{1+Y^2}} = 0.1118$.

$$F_{dA-r} = \frac{1}{2\pi}[0.124\tan^{-1}(0.4961) + 0.4472\tan^{-1}(0.1118)] = 0.017.$$

Since there are two equal rectangles, we use the summation rule and find that the total configuration factor from the two rectangular sheets is 0.034. This also gives a slightly lesser value than the exact equation for the cylinder in Table 7.1.

Absorbing, emitting, and scattering media: For media in which the intensity can be absorbed, scattered, and increased by emission, the intensity must be computed. For simplicity, we shall consider "gray" media in which the properties are independent of wavelength. As we have pointed out, particular problems bear close attention to spectral effects. For enclosure fire radiative exchanges gray gas and gray or blackbody surface approximations are generally satisfactory. For example, the emissivity of most nonmetal enclosure surfaces in the temperature range of fires can be estimated as 0.7 or greater. Such properties are difficult to specify accurately, but the prospect of soot deposition tends to make their pristine surface estimations moot. The governing equation for the intensity is known as the *Transfer Equation,* which applies along a path length r,

$$\frac{dI}{dr} = -\kappa I + \kappa I_b \tag{7.25}$$

where scattering effects have been ignored, and κ is the absorption coefficient, and $I_b = \dfrac{\sigma T^4}{\pi}$ where T is the temperature of the medium.

For a path length of distance, L, over a medium of uniform temperature, T_g, Eq. (7.25) can be solved as

$$I(L) - I_b = (I(0) - I_b)e^{-\kappa L} \tag{7.26}$$

We can consider this equation to be in spectral terms, applying to each wavelength. It can be used to define emissivity and absorptivity for the medium or, in our application, the gas.

Spectral absorptivity, α_λ: To define α_λ consider the pure absorbing (transmitting) case where $I_b = 0$. By definition,

$$\alpha_\lambda = \frac{I_\lambda(0) - I_\lambda(L)}{I_\lambda(0)} \sim \frac{\text{absorbed}}{\text{sent}} \tag{7.27a}$$

From Eq. (7.26),

$$\alpha_\lambda = 1 - e^{-\kappa_\lambda L} \qquad (7.27b)$$

Note the corresponding transmissivity τ_λ is $1 - \alpha_\lambda$, since no scattering is considered here.

Spectral emissivity, ε_λ: To define ε_λ consider the case of emission due to temperature T_g with $I(0) = 0$. By definition

$$\varepsilon_\lambda = \frac{I_\lambda(L)}{I_{b,\lambda}} \sim \frac{\text{emitted}}{\text{maximum possible emission}} \qquad (7.28a)$$

From Eq. (7.26)

$$\varepsilon_\lambda = 1 - e^{-\kappa_\lambda L} \qquad (7.28b)$$

Hence, we see that $\varepsilon_\lambda = \alpha_\lambda$ as followed by Kirchhoff's Law for surfaces.

Gray gas assumption: Similarly, if we regard κ independent of wavelength, which constitutes the "gray gas" assumption, we can show from the total properties that $\varepsilon = \alpha$. This follows from the definitions for the total (integrated over $0 \le \lambda \le \infty$) properties:

$$\alpha \equiv \frac{\int_0^\infty \alpha_\lambda E_{b,\lambda}(T) d\lambda}{\sigma T^4} \qquad \text{where } I_\lambda(o) = \frac{E_{b,\lambda}(T)}{\pi} \qquad (7.29a)$$

and

$$\varepsilon \equiv \frac{\int_0^\infty \varepsilon_\lambda E_{b,\lambda}(T_g) d\lambda}{\sigma T_g^4} \qquad \text{where } I_{b,\lambda} = \frac{E_{b,\lambda}(T_g)}{\pi} \qquad (7.29b)$$

It is readily seen that if κ_λ is independent of λ, by Planck's Law, $\alpha = \varepsilon$. This assumption will be used in our application to compartment fires, and has already been used for flames in Eq. (7.17).

Real gas properties: Combustion gases are complex in terms of radiative properties, since the gases (principally H_2O and CO_2) radiate discretely over wavelength bands that can overlap, and soot radiates continuously over wavelength. The most practical approach to computing the total emissivity of H_2O or CO_2 comes from the charts developed by Hottel, as shown in Figures 7.20 and 7.21.

The graphs are given in terms of temperature over the path length L, and concentration expressed in partial pressure P_a. Recall (P_a/mixture pressure) is the molar concentration. Each graph gives the emissivity of the species acting alone. The emissivity due to both gases is roughly

$$\varepsilon = \varepsilon_{H_2O} + \varepsilon_{CO_2} + \begin{bmatrix} \text{correction factor} \\ \text{due to overlap} \\ \text{wavelength} \end{bmatrix} \qquad (7.30)$$

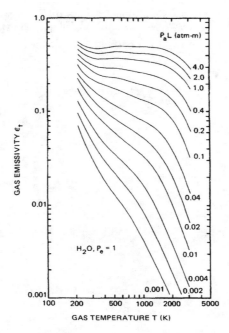

FIGURE 7.20   Total emittance of water vapor. (From Tien et al.[3])

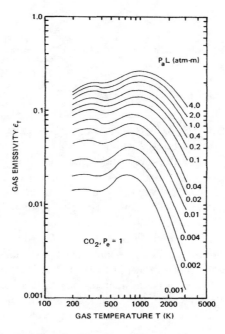

FIGURE 7.21   Total emittance of carbon dioxide. (From Tien et al.[3])

Soot emissivity can be included also. Soot absorption coefficient is found to depend on wavelength and on soot volume fraction (f_v), usually of the order of 10^{-6},

$$\kappa_{soot} \propto \frac{f_v}{\lambda} \qquad (7.31)$$

An illustration of how these various components with different wavelength characteristics can be combined is given as follows. Consider two species (1 and 2). The total emissivity would be given by Eq. (7.29b), where the absorption coefficients are added.

$$\varepsilon_{1+2} = \frac{1}{\sigma T^4} \int_0^\infty \left(1 - e^{-(\kappa_{\lambda,1} + \kappa_{\lambda,2})L}\right) E_{b,\lambda} d\lambda$$

Rearranging, we get

$$\varepsilon_{1+2} = \frac{1}{\sigma T^4} \left\{ \int_0^\infty E_{b,\lambda} \left[(1 - e^{-\kappa_{\lambda,1}L}) + (1 - e^{-\kappa_{\lambda,2}L}) - (1 - e^{-\kappa_{\lambda,1}L})(1 - e^{-\kappa_{\lambda,2}L}) \right] d\lambda \right\}$$

If gray gas approximations are used for each species, e.g.,

$$\varepsilon_1 \equiv \frac{\int_0^\infty (1 - e^{-\kappa_{\lambda,1}L}) E_{b,\lambda} d\lambda}{\sigma T^4} \approx 1 - e^{-\kappa_1 L}$$

where κ_1 is the gray gas approximation to $\kappa_{\lambda,1}$, then

$$\varepsilon_{1+2} = \varepsilon_1 + \varepsilon_2 - \varepsilon_1 \varepsilon_2 \tag{7.32}$$

This is the basis of Eq. (7.30), and can serve to combine the soot (ε_s) and gas (ε_g) contributions to flame and smoke radiation, i.e.,

$$\varepsilon \approx \varepsilon_g + \varepsilon_s - \varepsilon_g \varepsilon_s \tag{7.33}$$

Typically, we might expect the products H_2O and CO_2 to contribute about $\varepsilon_g \sim 0.3$ in a room smoke layer, and ε_s to range from nil to 0.7 taking ε up to 1. But note that the path length, if long enough, can take ε up to 1. In flames, this critical path length is nominally 1 to 2 m for most hydrocarbon fuels.

A few empirical values for an effective absorption coefficient, κ_m, are available in the literature for a number of fuel types and have been summarized by Drysdale.[16] Table 7.2 gives these values and their references.

Radiative exchange with an intervening gray gas: Consider two area elements A_i and A_j communicating by radiation through a homogeneous medium at temperature T_g. Each area element is at a uniform temperature and has uniform properties. The surfaces and the medium (gas) are considered to be "gray." No scattering is considered. This case is illustrated in Figure 7.22.

By Eq. (7.26) for a gray gas medium with emissivity ε_g. The intensity at A_i from A_j is

$$I_j(r_{ij}) = I_j(0)(1 - \varepsilon_g) + \frac{\sigma T_g^4}{\pi} \varepsilon_g \tag{7.34}$$

For diffuse surfaces, the rate of energy per unit area leaving A_j is

$$\left(\frac{d\dot{q}_0}{dA}\right)_j = \pi I_j(0) \equiv J_j \tag{7.35}$$

TABLE 7.2
Effective Absorption Coefficient κ_m for Various Fuels

Material	κ_m (m^{-1})	Reference
Diesel oil	0.43	Sato and Kunimoto[17]
Polymethylmethacrylate	0.5	Yuen and Tien[18]
Polystyrene	1.2	Yuen and Tien[18]
Wood cribs	0.8	Hägglund and Persson[19]
Wood cribs	0.51	Beyreis et al.[20]
Assorted furniture	1.13	Fang[21]

Source: Drysdale, D., *An Introduction to Fire Dynamics*, Wiley-Interscience, 1992. With permission.

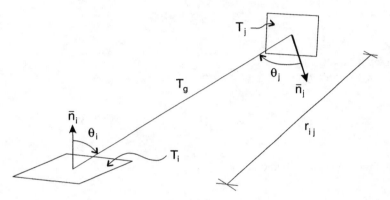

FIGURE 7.22   Radiation exchange with an intervening medium.

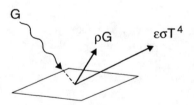

FIGURE 7.23   Irradiance (G) and radiosity (J).

and is typically called the "radiosity" of surface A_j denoted by J_j.

For any gray diffuse surface the radiosity J can be related to the emissive power E, and the total rate of energy received per unit area (called the irradiance, G) by (see Figure 7.23)

$$J = \rho G + \varepsilon \sigma T^4 \tag{7.36}$$

The subscript designating the surface has been dropped in Eq. (7.36) for convenience.

Now we have enough tools to consider the net radiative outward rate from any surface A_i. From all surfaces j to i, for N total uniform surfaces, from Eq. (7.34), (7.35), and (7.21) we find

$$\dot{q}_{j \to i} = \sum_{j=1}^{N} \left[\frac{J_j}{\pi}(1 - \varepsilon_g) + \frac{\sigma T_g^4}{\pi}\varepsilon_g \right] \iint_{A_i A_j} \frac{\cos\theta_i \cos\theta_j dA_j dA_i}{r_{ij}^2} \tag{7.37}$$

Since the configuration factor relationship, from Eq. (7.24) is

$$A_i F_{ij} = A_j F_{ji} = \int\limits_{A_i} \int\limits_{A_j} \frac{\cos\theta_i \cos\theta_j}{\pi r_{ij}^2} dA_j dA_i \tag{7.38}$$

we find that

$$\dot{q}_{j\to i} = \sum_{j=1}^{N} [J_j(1-\varepsilon_g) + \varepsilon_g \sigma T_g^4] A_i F_{ij} \tag{7.39}$$

The outgoing rate of energy from A_i to all $j = 1, \ldots , N$ surfaces is

$$\dot{q}_{i\to j} = \sum_{j=1}^{N} J_i A_i F_{ij} \tag{7.40}$$

in a similar manner realizing $I_i(0) = J_i/\pi$.

Hence, the *net outgoing* energy rate *per unit area* from A_i is

$$\dot{q}_i'' = \frac{\dot{q}_{i\to j} - \dot{q}_{j\to i}}{A_i} \sum_{j=1}^{N} [J_i - J_j(1-\varepsilon_g) - \varepsilon_g \sigma T_g^4] F_{ij} \tag{7.41}$$

Alternatively, this can be rearranged as

$$\dot{q}_i'' = \sum_{j=1}^{N} [(J_i - J_j)(1-\varepsilon_g)F_{ij} + (J_i - \sigma T_g^4)\varepsilon_g F_{ij}] \tag{7.42}$$

This last form has some utility, since it can be related to an electric circuit analogy that can help to shape problem solutions. To complete this analysis we need to establish a solution for J_i. From our definitions of J_i and G_i (Eq. (7.36) and (7.37), since G is $\dot{q}_{j\to i}/A_i$):

$$\dot{q}_i'' = J_i - G_i \tag{7.43a}$$

or

$$\dot{q}_i'' = [\varepsilon_i E_{b_i} + (1-\varepsilon_i)G_i] - G_i \tag{7.43b}$$

where ε_i is the emissivity of the A_i surface (opaque).

Therefore,

$$\dot{q}_i'' = \varepsilon_i (E_{b_i} - G_i) \tag{7.44a}$$

and

$$\dot{q}_i'' = \varepsilon_i E_{b_i} - \varepsilon_i \sum_{j=1}^{N} \left(J_j(1-\varepsilon_g) + \varepsilon_g \sigma T_g^4 \right) F_{ij} \tag{7.44b}$$

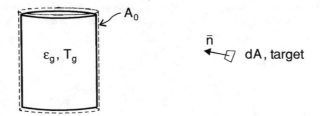

FIGURE 7.24 Gray gas analysis of flame radiation to a target.

Subtracting the ε_i times Eq. (7.41) from the above yields

$$\dot{q}_i'' = \frac{\left(\varepsilon_i E_{b_i} - \sum_{j=1}^{N} \varepsilon_i J_i F_{ij}\right)}{(1-\varepsilon_i)} \tag{7.45}$$

Equations (7.42) and (7.45) represent $2 \times N$ equations, two for each surface area considered. For each surface there are three variables, one of which must be known: $\dot{q}_i''$, J_i, or T_i. Usually T_i or $\dot{q}_i''$ are known in the problem. These equations represent the general tools in solving many radiation heat transfer problems. Sometimes fictitious surfaces can be used that represent open windows or the envelope of a flame.

EXAMPLE 7.8

Let us reconsider the cylindrical flame radiation heat transfer to a target. See Figure 7.24 where an "enclosure" surface is the cylindrical flame envelope. Consider the target at the ambient temperature, T_∞. The envelope is also considered at this temperature.

SUGGESTED SOLUTION

Here we apply Eq. (7.42) and (7.44) or (7.45) to the cylinder with only one enclosure A_0 at T_∞, $(\varepsilon_0 = 1)$. Therefore, from Eq. (7.42),

$$\dot{q}_0'' = (J_0 - J_0)(1 - \varepsilon_g)F_\infty + (J_0 - \sigma T_g^4)\varepsilon_g F_\infty.$$

But the configuration factor from the interior cylinder to the cylinder, $F_\infty = 1$, since this surface "sees" only itself. Then, $\dot{q}_0'' = (J_0 - \sigma T_g^4)\varepsilon_g$

From Eq. (7.45),

$$(1 - \varepsilon_0)\dot{q}_0'' = (1)\sigma T_\infty^4 - (1) \cdot J_0 \cdot (1) = 0$$

Combining the two equations, noting that $\varepsilon_0 = 1$, we get

$$\dot{q}_0'' = -\varepsilon_g \sigma (T_g^4 - T_\infty^4)$$

Note this is the net flux outward from A_0 but *within* the envelope. Hence, it is *out* of the flame as we expect for $T_g \gg T_\infty$.

This only gives half of the solution. We now need the heat flux to the target dA at T_∞. Consider now an exchange between A_o and dA with no intervening medium ($\varepsilon_g = 0$). From Eq. (7.44), and using Table 7.1 or such to find F_{dA-A_o} we obtain

$$\dot{q}''_{\substack{target \\ (net,out)}} = (1)\sigma T_\infty^4 - (1)\left(J_{dA} \overbrace{F_{dA-dA}}^{=0} + J_{A_0} F_{dA-A_0} \right)$$

F_{dA-dA} is zero, since it does not see itself.

Now J_{A_0} is the radiosity (apparent) from the flame cylinder envelope which is $(-\dot{q}''_o)$. Hence, the heat flux received by the target is

$$\dot{q}''_{target,received} = -\sigma T_\infty^4 + \varepsilon_g \sigma \left(T_g^4 - T_\infty^4 \right) F_{dA-A_0}$$

The first term on the right-hand side is the re-radiation flux from the target, the second term is the flame contribution.

The cylindrical flame emissivity ε_g from the above example must also be determined. For the gray gas model,

$$\varepsilon_g = 1 - e^{-\kappa_g L} \tag{7.46}$$

where κ_g must be a known property, and L is a characteristic length for the cylindrical volume. This is called a "mean beam length." Its value depends on the volumetric configuration and on the orientation of the flux direction. Such values, computed from theory, equate Eq. (7.46) to a full integration of the Transfer Equation for the geometry. Table 7.3 (taken from *SFPE Handbook of Fire Protection Engineering*[3]) gives the mean beam length computed for the optically thin limit ($\kappa L \rightarrow$ small), and a correction factor C to give L for the optically thick limit: $L = C L_o$. For many cases,

$$L_o = \frac{4V}{A} \tag{7.47}$$

is a good approximation with $C = 0.9$ where V is the volume of the medium, and A is the bounding surface area.

Notice from Table 7.3 that $L_o \rightarrow D$, the diameter of the base of a cylinder, as the cylinder becomes tall. This suggests, for a tall flame whose height is more than $2D$, that the emissivity becomes constant. If the average flame temperature does not change, the radiative heat flux to its base (fuel surface) can be constant. The Cone Calorimeter produces such a flame as the radiation from the cone heater is increased and the flame elongates.

7.5   ENCLOSURE APPLICATIONS

Quintiere and McCaffrey conducted a comprehensive experimental study of wood and plastic fires in an enclosure, where they compared experimental data with calculated values for a number of

TABLE 7.3
Mean Beam Lengths for Various Gas Body Shapes

Geometry of Gas Body	Radiating to	Geometric Mean Beam Length L_0	Correction Factor C
Sphere	Entire surface	0.66 D	0.97
Cylinder, $H = 0.5D$	Plane end surface	0.48 D	0.90
	Concave surface	0.52 D	0.88
	Entire surface	0.50 D	0.90
Cylinder, $H = D$	Center of base	0.77 D	0.92
	Entire surface	0.66 D	0.90
Cylinder, $H = 2D$	Plane end surface	0.73 D	0.82
	Concave surface	0.82 D	0.93
	Entire surface	0.80 D	0.91
Semi-infinite cylinder, $H \to \infty$	Center of base	1.00 D	0.90
	Entire base	0.81 D	0.80
Infinite slab	Surface element	2.00 D	0.90
	Both bounding planes	2.00 D	0.90
Cube $D \times D \times D$	Single face	0.66 D	0.90
Block $D \times D \times 4D$	1×4 face	0.90 D	0.91
	1×1 face	0.86 D	0.83
	Entire surface	0.89 D	0.91

Source: Tien, C.L., Lee, K.Y., Stretton, A.J., "Radiation Heat Transfer," *SFPE Handbook of Fire Protection Engineering*, 2nd ed., National Fire Protection Association, Quincy, MA, 1995. With permission.

environmental variables.[7] Among these were comparisons where the conductive heat transfer was calculated using Eq. (7.11) and the radiative transfer computed using the concepts and equations discussed in Section 7.4.

This section will first briefly discuss how the electrical circuit analogy can be practically used for enclosure applications. We then give two examples of how the previously derived equations for convective heat transfer and radiative heat transfer can be applied to enclosure fires and compare these to experimental data from Quintiere and McCaffrey.[7] The first example considers heat flux—from a gas layer, the upper walls, and the lower surfaces—to a sensor at ceiling level. The second example considers the similar case of heat transfer to a sensor at floor level.

7.5.1 ELECTRICAL CIRCUIT ANALOGY

Consider Eq. (7.42) and (7.45) where the surfaces form an *enclosure* surrounding the homogeneous gas medium at T_g. For an enclosure

$$\sum_{j=1}^{N} F_{ij} = 1 \tag{7.48}$$

because the fraction of energy leaving A_i received by all the surfaces must be 1. Then Eq. (7.45) can be written as

$$\dot{q}_i = \dot{q}_i'' A_i = \frac{\varepsilon_i E_{b_i} - \varepsilon_i J_i(1)}{\frac{1-\varepsilon_i}{A_i}} = \frac{E_{b_i} - J_i}{\left(\frac{1-\varepsilon_i}{A_i \varepsilon_i}\right)} \tag{7.49}$$

and from Eq. (7.42),

$$\dot{q}_i = \dot{q}_i'' A_i = \sum_{j=1}^{N} \frac{\left(J_i - J_j\right)}{\left[1/\left((1-\varepsilon_g)F_{ij}A_i\right)\right]} + \frac{J_i - \sigma T_g^4}{\left(1/(\varepsilon_g A_i)\right)} \tag{7.50}$$

The form of these two equations suggests an electrical circuit with "current" $\dot{q}_i$ and resistors and potentials as shown in Figure 7.25. The sum of the "current flows" in the circuits from J_i to $J_{j=1,N}$ and σT_g^4 is $\dot{q}_i$, the net flow rate. For example, Figure 7.26 represents two black surfaces separated by a participating gas, for example two infinite parallel plates.

Since the heat flow rates must balance at a node,

$$\dot{q}_i = \frac{\sigma T_1^4 - \sigma T_2^4}{\frac{1}{(1-\varepsilon_g)A_1 F_{1,2}}} + \frac{\sigma T_1^4 - \sigma T_g^4}{\frac{1}{\varepsilon_g A_1}}$$

Similarly, $\dot{q}_2$ can be found.

7.5.2 FIRST EXAMPLE: HEAT FLUX TO A SENSOR AT CEILING LEVEL

The ceiling heat flux arrangement for the treated scenario is shown in Figure 7.27. The subscript "u" corresponds to the measured average properties of the "upper" smoke layer. The surfaces were assumed to be black, so $J_i = E_{bi}$, and the result follows from Eq. (7.50).

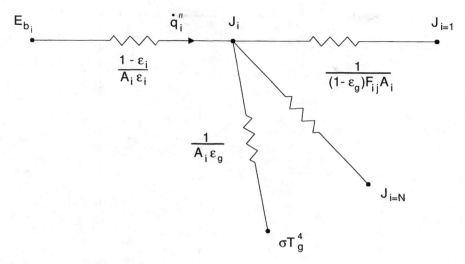

FIGURE 7.25   Enclosure analog circuit diagram.

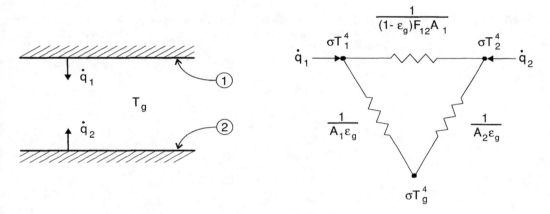

FIGURE 7.26   Circuit analog for two black surfaces separated by a gray gas.

Assuming an ambient lower region at T_a, the net radiative flux at the sensor at T_s is given as follows:

$$\dot{q}_r'' = \varepsilon_{g,u}\sigma T_{g,u}^4 + F_{12}\left(1 - \varepsilon_{g,u}\right)\sigma T_{w,u}^4 + F_{13}\left(1 - \varepsilon_{g,u}\right)\sigma T_a^4 - \sigma T_s^4 \tag{7.51a}$$

with

$$F_{12} = F_{13} = 1 \tag{7.51b}$$

$$F_{13} = \sum_{i=1}^{4} F_i \tag{7.51c}$$

$$F_i = \frac{1}{2\pi}\left[\left(\frac{a_i}{L_{a,i}}\right)\tan^{-1}\left(\frac{b_i}{L_{a,i}}\right) + \left(\frac{b_i}{L_{b,i}}\right)\tan^{-1}\left(\frac{a_i}{L_{b,i}}\right)\right] \tag{7.51d}$$

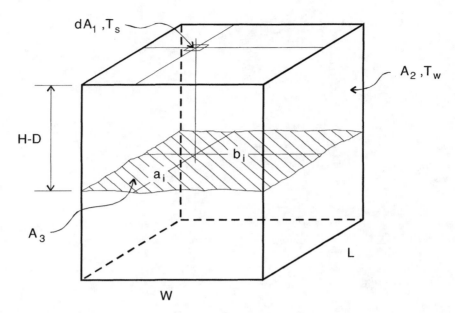

FIGURE 7.27 The model used to compute radiation heat transfer to a cooled sensor H1 at T_s from a gas layer of thickness $H - D$ and bounded by a cold surface 3 and hot walls 2 with temperature, T_w, based on measurements. (Adapted from Quintiere and McCaffrey[7].)

where

$$L_{a,i} = \sqrt{a_i^2 + (H - D)^2} \qquad (7.51e)$$

$$L_{b,i} = \sqrt{b_i^2 + (H - D)^2} \qquad (7.51f)$$

$$\varepsilon_{g,u} = 1 - \exp(-\kappa_{g,u} L_m) \qquad (7.51g)$$

and from Eq. (7.47), the mean beam length is

$$L_m = \frac{2WL(H - D)}{(H - D)(W + L) + WL} \qquad (7.51h)$$

Note that Eq. (7.51d, e, f) are for calculating the configuration factor and are equivalent with the first case in Table 7.1. The notation in Table 7.1 can be used for this example setting $X_i = a_i/(H - D)$ and $Y_i = b_i/(H - D)$. We will, however, use the above notation in order to combine the first and second cases from Table 7.1 in the next section.

The emissivity $\varepsilon_{g,u}$ was computed empirically in this analysis, but a more formal method is available from Modak that accounts for CO_2, H_2O, and soot.[15] The empirical method assumed that the smoke layer absorption coefficient κ_g was proportional to the flame absorption coefficient κ_f (0.8 m^{-1} for wood, 1.3 m^{-1} for polyurethane) by an overall combustion product mass fraction Y_p:

$$\kappa_g = \kappa_f Y_p \qquad (7.52a)$$

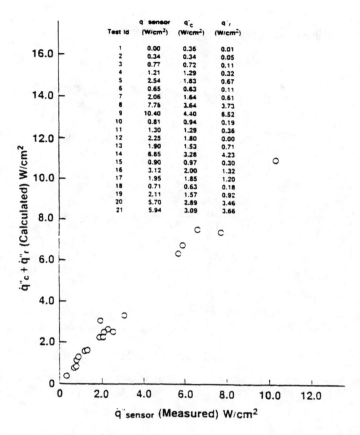

Test Id	q sensor (W/cm²)	q'c (W/cm²)	q'r (W/cm²)
1	0.00	0.36	0.01
2	0.34	0.34	0.05
3	0.77	0.72	0.11
4	1.21	1.29	0.32
5	2.54	1.83	0.67
6	0.65	0.63	0.11
7	2.06	1.64	0.61
8	7.78	1.64	3.73
9	10.40	4.40	6.52
10	0.81	0.94	0.19
11	1.30	1.29	0.36
12	2.25	1.80	0.00
13	1.90	1.53	0.71
14	6.65	3.28	4.23
15	0.90	0.97	0.30
16	3.12	2.00	1.32
17	1.95	1.85	1.20
18	0.71	0.63	0.18
19	2.11	1.57	0.92
20	5.70	2.89	3.46
21	5.94	3.09	3.66

FIGURE 7.28 A comparison of measured $\dot{q}''_{sensor}$ and computed convective $\dot{q}''_c$ and radiative $\dot{q}''_r$ total heat flux to a cooled ceiling sensor for the peak conditions of crib fire experiments in a room. (From Quintiere and McCaffrey[7].)

where

$$Y_p = \frac{(1+r)\dot{m}_f}{\dot{m}_a + \dot{m}_f} \qquad (7.52b)$$

where r = stoichiometric air to fuel ratio,
  $\dot{m}_f$ = mass flow rate of fuel,
  $\dot{m}_a$ = mass flow rate of air.

The results are shown in Figure 7.28, showing both the radiative and the convective computed values compared to the measured sensor values. The convective results were computed from Eq. (7.11). In these tests the smoke layer temperatures ranged from 200°C to 800°C for the polyurethane crib fuels and roughly 150°C to 500°C for the wood cribs. It is interesting that the ceiling sensor heat flux, despite varying the crib fuel and the doorway to the room, is mostly dependent on the average smoke layer temperature. These results are shown in Figure 7.29. It is interesting to note that the convective heat flux component at the ceiling heat flux is comparable to the radiative estimations, and the comparison with the measured values tends to confirm their accuracy. For this temperature range of the *developing* fire, they are both comparable to each other. Also, the convective

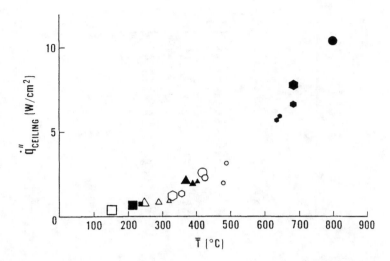

FIGURE 7.29 Ceiling incident heat flux vs. upper layer gas temperature. (From Quintiere and McCaffrey[7]. With permission.)

heat transfer coefficient can be as high as 55 W/m²°C for the highest case of 44 kW/m² at approximately 800°C for a 690 kW plastic crib fire. The high plume velocities contribute to this.

7.5.3 SECOND EXAMPLE: HEAT FLUX TO A SENSOR AT FLOOR LEVEL

The second example considers heat transfer to a floor sensor for the same set of experiments. The total flux measured by the sensor H2 (shown in Figure 7.30) can be considered as composed of several fluxes.

$$\dot{q}''_{sensor} = \dot{q}''_{f,r} + \dot{q}''_{e,r} + \dot{q}''_c - \sigma T_s^4 \tag{7.53}$$

where $\dot{q}''_{f,r}$ is the flame incident radiation computed by Eq. (7.17) but modified by a factor $\tau_{g,l}$ to account for attenuation by the lower gas layer; $\dot{q}''_{e,r}$ is the enclosure incident radiative flux; $\dot{q}''_c$ is the convective heat flux from the lower gas layer. For the data considered, the convective flux was found negligible. The radiative heat flux to the floor element from the upper smoke layer was found assuming black surfaces from Eq. (7.50) for each quadrant, $\dot{q}''_{i=1,4}$, the unattenuated heat flux. The transmitted portion is given by the transmissivity, $\tau_{g,l}$.

$$\dot{q}''_{e,r} = \tau_{g,l} \sum_{i=1}^{4} \dot{q}''_i \tag{7.54a}$$

$$\dot{q}''_i = \frac{1}{2\pi} \left[\left(\frac{a_i}{L_{a,i}} \right) \tan^{-1} \left(\frac{b_i}{L_{a,i}} \right) + \left(\frac{b_i}{L_{b,i}} \right) \tan^{-1} \left(\frac{a_i}{L_{b,i}} \right) \right]$$

$$\times \left\{ \left[1 - \exp\left(-\kappa_{g,u} L_{m,i} \right) \right] \sigma T_{g,u}^4 + \exp\left(-\kappa_{g,u} L_{m,i} \right) \sigma T_c^4 \right\}$$

$$+ \frac{\sigma T_a^4}{2\pi} \left[\tan^{-1} \left(\frac{a_i}{b_i} \right) - \left(\frac{b_i}{L_{b,i}} \right) \tan^{-1} \left(\frac{a_i}{L_{b,i}} \right) \right]$$

$$+ \frac{\sigma T_b^4}{2\pi} \left[\tan^{-1} \left(\frac{b_i}{a_i} \right) - \left(\frac{a_i}{L_{a,i}} \right) \tan^{-1} \left(\frac{b_i}{L_{a,i}} \right) \right] \tag{7.54b}$$

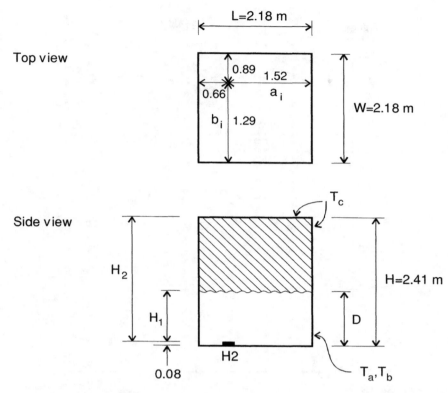

FIGURE 7.30 A description of the arrangement and model used in determining the incident heat flux to a floor sensor, H2. (Adapted from Quintiere and McCaffrey[7].)

$$L_{a,i} = \sqrt{(H_1^2 + a_i^2)} \tag{7.54c}$$

$$L_{b,i} = \sqrt{(H_1^2 + b_i^2)} \tag{7.54d}$$

$$L_{m,i} = \frac{2a_i b_i (H_2 - H_1)}{(H_2 - H_1)(a_i - b_i) + a_i b_i} \tag{7.54e}$$

where $T_{g,u}$ is the upper gas temperature,
T_c is the upper surface temperature,
T_a and T_b are the lower enclosure surface temperatures, and
a_i and b_i are the quadrant dimensions in Figure 7.30.

Note that Eq. (7.54b) calculates the configuration factor and is equivalent to a combination of the first and second cases shown in Table 7.1. The notation here is slightly different to that used in Table 7.1 in order to allow the two cases to be expressed in a single notation.

The lower gas layer is considered cool so that its emission is negligible, but sufficiently smoky so that it attenuates. Its transmissivity can be estimated as

$$\tau_{g,l} = \exp(-\kappa_{g,l}D) \tag{7.55a}$$

$$\kappa_{g,l} = C_k Y_{p,l} \tag{7.55b}$$

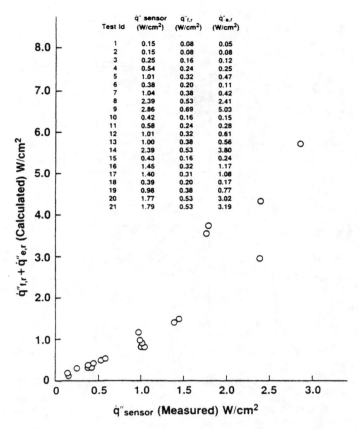

Test Id	$\dot{q}''$ sensor (W/cm^2)	$\dot{q}''_{t,r}$ (W/cm^2)	$\dot{q}''_{e,r}$ (W/cm^2)
1	0.15	0.08	0.05
2	0.15	0.08	0.08
3	0.25	0.16	0.12
4	0.54	0.24	0.25
5	1.01	0.32	0.47
6	0.38	0.20	0.11
7	1.04	0.38	0.42
8	2.39	0.53	2.41
9	2.86	0.69	5.03
10	0.42	0.16	0.15
11	0.58	0.24	0.28
12	1.01	0.32	0.61
13	1.00	0.38	0.56
14	2.39	0.53	3.80
15	0.43	0.16	0.24
16	1.45	0.32	1.17
17	1.40	0.31	1.08
18	0.39	0.20	0.17
19	0.98	0.38	0.77
20	1.77	0.53	3.02
21	1.79	0.53	3.19

FIGURE 7.31 A comparison of measured and computed heat flux to a cooled floor sensor during peak burning conditions of crib fires in room. (From Quintiere and McCaffrey[7]. With permission.)

and

$$Y_{p,l} = \left(\frac{\dot{m}_e / \dot{m}_a}{1 + \dot{m}_e / \dot{m}_a} \right) Y_{p,u} \qquad (7.55c)$$

where $\dot{m}_e$ is the mixing rate at the vent that contaminates the lower region with products, and $Y_{p,u}$ is given by Eq. (7.52b). For the data considered, $\tau_{g,l}$ was found to be 0.93 or greater.

Data compared for the 21 room crib fire experiments show the accuracy of the analysis; see Figure 7.31.[7] The analysis shows that as the smoke layer gets hotter, the radiation from the layer becomes much more significant than the radiation from the flame. Also, the computed smoke layer contribution appears too high at the larger fire conditions, suggesting that the estimates for $\tau_{g,l}$ may have been too high. Cold smoke may be more significant in fire heat transfer than we think. An interesting result from this study shows that the floor heat flux (to H2) is principally dependent on the average smoke layer temperature for the different fuels, fuel loads, and ventilation conditions. Hence, effects of emissivity appear to be accounted for by the layer temperature.

7.6 SUMMARY

This chapter has discussed convective and radiative heat transfer in enclosure fires, giving a brief discussion on the fundamental processes and describing methods for practical applications. Examples are given and calculations are compared to experimental data.

Section 7.2 discusses the different modes of heat transfer and offers a brief introduction to the techniques used to measure temperatures and heat transfer. Equation (7.9) can be used as a basis for estimating errors inherent in such measurements.

Section 7.3 discusses convective heat transfer, which in some instances can be the dominant heat transfer mechanism for heating up small devices such as detectors and temperature measuring devices. Both Eq. (7.11) and (7.15) can be used for estimating the convective heat flux at a distance r from the stagnation point. At this point and in the stagnation region ($r/H < 0.2$), Eq. (7.14) can be used to estimate the convective heat flux.

In Section 7.4 we described a number of methods to estimate heat flux received at a target. We started in Section 7.4.1 by describing two simple methods for this purpose. Neither method takes account of the influence of an intervening media, such as a gas layer. Modak's method (Eq. (7.16)) is suitable when the distance from the flame to the differential element, L, is far greater than the radius of the fire, r. A limiting value of $L/r > 10$ can be recommended. The second method, Eq. (7.17), considers radiation from a flame cylinder toward a differential element that can be at any angle to the cylinder axis. This approximate method is therefore very versatile, and errors are small for $L/r \geq 3$.

In Section 7.4.2 we considered some of the basic principles of radiative transfer, such as configuration factors, gray gases, and radiation through intervening medium.

Table 7.1 can be very useful for calculating configuration factors, and this eliminates some of the errors inherent in the methods described by Eq. (7.16) and (7.17), as Examples 7.7a, b, and c show.

The gray gas assumption is often used in fire protection engineering applications. By assuming that the emissivity and absorptivity of a surface or a medium are equal, the Transfer Equation can be greatly simplified. The effective absorption coefficient of the medium can be estimated or taken from Table 7.2 and the mean beam length from Table 7.3, which allows the emissivity and the absorptivity of the medium the to be estimated.

Having established average emissivities and absorptivities, the electric circuit analogy is a very useful tool for calculating radiant heat fluxes from several different sources towards a target, some simple examples are given.

Section 7.4.3 gives two examples where the methods above for calculating radiative heat fluxes are combined with methods for calculating convective heat fluxes. The calculations are performed and compared to experiments conducted in a compartment. Figures 7.28 and 7.31 show a good agreement with calculations of heat flux to the ceiling and heat flux to the floor.

REFERENCES

1. Rockett, J.A. and Milke, J.A., "Conduction of Heat in Solids," *SFPE Handbook of Fire Protection Engineering*, 2nd ed., National Fire Protection Association, Quincy, MA, 1995.
2. Atreya, A., "Convection Heat Transfer," *SFPE Handbook of Fire Protection Engineering*, 2nd ed., National Fire Protection Association, Quincy, MA, 1995.
3. Tien, C.L., Lee, K.Y., and Stretton, A.J., "Radiation Heat Transfer," *SFPE Handbook of Fire Protection Engineering*, 2nd ed., National Fire Protection Association, Quincy, MA, 1995.
4. Cooper, L.Y., "Heat Transfer from a Buoyant Plume to an Unconfined Ceiling," *Journal of Heat Transfer*, Vol. 104, No. 3, pp. 446–451, 1982.
5. Alpert, R.L., "Convective Heat Transfer in the Impingement Region of a Buoyant Plume," *Journal of Heat Transfer*, Vol. 109, pp. 120–124, Feb. 1987.
6. Kokkala, M., "Heat Transfer to and Ignition of Ceiling by an Impinging Diffusion Flame," Research Report 586, Fire Technology Laboratory, Technical Research Center of Finland, Espoo, February 1989.
7. Quintiere, J.G. and McCaffrey, B.J., "The Burning of Wood and Plastic Cribs in an Enclosure: Volume I," NBSIR 80-2054, National Bureau of Standards, Washington, D.C., November 1980.
8. Veldman, C.C., Kubota, T., and Zukoski, E.E., "An Experimental Investigation of the Heat Transfer from a Buoyant Gas Plume to a Horizontal Ceiling, Part 1: Unobstructed Ceiling," NBS-GCR-77-97, National Bureau of Standards, Washington, D.C., 1975.

9. You, H.Z. and Faeth, G.M., "Ceiling Heat Transfer during Fire Plume and Fire Impingement," *Fire and Materials*, Vol. 3, p. 140, 1979.
10. Alpert, R.L., "Convective Heat Transfer in the Impingement Region of a Buoyant Plume," *Journal of Heat Transfer*, Vol. 109, pp. 120–124, Feb. 1987.
11. Burgess, D.S., Grumer, J., and Wolfhard, H.G., "Burning Rates of Liquid Fuels," in Berl, W.G. (Ed.), *The Use of Models in Fire Research*, National Academy of Sciences, Washington, D.C., p. 68, 1961.
12. Orloff, L., Modak, A.T., and Alpert, R.L., "Burning of Large Scale Vertical Surfaces," Sixteenth Symposium (International) on Combustion, The Combustion Institute, Pittsburgh, PA, p. 1345, 1977.
13. Modak, A.T., "Thermal Radiation from Pool Fires," *Combustion and Flame*, Vol. 29, p. 177, 1977.
14. Dayan, A. and Tien, C.L., "Radiant Heating from a Cylindrical Fire Column," *Combustion Science and Technology*, Vol. 9, p. 41, 1974.
15. Modak, A.T., "Radiation from Products of Combustion," FMRC No. OAOE6.BU-1, Factory Mutual Research Corp., Norwood, MA, 1978.
16. Drysdale, D., *An Introduction to Fire Dynamics*, Wiley-Interscience, New York, 1992.
17. Sato, T. and Kunimoto, T., *Mem. Faculty of Engineering*, Kyoto University, Vol. 31, p. 47, 1969.
18. Yuen, W.W. and Tien, C.L., "A Simple Calculation Scheme for the Luminous Flame Emissivity," Sixteenth Symposium (International) on Combustion, The Combustion Institute, Pittsburgh, PA, 1977.
19. Hägglund, B. and Persson, L.E., "The Heat Radiation from Petroleum Fires," FOA Report C20126-D6 (A3), Forsvarets Forskningsanstalt, Stockholm, 1976.
20. Beyreis, J.R., Monsen, H.W., and Abbasi, A.F., "Properties of Wood Crib Flames," *Fire Technology*, Vol. 7, pp. 145–155, 1971.
21. Fang, J.B., "Measurement of the Behaviour of Incidental Fires in a Compartment," NBSIR 75-679, National Bureau of Standards, Washington, D.C., 1975.
22. Quintiere, J.G., "Radiative Characteristics of Fire Fighters' Coat Fabrics," *Fire Technology*, Vol. 10, p. 153, 1974.

PROBLEMS AND SUGGESTED ANSWERS

7.1 An oil leak from a pump ignites and causes cylindrical flame to be established on a factory floor. The flame cylinder has a height of 1.9 m and a radius of 0.25 m. Calculate the maximum view factor from the cylindrical flame at a distance of 1.72 m from the flame cylinder axis. Assume that the receiver is a differential element with its normal vector perpendicular to the cylinder axis. Use
(a) the view factor calculated by Eq. (7.17).
(b) the view factor calculated using Table 7.1 (fourth case from the top).

Suggested answer: (a) The maximum radiation is at the mid-height of the cylinder. For a half-cylinder we find that $F_1 = F_3 = 0$ and $F_2 = 0.0429$. The total view factor at mid-height is therefore $2 \cdot 0.0429 = 0.0859$. (b) Using the fourth case from top in Table 7.1 we find for a half-cylinder, $F = 0.0466$. The total view factor is thus 0.0932.

7.2 Assume that the cylinder in Problem 7.1 is now replaced with a rectangular sheet of dimensions 0.5 wide and 1.9 high. Assume that the distance to the differential element now is $1.72 - 0.25 = 1.47$ m. Calculate the view factor from the center of the sheet the differential element, using Table 7.1 (top case).

Suggested answer: One quarter of the sheet gives $F = 0.028$. The view factor from the whole sheet is therefore $4 \cdot 0.028 = 0.112$.

8 Conservation Equations and Smoke Filling

Most fire deaths are due to the inhalation of smoke and toxic gases. It is therefore important for the fire protection engineer to be well acquainted with the design methods used to control the flow of smoke in a building and to know how these are arrived at. In this chapter we state the conservation laws for mass and energy and apply these to a number of fire protection problems. The conservation equations are often presented as coupled differential equations that must be solved simultaneously by computer. We introduce a number of commonly applied assumptions that allow these equations to be considered separately and thus derive analytical solutions and iterative methods that can be applied to problems to do with the smoke filling process. We consider compartments under two types of ventilation conditions: closed compartments with only small leakage vents, which results in a dynamic pressure build-up, and compartments with openings large enough to prevent the build-up of pressures due to gas expansion. We apply the conservation equations to calculate smoke filling time and derive smoke-control methodologies for several cases.

CONTENTS

8.1   Terminology 182
8.2   Introduction 182
8.3   Conservation Equations for a Control Volume...183
      8.3.1   The Conservation of Mass..184
      8.3.2   Some Thermodynamic Properties...184
      8.3.3   The Conservation of Energy...187
8.4   Pressure Rise in Closed Rooms..190
      8.4.1   Pressure Rise in a Closed Volume..190
      8.4.2   Pressure Rise in a Leaky Compartment 192
8.5   Smoke Filling of an Enclosure with Leaks...196
      8.5.1   Small Leakage Areas at Floor Level 196
      8.5.2   Small Leakage Areas at Ceiling Level 200
      8.5.3   Estimating Gas Temperatures for the Floor Leak Case 202
      8.5.4   Limitations 203
8.6   Smoke Control in Large Spaces 204
      8.6.1   Smoke Filling: The Non-Steady Problem 204
      8.6.2   Smoke Control: The Steady-State Problem.......................................209
      8.6.3   Case 1: Natural Ventilation from Upper Layer 212
      8.6.4   Case 2: Mechanical Ventilation from Upper Layer 215
      8.6.5   Case 3: Lower Layer Pressurization by Mechanical Ventilation 216
8.7   Summary 218
References 221
Problems and Suggested Answers 221

8.1   TERMINOLOGY

Control volume — The first law of thermodynamics and the continuity equation for mass are
    sometimes applied to a volume in space into which, or out from which, a substance flows.
    This volume is called a control volume.

Control surface — The surface that completely surrounds the control volume is called the
    control surface.

Total energy — The sum of the internal energy, the kinetic energy, and the potential energy
    of a system, or $E = U + KE + PE$, where U stands for the internal energy at constant
    volume. We will only be interested in the increase or decrease in total energy as the system
    goes from one stage to another. In our applications the changes in kinetic and potential
    energy are very small compared to the changes in internal energy; we shall therefore focus
    our attention on changes in internal energy.

Internal energy — Associated with the translation, rotation, and vibration of the molecules
    and the chemical energy due to bonding between atoms. Our attention will be focused on
    the influence that temperature and pressure have on the internal energy. Internal energy
    is assigned the symbol U.

Enthalpy — The sum of the internal energy and the pressure work performed by or on a
    system. The term arises since the combination of the two properties often appear when
    analyzing thermodynamic processes. Enthalpy is assigned the symbol H.

Specific properties — When a property of a system is divided by unit mass, it is called a
    specific property. Specific properties are assigned lowercase letters, so that specific total
    energy is $e = E/m$, specific internal energy is $u = U/m$, and specific enthalpy is $h = H/m$.

Specific heat — The energy required to raise the temperature of a unit mass of a substance
    by 1 degree; this depends on the substance. We are interested in two kinds of specific
    heats: specific heat at constant volume, c_v, and specific heat at constant pressure, c_p.

8.2   INTRODUCTION

General: The conservation equations of mass, momentum, and energy can be used in many ways
for assessing the environmental consequences of a fire in an enclosure. In this chapter we use the
conservation of mass and energy; the momentum equation will not be explicitly applied, since
information needed to calculate velocities and pressures across openings will come from assumptions and specific applications discussed in earlier chapters.

Controlling the smoke extraction from a compartment can save lives, aid firefighting, and protect
property. Smoke control design must be based on the conservation of mass and the conservation
of energy; the first can be written to allow the mass flow rates of gases through openings to be
expressed in terms of pressure differences, the second allows the gas temperatures to be calculated
so that the pressure differences can be assessed.

Since the mass flows are dependent on temperature and the energy content of the compartment
is dependent on mass flows, the conservation equations must be coupled and solved simultaneously
by computer. However, some simple assumptions will allow the equations to be solved separately
so that analytical solutions or iteration schemes can be derived. In this chapter we state the
conservation equations for a specific volume in space and use these to derive equations for the
prediction of a number of environmental parameters of interest.

Ventilation conditions: We direct our attention to two ventilation conditions. In a closed
compartment, or a compartment with small leakages, the release of heat will cause an increase in
pressure. This pressure drives the mass flow out and there will be no mass flow into the compartment.
In Chapter 5, we referred to this stage of the fire as the first stage.

In a compartment with larger openings there will be little or no build-up of dynamic pressure. The opening flows are determined by the hydrostatic pressure differences across the openings, and there will be mass flow out of and into the compartment. In Chapter 5, we referred to this stage of the fire as the third stage.

Overview: We start by discussing the conservation laws for a specific volume in space and thus arrive at expressions that we will use for our applications. The applications include calculating dynamic pressures and smoke filling in a compartment with small leaks as well as transient smoke filling and steady-state smoke control for compartments with larger openings. A number of common smoke-control methods will be discussed and the employed equations will be derived.

The aim of this chapter is not to give a thorough presentation of the various smoke control design methodologies available to the fire protection engineer. Klote and Milke have provided an excellent and thorough guide to the use of design methodologies for smoke management in atria, for stairwell pressurization, for mechanical smoke removal, and for many more smoke-control applications.[1] Our aim is rather to show how the conservation equations can be used to derive such design equations from first principles and to enhance the readers' awareness of the multitude of assumptions commonly made along the way.

8.3 CONSERVATION EQUATIONS FOR A CONTROL VOLUME

The law of conservation of energy is commonly called the *first law of thermodynamics* when it is applied to problems where the effects of heat transfer and internal energy changes are included. The first law and the continuity equation for mass can be applied to a system or to a finite *control volume*.

A thermodynamical system is defined as a definite quantity of matter contained within a closed surface where the system can move in time and space. This approach is often applied when analyzing velocity distributions, diffusion processes, etc., where the state can change from one point to another. The resulting equations are in differential form and are sometimes referred to as the *point-wise equations*.

In many cases an analysis is simplified if attention is focused on a volume in space into which, or out from which, a substance flows. Such a volume is a control volume. The surface that completely surrounds the control volume is the *control surface*. This way of expressing the conservation equations is used where knowledge of the inner structure of the flow is not necessary. We shall be using the conservation equations expressed for a control volume and apply these to fire compartments.

The derivation of the equations for control volumes is given in many elementary textbooks on fluid mechanics and thermodynamics (see for example Welty et al.[2]), and it is assumed that the reader is acquainted with these. We use Figure 8.1 to set up the conservation equations for mass and energy for a control volume.

The following definitions apply:

CV   is a control volume that is selected as a particular region in space
CS   is the control surface, the boundary of the control volume
V is the volume of the control volume
ρ is the density of the matter in the control volume
$\dot{m}_j$ is mass flow rate of fluid mixture through the control surface at stream j (out of CV is positive)
m is the mass of the control volume
$\bar{v}$ is the velocity of the matter
$\bar{n}$ is the unit normal vector (outward) on the control surface

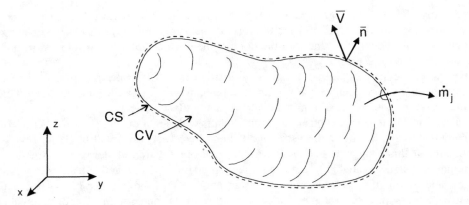

FIGURE 8.1   Control volume.

8.3.1   THE CONSERVATION OF MASS

The conservation of mass can be written as

$$
\begin{bmatrix} \text{Rate of change} \\ \text{mass in the CV} \end{bmatrix} + \begin{bmatrix} \text{Net rate of mass flow} \\ \text{leaving the CV or } \sum_{\text{out}} - \sum_{\text{in}} \end{bmatrix} = 0
$$

or

$$
\frac{dm}{dt} + \sum_{j=1}^{n} \dot{m}_j = 0 \tag{8.1}
$$

Since the mass can be written as $m = \rho\, V$, this can be written in integral form as

$$
\frac{d}{dt} \iiint_{CV} \rho dV + \iint_{CS} \rho \bar{v} \cdot \bar{n} dS = 0 \tag{8.2}
$$

Figure 8.2 shows the relationship between the velocity of the matter and the unit normal vector.
    The quantity $\bar{v} \cdot \bar{n}$ is the fluid velocity normal to the CS, which gives the rate at which fluid enters (–) or leaves (+) the CS. Hence $\rho \bar{v} \cdot \bar{n} dS$ is the rate of mass flow across the surface dS.
    For simplicity we denote the component of the velocity of matter normal to the CS as v_n so that $v_n = \bar{v} \cdot \bar{n}$. We can then drop the vector notation and the unit normal vector, $\bar{n}$.
    Equation (8.2) can then be written

$$
\frac{d}{dt} \iiint_{CV} \rho dV + \iint_{CS} \rho v_n dS = 0 \tag{8.3}
$$

8.3.2   SOME THERMODYNAMIC PROPERTIES

Before we set up the law of conservation of energy for a control volume, it is useful to discuss some thermodynamic properties. In this section we state the law for a closed system and discuss

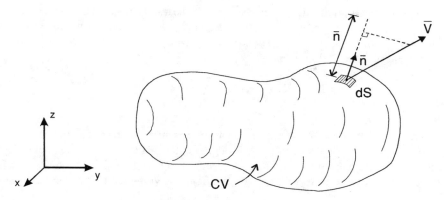

FIGURE 8.2 Mass flow rate at the control surface.

some properties of the system. This will be useful when we set up the energy conservation law for a control volume in Section 8.3.3.

For a closed system the conservation of energy can be written as

$$
\begin{bmatrix} \text{Rate of change of} \\ \text{energy in the CV} \end{bmatrix} = \begin{bmatrix} \text{Rate of heat} \\ \text{added to the CV} \end{bmatrix} - \begin{bmatrix} \text{Rate of work done} \\ \text{by fluid in the CV} \end{bmatrix}
$$

or

$$
\frac{dE}{dt} = \dot{Q} - \dot{W} \tag{8.4}
$$

Before we expand on this and formulate the first law for a control volume, where mass entering and leaving the CV contains energy and also influences the energy conservation, we shall define some thermodynamic properties.

Internal energy: The total energy of the system, E, is the sum of the internal energy, the kinetic energy, and the potential energy, or $E = U + KE + PE$, where U stands for the internal energy at constant volume. The internal energy is associated with the translation, rotation, and vibration of the molecules and the chemical energy due to bonding between atoms. Our attention will be focused on the influence that temperature and pressure have on the internal energy. Since we will be interested only in the increase or decrease in internal energy, we will not need to know its absolute value, only the change.

Further, in our applications, the changes in kinetic and potential energy are very small compared to the changes in internal energy. The changes in kinetic and potential energy are traditionally neglected when solving the types of engineering problems we are interested in, and we shall follow this in our treatment.

Equation (8.4) can therefore be rewritten as

$$
\frac{dU}{dt} = \dot{Q} - \dot{W} \tag{8.5}
$$

Specific internal energy: When a property of the system is divided by unit mass, it is called a *specific property*. The specific internal energy is thus $u = U/m$. Similarly, the specific total energy is $e = E/m$.

Enthalpy and specific enthalpy: When analyzing certain types of thermodynamic processes we frequently encounter the combination of properties $U + PV$, the sum of the internal energy and the pressure work. This sum is termed *enthalpy* and given the symbol H. When dividing by the mass we get the specific enthalpy, h, as

$$h = u + P/\rho \qquad (8.6)$$

since $m/V = \rho$.

Specific heat at constant volume and constant pressure: The energy required to raise the temperature of a unit mass of a substance by 1 degree is termed the specific heat, and depends on the substance. We are interested in two kinds of specific heats: specific heat at constant volume, c_v, and specific heat at constant pressure, c_p. These are defined symbolically through the relationships

$$du = c_v dT \qquad (8.7)$$

and

$$dh = c_p dT \qquad (8.8)$$

so

$$dh - du = (c_p - c_v)dT \qquad (8.9)$$

Note that when integrating the above equations we get $u = \int^T c_v dT$ and $h = \int^T c_p dT$. Applying the lower integration limit as zero degrees Kelvin (0 K) will result in the total specific internal energy and enthalpy. Usually, however, we are interested only in the change in these properties, and the lower limit therefore is chosen as the initial temperature of the system.

Further, the control volumes we will be looking at in many of our applications contain smoke, which in turn consists mainly of air (or, rather, nitrogen). It is commonly assumed that the specific heat of air is constant through the temperature range we are interested in, and we can sometimes write the above as

$$u = c_v(T - T_a) \qquad (8.10)$$

and

$$h = c_p(T - T_a) \qquad (8.11)$$

The ideal gas law: We will use the ideal gas law in many of our applications. This law can be stated in many different ways, and we will derive some expressions that will be of use to us later. A common expression for the ideal gas law is

$$PV = nR_0T$$

or

$$PV = \frac{m}{M}R_0T$$

where n is the number of molecules in the system, M is the molecular mass of the substance, and R_0 is the universal gas constant, which is the same for all substances $R_0 = 8.314$ J/(mol K). For each substance there is a unique gas constant, R, given by

$$R = R_0/M$$

and for air $R = 287$ J/(kg K).

Since our applications often deal with air (or smoke), we can rewrite the ideal gas law, by noting that $V = m/\rho$, as

$$P = \rho RT \tag{8.12}$$

Relationship between specific heat and the gas constant: We can now derive a relationship between the gas constant and the specific heats of a substance, which will be of use to us in our applications. Derivating Eq. (8.6), we get

$$dh = du + d(P/\rho).$$

Rearranging and derivating Eq. (8.12) we get

$$d(P/\rho) = d(RT) = RdT$$

since R is constant. Combining these leads to $dh - du = R\,dT$. Substituting into Eq. (8.9) we arrive at

$$R = c_p - c_v \tag{8.13}$$

which is a result we will use later.

8.3.3 THE CONSERVATION OF ENERGY

When the law of the conservation of energy is applied to problems where the effects of heat transfer and internal energy changes are accounted for, it is commonly called the *first law of thermodynamics*. Equation (8.5) states the law for a closed system, where the common assumption is made that changes in kinetic and potential energy are negligible.

Consider the control volume given in Figure 8.3. Heat is being added to the CV, work is being carried out by the CV, and mass is flowing in and out through the CS.

For a control volume where fluid flows through the control surface, account must be taken of the energy content of the fluid entering or leaving the control volume. We can restate the first law of thermodynamics, ignoring changes in kinetic and potential energy, as

$$\begin{bmatrix} \text{Rate of increase} \\ \text{of internal energy} \\ \text{in the CV} \end{bmatrix} + \begin{bmatrix} \text{Net rate of energy} \\ \text{out of CV due to} \\ \text{fluid flow} \end{bmatrix} = \begin{bmatrix} \text{Net rate of heat} \\ \text{added to} \\ \text{the CV} \end{bmatrix} - \begin{bmatrix} \text{Rate of work} \\ \text{done by fluid} \\ \text{in the CV} \end{bmatrix}$$

or in integral form, using the notation and assumptions made for Eq. (8.3),

$$\frac{d}{dt} \iiint_{CV} \rho \cdot u \cdot dV + \iint_{CS} \rho \cdot u \cdot v_n \cdot dS = \dot{Q} - \dot{W} \tag{8.14}$$

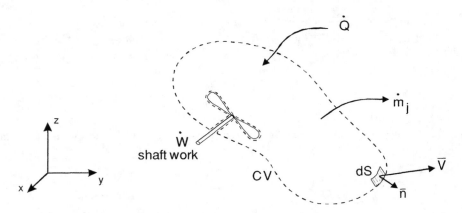

FIGURE 8.3   Control volume for the conservation of energy.

Energy can cross the control surface in three forms: as work, as heat, and as energy contained in the mass entering or leaving the control volume. It is important to distinguish between these, and we will expand on this by examining the two last terms more closely and thereby arrive at an equation we will use to solve practical problems.

Work: The work done by a constant force F on a body that is displaced a distance x is given by

$$W = Fx$$

where the work is measured in joules. Work done per unit time is denoted $\dot{W}$ and can be written (measured in watts) as

$$\dot{W} = F\frac{dx}{dt} \tag{8.15}$$

There are several different ways of doing work, each in some way related to a force acting through a distance. In engineering problems we often distinguish three types of work. The first is the *shaft work*, W_s, the work done by the control volume on its surroundings that could cause a shaft to rotate or a weight to be raised. The second is the *pressure work*, W_p, which is done on the surroundings to overcome normal stresses on the control surface where there is fluid flow (also called flow work). The third is the *shear work*, W_τ, which is performed on the surroundings to overcome shear stresses at the control surface. We can therefore write the rate at which work is performed on the control volume surroundings as

$$\dot{W} = \dot{W}_s + \dot{W}_p + \dot{W}_\tau \tag{8.16}$$

In our applications there are no mechanical devices, such as shafts, present. The shaft work is therefore zero. The shear force at a solid surface is zero since the fluid velocity there is zero (for a fixed control volume). At openings in the control surface, the shear work can be made zero by choosing a control surface that cuts across the opening perpendicular to the flow, and we shall only be considering such control volumes. No component of the velocity vector is then in the direction of the shear force vector, and the shear work is therefore zero.

Consider Figure 8.4. The pressure work can be formulated by observing that the force, F, acting on a surface dS is given as $F = P\,dS$.

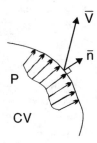

FIGURE 8.4   Force acting on the control surface.

The total force is obtained by integrating over the surface dS. Using Eq. (8.15) where dx/dt is the velocity normal to the surface dS, we can write the following equation expressing the total rate of work done on the entire control surface by normal stresses as

$$\dot{W}_p = + \iint_{CS} Pv_n dS \qquad (8.17)$$

The plus sign indicates that the work is being carried out by the control volume on the control surface and not vice-versa.

Heat: Heat is the form of energy transferred to or from the control volume, through the control surface, due to a temperature difference. It is important to distinguish this from the heat transferred with mass into and out of the control volume. In our applications the heat added to the control volume is due to the energy released when chemical reactions take place and heat transfer to the control volume. Here we call this term $\dot{Q}_{ch}$ (in earlier chapters, we simply called this term $\dot{Q}$). Some of this heat will be transferred to the control surface; we call this $\dot{q}_{loss}$. We can therefore write

$$\dot{Q} = \dot{Q}_{ch} - \dot{q}_{loss} \qquad (8.18)$$

We will continue to use the term $\dot{Q}$ to describe the net rate of heat added to the control volume, but will keep in mind that this consists of the chemical energy released and the heat transfer losses to the boundaries.

Resulting equation for our applications: We can now rewrite Eq. (8.14) into a form that we will use for our applications. Replacing the work term with Eq. (8.17), we get

$$\frac{d}{dt} \iiint_{CV} \rho u dV + \iint_{CS} \rho u v_n dS = \dot{Q} - \iint_{CS} Pv_n dS \qquad (8.19)$$

We can now combine the second and the last term in the above equation and get

$$\frac{d}{dt} \iiint_{CV} \rho u dV + \iint_{CS} \rho \left(u + \frac{P}{\rho} \right) v_n dS = \dot{Q} \qquad (8.20)$$

From the definition of enthalpy as $h = u + P/\rho$, we can use Eq. (8.6) to rewrite the second term as

$$\frac{d}{dt} \iiint_{CV} \rho u dV + \iint_{CS} \rho h v_n dS = \dot{Q} \qquad (8.21)$$

We will be using this equation extensively in our applications. The first term is the time rate of change of internal energy within the control volume. The second term is the net rate of enthalpy flow out of the control volume through the control surface. The third term is the net rate of heat added to the control volume.

The main assumptions we have made so far are neglecting the changes in kinetic and potential energy and assuming that the control volume can be set up such that there will be no work carried out due to shear forces. These are commonly used engineering assumptions.

We can further simplify Eq. (8.21) by assuming that the state of the control volume is uniform: the state may change with time, but it will do so uniformly. This means that the density in Eq. (8.21) is uniform over the entire control volume and can be moved out of the integrals. Finally, by assuming that c_v and c_p are constant for the temperature ranges we are interested in, Eq. (8.10) and (8.11) will allow us to express the internal energy and the enthalpy changes of the control volume. We use these assumptions in the following sections.

8.4 PRESSURE RISE IN CLOSED ROOMS

In a closed compartment, or a compartment with small leakages, the release of heat will cause volumetric expansion of gases and an increase in compartment pressure. It is this pressure that drives the mass flow out, and there will be no mass flow into the compartment. In Chapter 5 we referred to this stage of the fire as the *first stage*.

When there is rapid accumulation of mass or energy, or when the compartment has small openings to the surroundings, this pressure rise is very rapid and any hydrostatic pressure differences with height are negligible. For example, an addition of 100 kW to a 60 m^3 enclosure with a 0.01 m^2 opening will cause a steady-state pressure increase of ≈ 1000 Pa in a number of seconds. The hydrostatic pressure difference decreases at the rate of 10 Pa per meter as the height increases. For this case we see that the hydrostatic pressure difference is negligible and the opening flow will be determined by the pressure caused by the volumetric expansion of gases.

In this section we follow the work of Zukoski,[3] who set up the conservation equations for mass and energy for the closed room case and the case where there are small leakage areas to the surroundings, and derived expressions for the pressure rise. Zukoski also considered the smoke filling process for rooms with a leakage, and we consider this in Section 8.5.

8.4.1 PRESSURE RISE IN A CLOSED VOLUME

When heat is added to an ideal gas in a fixed volume, the pressure must increase in response to the temperature according to the ideal gas law. In a building fire situation the resulting pressure and the rate of pressure rise are often kept very small by gas leaks through openings in the walls of the buildings such as cracks around windows and doors.

However, situations may arise where the enclosure can be considered to be very well sealed, such as certain compartments on ships. The purpose of this section is to derive simple equations for calculating the dynamic pressure build-up in a hermetically closed compartment. We will then use the results to show that the pressure rise is very rapid. This result can be used to justify the so-called constant pressure assumption, used when examining a leaky room fire.

We consider this problem for a very simple example; see Figure 8.5. Consider a room of volume V with gases at an initial temperature, density, and pressure of T_a, ρ_a, and P_a. A small fire of constant heat output $\dot{Q}$ is treated as a point source of heat, and any heat losses to the surrounding structure are ignored. The room is hermetically closed.

Conservation of mass: Consider Eq. (8.3). We assume that the fire is a source of heat only and the mass release rate of the fuel is neglected. The second term in Eq. (8.3) is thus zero.

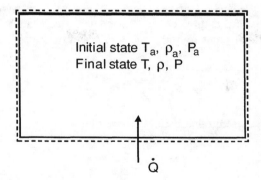

FIGURE 8.5 Hermetically closed room.

The total mass of gases in the volume will remain constant for all times and can be written as $\iiint_{CV} \rho_a dV$, where r_a is the initial density.

Conservation of energy: Since the mass can be written in terms of r_a and since there is no fluid flow into or out of the control volume, the second term in Eq. (8.21) is zero, and the equation can be rewritten as

$$\frac{d}{dt} \iiint_{CV} \rho_a u dV = \dot{Q} \tag{8.22}$$

We can assume that the gas in the volume consists mainly of air and that the specific heat is constant over the temperature range in which we are interested. This is quite a reasonable assumption, since for the temperature range 300 to 1000 K the value of c_p for air ranges from 1.0 to 1.14 kJ/(kg K) and c_v ranges from 0.71 to 0.85 kJ/(kg K). For simplicity, these values are often taken to be those at ambient temperature, and for many of our calculations we shall use $c_p = 1.0$ kJ/(kg K) and $c_v = 0.7$ kJ/(kg K).

We can now use Eq. (8.10) and write $u = c_v (T - T_a)$. Equation (8.22) becomes

$$\frac{d}{dt} \iiint_{CV} \rho_a c_v (T - T_a) dV = \dot{Q}$$

The terms within the integral sign are constants, except dV, and the term $\iiint_{CV} dV$ is simply the volume, V. This results in

$$\frac{d}{dt} (V \rho_a c_v (T - T_a)) = \dot{Q}$$

Integrating both sides of this equation from time 0 to time t gives us

$$V \rho_a c_v (T - T_a) = \dot{Q} t \tag{8.23}$$

Resulting equation: We wish to arrive at an equation for the pressure rise. We must therefore rewrite Eq. (8.23) in terms of pressure. The ideal gas law for constant volume (Eq. (8.12)) can be applied at the initial and final states to find $P_a/T_a = P/T$, and by simple manipulation this can give

$$\frac{P - P_a}{P_a} = \frac{T - T_a}{T_a}$$

Dividing Eq. (8.23) by T_a, we find that it can be rewritten as

$$\frac{P - P_a}{P_a} = \frac{\dot{Q}t}{V\rho_a c_v T_a} \tag{8.24}$$

We have thus arrived at an expression that can be used to estimate the dynamic pressure build-up due to thermal expansion in a hermetically closed compartment. In addition to the assumptions made when deriving Eq. (8.21), some further assumptions have been made:

- The energy release rate is constant.
- The mass loss rate of the fuel is neglected in the conservation of mass.
- The specific heat does not change with temperature.
- The hydrostatic pressure difference over the height of the compartment is ignored and assumed to be negligible compared to the dynamic pressure.

EXAMPLE 8.1

A hermetically closed machine room in a ship has a volume of 60 m³. A fire starts with a constant effect of 100 kW. Estimate the pressure rise due to the expansion of the gases after 10 seconds.

SUGGESTED SOLUTION

Ignoring heat losses to the compartment boundaries, assuming $\rho_a = 1.2$ kg/m³, we use Eq. (8.24) to find $\dfrac{P - P_a}{P_a} = \dfrac{100 \cdot 10}{60 \cdot 1.2 \cdot 0.7 \cdot 293} = 0.068$ atm. Multiplying by the atmospheric pressure 101 kPa gives a pressure difference of 6.8 kPa, which is a considerably high value.

Example 8.1 shows that in a very short time the pressure in a hermetically closed room rises to quite large values. A pressure difference of 6.8 kPa across a window of 0.6 m² will produce a total load of 4100 Newtons, which will probably be enough to destroy the window.

Most buildings have leaks of some sort. The above example indicates that even though a fire room may be closed, the pressure rise is very rapid and would presumably lead to sufficient leaks to prevent further pressure rise from occurring. We will use this conclusion when dealing with pressure rises in enclosures with small leaks.

8.4.2 PRESSURE RISE IN A LEAKY COMPARTMENT

Consider again a fixed volume, but this time with a small opening at floor level. The fire is considered as a source of heat only; see Figure 8.6. Again we use the conservation of energy as given in Eq. (8.21), reproduced here for clarity:

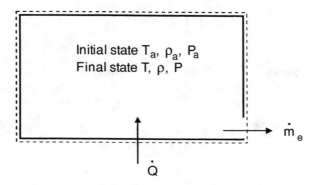

FIGURE 8.6 Control volume for a leaky compartment.

$$\frac{d}{dt}\underbrace{\iiint_{CV}\rho u dV}_{\text{first term}} + \underbrace{\iint_{CS}\rho h v_n dS}_{\text{second term}} = \dot{Q} \qquad (8.21)$$

Equations (8.10) and (8.11) define u and h as $u = c_v(T - T_a)$ and $h = c_p(T - T_a)$ where T_a is some reference temperature. Here, it will be useful for us to express these in terms of total internal energy per unit mass and total enthalpy per unit mass. This will simplify our treatment. We therefore take the reference temperature T_a to be zero degrees Kelvin (0 K) and write

$$u = c_v T \qquad (8.25)$$

and

$$h = c_p T \qquad (8.26)$$

thus expressing the total internal energy and total enthalpy per unit mass.

First term in Equation (8.21): We are interested in arriving at an expression that allows us to evaluate the dynamic pressure in an enclosure with a leakage opening, and we must therefore write the first term in Eq. (8.21) in terms of pressure. We can do this by using the ideal gas law as given by Eq. (8.12) to express the density as

$$\rho = \frac{P}{RT} \qquad (8.27)$$

Using Eq. (8.25) we can write the first term as

$$\frac{d}{dt}\iiint_{CV}\frac{P}{R \cdot T}c_v T dV \text{ which becomes } \frac{d}{dt}\iiint_{CV}P\frac{c_v}{R}dV$$

Performing the integration results in $\frac{d}{dt}\left(\frac{Pc_v}{R}V\right)$, since P, c_v, and R are not dependent on volume. This expression in turn results in $\frac{dP}{dt}\frac{Vc_v}{R}$, since V, R, and c_v are not dependent on time. To summarize, we have found that

$$\frac{d}{dt}\iiint_{CV}\rho u dV = \frac{dP}{dt}\frac{Vc_v}{R}$$ (8.28)

We shall use this result in our final expression.

Second term in Equation (8.21): Using Eq. (8.26) to express the total enthalpy per unit mass of the gases flowing out of the compartment, we can rewrite the second term in Eq. (8.21) as

$$\iint_A \rho v c_p T dA$$

since the area A is the only part of the control surface allowing mass to exit. Further, we can write v for velocity instead of v_n, since the direction of the flow is perpendicular to the surface of the opening. We know that the mass flow rate through an opening of area A can be written $\dot{m} = \rho \, v \, A$ if the density and velocity are constant over the area and the velocity normal to the surface of A.

Adopting the suffix "e" to denote "exit," we can express the mass flow rate exiting the opening as $\dot{m}_e = \rho_e v_e A_e$. Performing the integration over the opening area we find that the second term in Eq. (8.21) can be written as $\rho_e v_e A_e c_p T_e$ or $\dot{m}_e c_p T_e$.

To summarize, we have found that

$$\iint_{CS} \rho \cdot h \cdot v_n \cdot dS = \rho_e \cdot v_e \cdot A_e \cdot c_p \cdot T_e = \dot{m}_e \cdot c_p \cdot T_e$$ (8.29)

Third term in Equation (8.21): The third term consists of the chemical heat release rate (assumed to be released as a point source) minus the heat losses to the boundary or, as expressed by Eq. (8.18), $\dot{Q} = \dot{Q}_{ch} - \dot{q}_{loss}$.

Resulting equation: Combining the three terms in Eq. (8.21), we can rewrite the equation as

$$\frac{c_v \cdot V}{R}\frac{dP}{dt} + \dot{m}_e c_p T_e = \dot{Q}$$ (8.30)

In Section 8.4.1 and Example 8.1 we found that the rate of pressure rise, dP/dt, was very high for the first few seconds in a closed compartment. Thus, leakage areas would be established relatively quickly, resulting in a constant level of pressure as shown in Figure 8.7. Example 8.1 therefore suggests that the constant pressure assumption is reasonable. To simplify our application of Eq. (8.30) we can now assume that $dP/dt \approx 0$ and therefore arrive at the expression

$$\dot{m}_e c_p T_e = \dot{Q}$$ (8.31)

If T_e and $\dot{Q}$ are known, we can now calculate the mass flow rate out of the opening.

Equation (8.31) shows that the enthalpy flux due to mass flow out of the opening is equal to the heat addition rate. We can express this in terms of pressure by expressing the pressure difference over the opening in the well-known form

$$\Delta P = \frac{1}{2}\rho_e v_e^2 \text{ and therefore } v_e = \sqrt{2\Delta P/\rho_e}$$

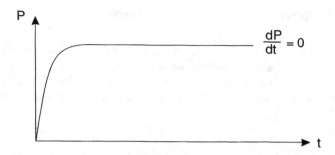

FIGURE 8.7 Rate of pressure rise in a leaky compartment.

Writing the mass flow rate as $\dot{m}_e = A_e \rho_e v_e$ and using the above expression for v_e, we can rewrite Eq. (8.31) as $A_e \sqrt{2 \Delta P \rho_e} \, c_p T_e = \dot{Q}$. Solving for the pressure difference we get

$$\Delta P = \left(\frac{\dot{Q}}{c_p T_e A_e} \right)^2 \frac{1}{2 \rho_e}$$

In order to improve our estimates of the pressure difference, we can include the flow coefficient C_d in the above expression. The value of the flow coefficient is discussed in Chapter 5 (Section 5.2.3), and for most openings the value of C_d is between 0.6 and 0.7. Including C_d in our expression leads to

$$\Delta P = \frac{1}{2 \rho_e} \left(\frac{\dot{Q}}{c_p T_e A_e C_d} \right)^2 \qquad (8.32)$$

We have thus arrived at two equivalent expressions to estimate mass flow rate out through an opening (Eq. (8.31)) and the pressure difference across the opening (Eq. (8.32)). The expressions are approximations, since many assumptions have been made along the way. It is advisable to keep the following assumptions in mind:

- The energy release rate is constant.
- The mass loss rate of the fuel is neglected in the conservation of mass.
- The specific heat does not change with temperature.
- The hydrostatic pressure difference over the height of the compartment is ignored and assumed to be negligible compared to the pressure due to expansion of hot gases.
- Constant pressure is assumed, neglecting the initial rate of pressure rise.
- The opening cannot be taken to be a vertically oriented slit, since the properties of the exiting gas must be uniform over the opening.
- The opening can be taken to be at floor level so the exiting gas temperature $T_e \approx T_a$. If the opening is at ceiling level, then the hot gas temperature must be known and assumed to be a constant value.
- The area of the leakage opening is assumed to be known but this is difficult to assess in real buildings.

Due to these and other assumptions made, Eq. (8.31) and (8.32) can only be expected to give answers that can be used as order of magnitude estimates.

EXAMPLE 8.2

A relatively air-tight room has a floor area of 6 m by 4 m and a height of 3 m. It has a door of width 1 m with a 1 cm high slit at the floor level. A hydraulic oil leak has ignited, causing a fire with an effect of 100 kW. Calculate the pressure difference that arises, the velocity in the opening, and the mass flow through the slit. What happens if the slit is made 10 times higher?

SUGGESTED SOLUTION

Assuming c_p = 1.0 kJ/(kg K), T_e = 300 K, and C_d = 0.7, we find (using Eq. (5.9)) that $\rho_e = 353/300 = 1.18$ kg/m^3. Ignoring heat losses to the boundaries and using Eq. (8.32) we find

$$\Delta P = \frac{1}{2 \cdot 1.18}\left(\frac{100}{1.0 \cdot 300 \cdot 1 \cdot 0.01 \cdot 0.7}\right)^2 = 960 \text{Pa}.$$

The velocity is $v_e = \sqrt{2\Delta P/\rho_e} = \sqrt{2 \cdot 960/1.18} = 40$ m/s (an unreasonably high value). Using Eq. (8.31) we find $\dot{m}_e = \frac{100}{1.0 \cdot 300} = 0.33$ kg/s.

Making the slit 10 times larger gives a pressure difference that is 100 times smaller (≈ 10 Pa) and a velocity 10 times smaller, but the mass flow rate is the same.

8.5   SMOKE FILLING OF AN ENCLOSURE WITH LEAKS

In this section we expand on the previous sections in Chapter 8 and use our conclusions to present simple models for calculating the time it takes to fill a compartment with smoke. We examine a volume composed of a single room where a fire of a constant heat output will cause smoke to rise and form a horizontal ceiling layer of hot gas. The room then contains two layers: the upper hot layer and the lower cold (ambient) layer, both of which are assumed to have a uniform temperature of T_g and T_a, respectively.

We follow Zukoski and assume that the opening is a leakage, located either at the floor level or at the ceiling level.[3] We assume that the energy release rate results in a pressure increase due to the thermal expansion of the gases and air is pressed out through the opening. Another way of stating this limitation is to say that we assume that there is no mass flow into the room (pressure across the leak is always positive) and that the mass flow out is either cold gases (leakage at floor level) or hot gases (leakage at ceiling level) of temperature T_g or T_a.

We consider the two cases below, and in a third section we consider how the upper layer temperature can be calculated for the case where there is no mass flow into the room.

8.5.1   SMALL LEAKAGE AREAS AT FLOOR LEVEL

Figure 8.8 shows a schematic of the case we will consider. A room with a leak opening at floor level has a height H and a lower layer height z. The fire is treated as a point source of heat $\dot{Q}$, and no account is taken of the fuel mass flow rate. The mass flow rate out through the leakage opening is due to the expansion of the hot gases and is given by $\dot{m}_e$. The mass flow rate from the lower layer to the upper layer is given by $\dot{m}_p$, the plume mass flow rate. The control volume is chosen such that the volume of the plume is ignored; the plume is considered only as a means of transporting mass from the lower to the upper layer.

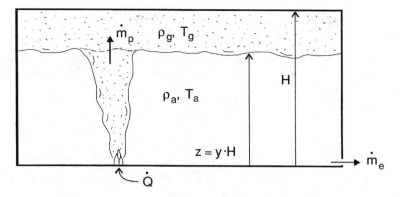

FIGURE 8.8   Simple smoke-filling model, leakage at floor level.

The conservation of mass and energy can be applied to this case in many different ways. Zukoski analyzed this case and found that applying the conservation of mass to the lower layer only, and using results from the conservation of energy arrived at earlier, resulted in a simple differential equation.[3] The solution of the equation can be presented graphically and then used to solve practical smoke-filling problems for rooms with leaks. We summarize this simple smoke-filling model in the following.

We therefore choose our control volume as the lower layer and apply the conservation of mass to this volume.

The conservation of mass was earlier stated through Eq. (8.3), reproduced here for clarity:

$$\frac{d}{dt}\iiint_{CV}\rho\,dV + \iint_{CS}\rho v_n\,dS = 0 \tag{8.3}$$

In the first term, the volume of the lower layer is not constant. We must therefore express volume as $V = z\,S$ where z is the height of the lower layer (and is dependent on time) and S is the floor area. The second term is simply the rate of mass leaving the CV. From Figure 8.8 we see that mass leaves the lower layer through the plume as $\dot{m}_p$ and through the leakage opening as $\dot{m}_e$. Performing the integration in the first term simply results in $\rho_a\,z\,S$, and Eq. (8.3) can then be rewritten as

$$\frac{d}{dt}(\rho_a z S) + \dot{m}_e + \dot{m}_p = 0 \tag{8.33}$$

where z is a function of time.

In Chapter 4 we discussed the Zukoski plume and gave the plume mass flow rate (by Eq. (4.21)) as

$$\dot{m}_p = 0.21\left(\frac{\rho_a^2 g}{c_p T_a}\right)^{1/3}\dot{Q}^{1/3}\cdot z^{5/3} \tag{4.21}$$

We can use this result and insert it into Eq. (8.33). But first we consider the conservation of energy to arrive at an expression for $\dot{m}_e$.

The conservation of energy was considered in Section 8.4.2, where we found it could be written as Eq. (8.30) (reproduced here for clarity):

$$\frac{c_v \cdot V}{R}\frac{dP}{dt} + \dot{m}_e c_p T_e = \dot{Q} \tag{8.30}$$

We argued in Section 8.3.2 that the rate of pressure rise could in many cases be ignored so that $dP/dt = 0$. This gave us Eq. (8.31). Noting that the mass flow rate out contains gases with temperature T_a we can rewrite Eq. (8.31) as

$$\dot{m}_e = \frac{\dot{Q}}{c_p T_a} \tag{8.34}$$

A differential equation for smoke-filling time can now be achieved by combining Eq. (8.33), (4.21), and (8.34), realizing that ρ_a and S are constants:

$$\frac{dz}{dt}\rho_a S + \frac{\dot{Q}}{c_p T_a} + 0.21\left(\frac{\rho_a^2 g}{c_p T_a}\right)^{1/3}\dot{Q}^{1/3}z^{5/3} = 0 \tag{8.35}$$

The above differential equation cannot be solved analytically, but for a constant $\dot{Q}$ it can very easily be solved using numerical techniques. A very convenient way to make the equation directly useful for us is to show the solution graphically. In order to make such a solution applicable to many different geometries and different heat release rates, we must first make Eq. (8.35) dimensionless, then solve it numerically and present the graphical solution in terms of nondimensional parameters. We therefore define three dimensionless parameters.

Dimensionless height is given the symbol y and is simply defined as the height of the lower layer, z, normalized by the height of the room, H, or

$$y = \frac{z}{H} \tag{8.36}$$

The dimensionless height varies from 0 to 1 and expresses the percentage of the room height below the smoke layer.

Dimensionless heat release rate is given the symbol $\dot{Q}^*$, and Zukoski[3] defined this as

$$\dot{Q}^* = \frac{\dot{Q}}{\rho_a c_p T_a \sqrt{g}H^{5/2}} \tag{8.37}$$

This way of expressing dimensionless heat release rate has been found very useful for many other applications. When discussing flame heights in Chapter 4 we gave the dimensionless heat release rate in Eq. (4.1) where the length scale is presented by the burner diameter. For smoke-filling applications the length scale is presented by the room height.

For normal conditions, taking $\rho_a = 1.2$ kg/m^3, $T_a = 293$K, $g = 9.81$ m/s^2, and $c_p = 1.0$ kJ/(kg K), we find that Eq. (8.37) can be expressed as

$$\dot{Q}^* = \frac{\dot{Q}}{1100H^{5/2}}$$

Note that the units of c_p in the above expression are given in terms of [kJ], and therefore the heat release rate must be given in [kW] if this simplified expression is to be used.

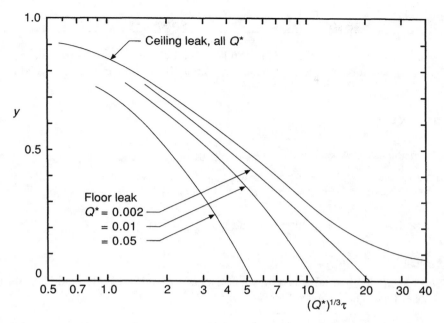

FIGURE 8.9 Dependence of ceiling layer height on time and heat release rate: solution of Eq. (8.39). (Adapted from Zukoski[3].)

Dimensionless time is given the symbol τ and is defined as

$$\tau = t\sqrt{\frac{g}{H}}\frac{H^2}{S} \tag{8.38}$$

where t is time in seconds, $g = 9.81$ m/s^2, H is the room height, and S the floor area.

Differential equation for smoke-filling time in dimensionless form: Equations (8.36), (8.37), and (8.38) can now be used to rewrite Eq. (8.35). After some manipulations we find that Eq. (8.35) can be rewritten in dimensionless form as

$$\frac{dy}{d\tau} + \dot{Q}^* + 0.21\left(\dot{Q}^*\right)^{1/3} y^{5/3} = 0 \tag{8.39}$$

Equation (8.39) cannot be solved analytically, but a numerical solution for a range of values of the dimensionless heat release rate, $\dot{Q}^*$, is shown in Figure 8.9, which gives the dimensionless height to the smoke layer vs. the parameter $(\dot{Q}^*)^{1/3}\tau$.

Calculational procedure: Assume that we are interested in knowing how much time it will take for the smoke layer to reach a certain height z in a room with given dimensions and a fire with a constant heat release rate. We must then

1. Calculate the dimensionless heat release rate, $\dot{Q}^*$, using Eq. (8.37).
2. Calculate the dimensionless height, y, using Eq. (8.36).
3. Read the value of the parameter $(\dot{Q}^*)^{1/3}\tau$ from Figure 8.9 for a given y and $\dot{Q}^*$ and solve for τ.
4. Use this value of τ to calculate the time t using the definition of dimensionless time, τ, in Eq. (8.38).

EXAMPLE 8.3

A pool of kerosene is ignited, releasing 186 kW, in a building with floor area of 5.62 m by 5.62 m and a height of 5.95 m. Calculate the time until the smoke layer has filled half the room.

SUGGESTED SOLUTION

The dimensionless heat release rate is found using Eq. (8.37), $\dot{Q}^* = \dfrac{186}{1100 \cdot 5.95^{5/2}} = 0.002$. The dimensionless height we are interested in is half the room height ($z = 2.975$ m), and Eq. (8.36) gives $y = 2.975/5.95 = 0.5$ (which of course is half the room height).

Figure 8.9 gives the factor $(\dot{Q}^*)^{1/3}\tau \approx 4$, and we can then calculate $\tau = \dfrac{4}{0.002^{1/3}} = 31.7$. The time t can be solved from Eq. (8.38) as $\tau = t \cdot \sqrt{\dfrac{9.81}{5.95}} \cdot \dfrac{5.95^2}{5.62^2} = 1.44 \cdot t$. Therefore, $t = 31.7/1.44$ $= 22$ seconds. It therefore takes less than half a minute for a 186 kW fire to fill half the room with smoke. Figure 8.10 shows the results for other heights.

Comparison with experiments: Hägglund et al. conducted several experiments in a room of floor area of 5.62 m by 5.62 m and a height of 6.15 m.[4] The room was closed except for a 0.25 m high by 0.35 m wide opening near the floor. The fire source was placed 0.2 m above the floor and we therefore use an effective room height of 5.95 m. The fuel was kerosene burnt in pans of various sizes with heat releases varying from 30 to 390 kW. Figure 8.10 shows the results from one such experiment where the steady-state heat release rate was given as 186 kW.

The figure shows very typical results from such comparisons. It shows that the simple Zukoski smoke-filling model[3] overestimates the rate at which smoke fills the room. The main reasons for this are as follows:

- In the experiments it took up to 1 minute for the heat release rate to reach the steady-state value of 186 kW.
- Additionally, it takes time for the smoke to travel to the ceiling and spread out over it. The model treats the plume simply as a pipeline where smoke is instantaneously transported to the ceiling and across the ceiling area.
- Using the Zukoski model in this comparison we have ignored heat losses to the surrounding structure. This can be taken into account by reducing the heat release rate by some fraction of the heat release rate in the term for the enthalpy flow.

8.5.2 SMALL LEAKAGE AREAS AT CEILING LEVEL

If the leakage area is at the ceiling level, the problem is even simpler. Since we have chosen our control volume as the lower layer, the mass balance is simplified and the only mass leaving the lower layer does so through the plume.

This means that the second term in the equation for the conservation of mass is canceled, and Eq. (8.33) can be rewritten as

$$\frac{d}{dt}(\rho_a zS) + \dot{m}_p = 0$$

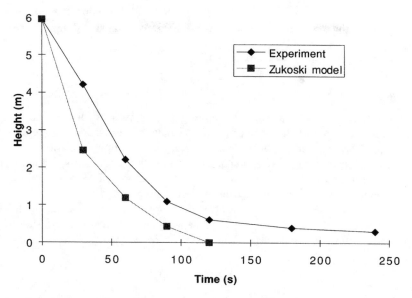

FIGURE 8.10 Comparison of experimentally observed and calculated smoke layer heights.

We can show that this also simplifies the differential equation for smoke filling (in dimensionless form) so that the second term in Eq. (8.39) is canceled and the equation can be rewritten as

$$\frac{dy}{d\tau} + 0.21(\dot{Q}^*)^{1/3} y^{5/3} = 0 \tag{8.40}$$

This differential equation can be solved immediately to give

$$y = \left[1 + \frac{2 \cdot 0.21}{3}(\dot{Q}^*)^{1/3}\tau\right]^{-3/2} \tag{8.41}$$

However, Figure 8.9 also provides the solution for this case, and the calculational procedure given for the floor leak case can also be applied here.

EXAMPLE 8.4

Consider Example 8.3, but assume that the leakage opening is near the ceiling. Calculate the time taken to fill half the room volume.

SUGGESTED SOLUTION

As before, $\dot{Q}^* = 0.002$, $y = 0.5$, and $\tau = 1.44 \cdot t$. For $y = 0.5$, Figure 8.9 gives a value of $(\dot{Q}^*)^{1/3}\tau \approx 4.5$. This gives $\tau = \dfrac{4.5}{0.002^{1/3}} = 35.6$, and therefore $t = 35.6/1.44 = 25$ seconds.

Using Eq. (8.41) instead gives $\tau = \dfrac{0.5^{-2/3} - 1}{0.14 \cdot 0.002^{1/3}} = 33.3$ and $t = 33.3/1.44 = 23$ seconds.

The reason for the slight difference in using Eq. (8.41) and using the graphical solution is that we have used the plume constant 0.21 in the equation, whereas Zukoski used the slightly lower constant of 0.19 when constructing Figure 8.9.[3] Considering the crudeness of this method in general and the purposes to which we shall apply it, these differences are considered to be negligible.

8.5.3 ESTIMATING GAS TEMPERATURES FOR THE FLOOR LEAK CASE

In Chapter 6 we discussed methods for calculating gas temperatures for the case where there was a flow of gases into and out of the compartment. We have been considering the case where there is no flow of upper layer gases out of the compartment and no flow of air into the compartment. This is referred to as the first stage in Section 5.3. Chapter 6 does not discuss methods for calculating gas temperatures for this case, and we present here a method of very roughly estimating the maximum temperature that can be attained in the upper layer.

Consider Figure 8.8, where at some time t, the upper layer is assumed to have a uniform temperature and density of T_g and ρ_g, respectively, and the lower layer has temperature and density of T_a and ρ_a. At this time the total mass of the upper layer, m, is

$$m = \rho_g V = \rho_g S(H - z) = \rho_g SH(1 - y) \tag{8.42}$$

We wish to obtain an expression for T_g or ρ_g. Since the temperature and mass in the upper layer change with time, we find it convenient to express the conservation of energy in terms of total energy added to the mass in the upper layer and the change in enthalpy of this mass over time 0 to t.

The first law of thermodynamics for this case can be stated as

$$Q - W = m\Delta h \tag{8.43}$$

where Q is the total heat added [in J] to the mass in the upper layer over time 0 to t and can be expressed as $\int_0^t \dot{Q}dt$, where $\dot{Q}$ is in [W]. This includes any heat transfer losses to the boundaries; here we make the crude assumption that $\dot{q}_{loss} = 0$. If the heat release rate is constant then this term simply becomes $\dot{Q}_t$.

W is the work carried out by the mass. The work required to push the mass into the control volume is taken care of by using enthalpies instead of internal energy in the third term of Eq. (8.43). Since there is no shaft work, W can be taken to be zero.

m is the mass of the upper layer at time t and is given by Eq. (8.42).

Δh is the change in enthalpy of this mass from time 0 to t, so $\Delta h = c_p (T_g - T_a)$.

Assuming that the heat release rate is constant, we can rewrite Eq. (8.43) as

$$\dot{Q}t = \rho_g SH(1 - y)c_p(T_g - T_a) \tag{8.44}$$

We now have the thermal state of the mass expressed in terms of both T_g and ρ_g. We can use the ideal gas law; writing $T_g \rho_g = T_a \rho_a$ we can write $T_g = T_a \rho_a/\rho_g$. The temperature term in Eq. (8.44) can then be written as $T_g - T_a = \dfrac{T_a}{\rho_g}(\rho_a - \rho_g)$ and Eq. (8.44) as

$$\dot{Q}t = SH(1 - y)c_p T_a(\rho_a - \rho_g) \tag{8.45}$$

We can now calculate ρ_g and therefore T_g. However, we wish to express the first term in Eq. (8.45) in terms of $\dot{Q}^*\tau$ so we can use this in connection with the smoke-filling calculations discussed in the previous sections. Using Eq. (8.37) and (8.38) we find that $\dot{Q} \cdot t = \dot{Q}^*\tau\, \rho_a c_p T_a HS$, and substituting this into Eq. (8.45) gives

$$\dot{Q}^*\tau = (1-y)\left(1-\frac{\rho_g}{\rho_a}\right) \tag{8.46}$$

and this can be used to solve for ρ_g.

EXAMPLE 8.5

Consider the case given in Example 8.3. What is the average adiabatic temperature of the gas layer after 22 seconds, when half the room is filled with smoke?

SUGGESTED SOLUTION

From Example 8.3 we know that $\dot{Q}^* \cdot \tau = 0.002 \cdot 31.7 = 0.0634$. Since $y = 0.5$ we can solve Eq. (8.46) to give $\rho_g = 1.2 \cdot \left(1 - \dfrac{0.0634}{0.5}\right) = 1.048$. Using Eq. (5.9) we find

$$T_g = 353/1.048 = 337\ K = 64°C.$$

When comparing the results from Eq. (8.46) with experiments, we find that the calculated temperatures are grossly overestimated. Since Eq. (8.46) expresses the adiabatic temperature, all the energy supplied by the fire is assumed to be taken up by the mass in the smoke and no heat losses are accounted for. This is in addition to other assumptions made, discussed in the previous sections. Equation (8.46) therefore cannot be used for design calculations, but only for a very rough estimate of maximum possible gas temperature of the upper layer.

Further, Eq. (8.46) is valid only for the case where there is no outflow of smoke or air from a compartment. This is referred to as the first stage in Section 5.3. For other cases the methods for calculating temperatures discussed in Chapter 6 should be used.

8.5.4 LIMITATIONS

The methods discussed in this section have several limitations. During the process of developing simple solution methodologies we have had to make a number of limiting assumptions. We must keep the following in mind:

- The fire is considered as a point source of heat only.
- The mass rate of the fuel is ignored.
- The plume is modeled as a pipeline of no volume, where mass is instantaneously transferred from the lower layer to the upper.
- The plume equation used (Zukoski plume) is valid for a weak fire source.
- No account is taken of heat losses to walls and ceiling.
- The rate of pressure rise is ignored so $dP/dt = 0$.
- No account is taken of hydrodynamic pressure differences with height; the pressure is assumed to have a single value P in the whole compartment.

With respect to the weak fire source limit, we can postulate that the method loses validity if the flame reaches the ceiling. This can be expressed numerically by considering the flame length, expressed by Eq. (4.3) as $L = 0.235 \, \dot{Q}^{2/5} - 1.02 \, D$. Taking $D = 0$ allows us to express a minimum room height in terms of heat release rate as

$$H^{5/2} > 0.235^{5/3} \dot{Q}$$

This can be expressed in terms of $\dot{Q}^*$ since $\dot{Q}^* = \dot{Q}^* = \dfrac{\dot{Q}}{H^{5/2}} \cdot \dfrac{1}{\rho_a c_p T_a \sqrt{g}}$. Inserting this into the

above relation and using common properties for air we find that $\dot{Q}^* < 0.036$ if the flame height does not reach the ceiling. Taking into account the fact that $D \neq 0$ we can expand this limit somewhat.

Our conclusion is that the methodologies discussed in this section should be used only if

$$\dot{Q}^* < 0.05 \qquad\qquad\qquad (8.47)$$

The other limitations mentioned above should also be carefully considered.

8.6   SMOKE CONTROL IN LARGE SPACES

In a compartment with larger openings there will be little or no build-up of pressure due to the volumetric expansion of hot gases. Except for rapid accumulation of mass or energy, or for compartments with small openings, this pressure rise is small, and the pressure nominally remains at the ambient pressure. The opening flows are thus determined by the hydrostatic pressure differences across the openings, and there will be mass flow out of and into the compartment. In Chapter 5 we referred to this stage of the fire as the *third stage*.

In this section we assume that the openings and the leakage areas from the compartment to the surroundings are sufficiently large to prevent any build-up of pressure due to the volumetric expansion of hot gases.

The conservation equations for mass and energy will usually have to be solved simultaneously by computer. The equations can be solved separately if, for example, the temperature is assumed to be an average constant value throughout the process. This will allow us to express the smoke filling process as a function of time.

Alternatively, we can concentrate on smoke control as opposed to smoke filling, where the accumulation of smoke in the space is controlled by some means, such as by opening a ceiling vent to exhaust the smoke. This will allow us to examine the problem at long times, when steady state has been established. We can then solve the conservation equations by iteration.

We first consider the smoke filling problem and then study the steady-state problem for a number of smoke-control methods.

8.6.1   SMOKE FILLING: THE NON-STEADY PROBLEM

In this section we present a method for calculating the smoke-filling time for the case where no measures are taken to vent the smoke out of the compartment. This is therefore similar to the analysis presented in Section 8.5 with two main exceptions: the pressure differences causing mass flow in and out of the compartment are due to hydrostatic pressures, and we shall assume that the upper layer density, ρ_g, can be taken to be some average constant value throughout the smoke-filling process. The method can therefore be used only where the space is large with respect to the heat release rate, so that the temperature rise in the upper layer can be considered to be relatively small. We discuss further the appropriateness of this assumption below.

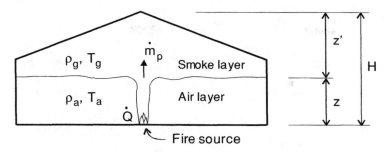

FIGURE 8.11   Schematic of the smoke filling process in a room with no venting of hot gases.

Assuming a constant average density in the upper layer for all times has the advantage that we can form an analytical solution for the smoke-filling rate, where the heat release rate does not need to be constant but can be allowed to change with time. We shall therefore stipulate a constant average value of ρ_g and use the conservation of mass to arrive at the expression for the smoke-filling rate. When this is done, the height of the smoke layer as a function of time is known and we can use the conservation of energy to check the stipulated value of ρ_g.

Consider Figure 8.11 where a fire of point source $\dot{Q}$ causes smoke filling of a compartment with height H and floor area S. At some time t, the lower layer has a thickness z and the upper layer has thickness z', volume V_g, temperature T_g, and density ρ_g. We use the schematic and symbols in Figure 8.11 to derive a simple equation for determining the upper layer interface height as a function of time.

Conservation of mass: In Section 8.5.1, where a similar case was considered, we chose our control volume as the lower layer, and therefore did not need to include ρ_g in our treatment. Here we choose our control volume as the upper layer and set up the law of the conservation of mass. The conservation of mass given by Eq. (8.3) can be written for the upper layer as

$$\frac{d}{dt}\left(\rho_g V_g\right) - \dot{m}_p = 0 \tag{8.48}$$

where the mass plume rate can be given by Eq. (4.21):

$$\dot{m}_p = 0.21\left(\frac{\rho_a^2 g}{c_p T_a}\right)^{1/3} \dot{Q}^{1/3} \cdot z^{5/3} \tag{4.21}$$

We now make the bold assumption that ρ_g is an average constant value for all times. The first term in Eq. (8.48) (since $V_g = S\,z'$) can be written as $\dfrac{dz'}{dt}\rho_g S$. Noting that the rate at which the upper layer descends must equal the rate at which the lower layer diminishes, we can write $\dfrac{dz'}{dt} = -\dfrac{dz}{dt}$. The mass balance for the upper layer can now be written from Eq. (8.48) as

$$\frac{dz}{dt}\rho_g S + 0.21\left(\frac{\rho_a^2 g}{c_p T_a}\right)^{1/3} \dot{Q}^{1/3} z^{5/3} = 0$$

This can be transformed to

$$\frac{dz}{z^{5/3}} = -\frac{k}{S}\dot{Q}^{1/3}dt \tag{8.49}$$

where k is a constant given by

$$k = \frac{0.21}{\rho_g}\left(\frac{\rho_a^2 g}{c_p T_a}\right)^{1/3} \tag{8.50}$$

We can now assume that the heat release rate changes as some function of time. We use the expression

$$\dot{Q} = \alpha t^n \tag{8.51}$$

where α is a growth rate factor (given in kW/s^2) and n is an exponent. When $n = 0$, $\dot{Q}$ is a constant with the numerical value of α; when $n = 2$, we make use of the t-squared fire discussed in Section 3.4.4 and the heat release rate is given by Eq. (3.7) as $\dot{Q} = \alpha t^2$.

By substituting Eq. (8.51) into Eq. (8.49) and by integrating both sides, we get an expression for the height of the layer interface, z, in terms of time as

$$z = \left(k\frac{\alpha^{1/3}}{S}\frac{2t^{(1+n/3)}}{n+3} + \frac{1}{H^{2/3}}\right)^{-3/2} \tag{8.52}$$

For the case where $n = 0$, the heat release rate is constant with $\dot{Q} = \alpha$, and Eq. (8.52) is written as $z = \left(\frac{2k\dot{Q}^{1/3}t}{3S} + \frac{1}{H^{2/3}}\right)^{-3/2}$.

EXAMPLE 8.6a

Derive an expression for the smoke layer height using Eq. (8.49) for the case where $n = 2$ in Eq. (8.51).

SUGGESTED SOLUTION

Substituting Eq. (8.51) into Eq. (8.49) and integrating both sides we get $\int_H^z \frac{dz}{z^{5/3}} = -\frac{k}{S}\alpha^{1/3}\int_0^t t^{2/3}dt$, since the position of the interface between the two layers goes from H to z during time 0 to t. The integral is solved to give $\frac{3}{2}\left(\frac{1}{z^{2/3}} - \frac{1}{H^{2/3}}\right) = \frac{3}{5}\frac{k}{S}\alpha^{1/3}t^{5/3}$. We can now solve for z to give $z = \left(\frac{k}{S}\alpha^{1/3}\frac{2}{5}t^{5/3} + \frac{1}{H^{2/3}}\right)^{-3/2}$.

We now have an expression for the layer interface height as a function of time, but in the expression for the constant k the upper layer density, ρ_g, must be known. We must therefore first guess a value of ρ_g and then use the law of the conservation of energy to check our guess, and iterate forth a solution.

Conservation of energy: In Section 8.3.3 we found the conservation of energy could be written as Eq. (8.21), reproduced here:

$$\frac{d}{dt}\iiint_{CV} \rho u dV + \iint_{CS} \rho h v_n dS = \dot{Q} \tag{8.21}$$

In Section 8.4.2 we found that the first term could be written as $\dfrac{c_v V}{R}\dfrac{dP}{dt}$ and that dP/dt could be considered to be zero.

The second term in Eq. (8.21) expresses the change in enthalpy due to mass flowing into and out of the control volume. Since we have chosen the upper layer as our control volume, this is simply the enthalpy of the mass that the plume pumps into the upper layer. From the mass balance given by Eq. (8.48) we know that the plume mass flow rate is $\dot{m}_p = \dfrac{d}{dt}(\rho_g V_g)$. The change in enthalpy is written as $h = c_p\,(T_g - T_a)$ and the second term can thus be expressed as $\dfrac{d}{dt}(\rho_g V_g)c_p(T_g - T_a)$.

The third term gives the net energy added to the control volume. For the moment we ignore the heat losses to the boundaries, but we will take into account the heat losses in the next section. The energy balance can now be written as

$$\frac{d}{dt}(\rho_g V_g)c_p(T_g - T_a) = \dot{Q} \tag{8.53}$$

When considering Eq. (8.48) we found that the term $\dfrac{d}{dt}(\rho_g V_g)$ could be written as $-\dfrac{dz}{dt}\cdot\rho_g\cdot S$. Expressing the heat release rate as $\dot{Q} = \alpha t^n$, substituting this into Eq. (8.53), and integrating from height H to z and from time 0 to t we find

$$(H - z)\rho_g S c_p(T_g - T_a) = \frac{\alpha t^{n+1}}{n+1} \tag{8.54}$$

In Section 8.5.3 we expressed the conservation of energy of the upper layer as Eq. (8.44), where the heat release rate was assumed to be constant. Setting $n = 0$ in Eq. (8.54) reduces the last term to $\dot{Q}$ and we see that Eq. (8.44) and (8.54) are equivalent.

We now have an expression in terms of z, ρ_g, and T_g. We can use the ideal gas law to express T_g in terms of density, using Eq. (5.9) so that $T_g = 353/\rho_g$. Substituting this into Eq. (8.54) results in

$$\rho_g = \rho_a\left(1 - \frac{\alpha t^{n+1}}{(n+1)(H-z)S c_p 353}\right) \tag{8.55}$$

The above energy balance does not take into account any heat losses to the boundaries, and Eq. (8.55) therefore gives maximum possible temperatures (lowest possible values of density) of the upper layer. All the released energy is assumed to be contained and stored in the upper layer gases. The same assumptions were made when deriving Eq. (8.46). When comparing these equations with experiments we find that they grossly overestimate the upper layer temperature, especially for long times. Equation (8.55) therefore can be used to give a rough (minimum) estimate of the upper layer density at relatively short times.

The constant densities assumption requires that the rate of temperature rise is not excessive, and this approach therefore gives best results for compartments that are large in relation to the heat being released.

Calculational procedure:

1. Give ρ_g a guess value. This should be in the vicinity of 1.0.
2. Calculate the constant k from Eq. (8.50).
3. Calculate z at the some time t from Eq. (8.52).
4. Check ρ_g from Eq. (8.55). Note the limitation in validity of Eq. (8.55) for long times.

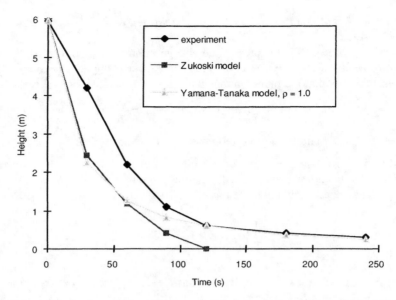

FIGURE 8.12 Comparison of experimentally measured interface height with the Zukoski[3] model and the Yamana–Tanaka[5] model.

EXAMPLE 8.6b

Use the information in Example 8.3 to calculate the smoke layer height at time $t = 60$ seconds using the Yamana–Tanaka model.

SUGGESTED SOLUTION

We give ρ_g a guess value of 1.0 kg/m³. Equation (8.50) gives $k = \dfrac{1}{1.0} \left(\dfrac{1.2^2 \cdot 9.81}{1.0 \cdot 293} \right)^{1/3} \cdot 0.21$
$= 0.0764$. For $\dot{Q} = 186$ kW, and $n = 0$, Eq. (8.52) gives z for $t = 60$ s as

$$z = \left(\frac{2 \cdot 0.0764 \cdot 186^{1/3} \cdot 60}{3 \cdot 5.62 \cdot 5.62} + \frac{1}{5.95^{2/3}} \right)^{-3/2} = 1.26 \text{ m.}$$

Using Eq. (8.55) to check ρ_g we get $\rho_g = 1.2 \left(1 - \dfrac{186 \cdot 60}{(5.95 - 1.26) \cdot 5.62^2 \cdot 1.0 \cdot 353} \right) = 0.94$ kg/m³.
This is the lowest possible value of the upper layer density at $t = 60$ s, since no account is taken of heat losses. Our assumption of $\rho_g = 1.0$ kg/m³ is therefore reasonable.

Comparison with experiments: In Section 8.5.1 (Figure 8.10) we compared the Zukoski model with an experiment in a room of size 5.62 by 5.62 floor area and a height of 5.95 m. Comparing the Yamana–Tanaka model with the same experiment gives the results presented in Figure 8.12.

Yamana and Tanaka also carried out a series of experiments in a large room with a floor area of 720 m² and a height of 26.3 m.[6] One of the experiments involved smoke filling with no outlet for the hot gases, with a heat release rate of $\dot{Q} \approx 1300$ kW. The smoke layer interface height was arrived at from three sources: measurements of temperature profile, optical smoke density profile, and observation by eye.

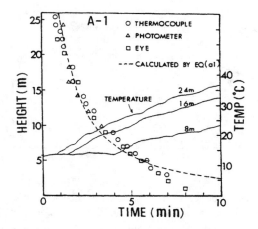

FIGURE 8.13 Height of the interface and temperature at a height of 24 m measured by Yamana and Tanaka[6] compared to calculated interface height, assuming $\rho_g = 1.0$ kg/m³, using Eq. (8.52), where the calculated results have been shifted by 1 minute. (From Yamana and Tanaka[6]. With permission.)

The results are shown in Figure 8.13, where the calculated values using the above equations are shown by the dotted line. The density of the smoke layer was assumed to be 1.0 kg/m³. There is one deviation from the procedure outlined above: the calculated values were shifted by 1 minute. This was to account for the time taken to transport the smoke to the upper layer and across the ceiling and the time it takes for the heat release rate to attain its steady value of 1300 kW.

Figure 8.13 also shows the temperature measured at a height of 24 m. The temperature increase is relatively low, and assuming a constant average value of ρ_g is reasonable. The smoke-filling curve shown by the dotted line is referred to as the "standard filling" curve by Yamana and Tanaka,[6] and we shall use the "standard filling" curve in later sections for comparisons.

8.6.2 SMOKE CONTROL: THE STEADY-STATE PROBLEM

In the previous section we presented the conservation of mass (Eq. (8.48)) and the conservation of energy (Eq. (8.53)) as two differential equations where the term $\frac{d}{dt}(\rho_g \cdot V_g)$ appears in both equations. For correct solutions the two equations should be coupled and solved numerically by computer. We achieved analytical solutions by assuming the density of the upper layer to be a constant average value throughout the smoke-filling process.

We now assume that some smoke-control measure has been taken where the hot gases are vented out through an opening and that the system reaches steady state at some point, i.e., that the mass rate of smoke being vented out equals the mass rate being pumped into the hot layer. We will therefore consider the mass and energy balances for the steady state, i.e., when time goes to infinity.

This will allow us to solve the mass balance and the energy balance separately, and we can then iterate forth a steady-state solution of the problem. This is what Yamana and Tanaka did and we will follow their methodology.[3]

Yamana and Tanaka considered a number of smoke-control measures, and we discuss three of the cases in this section:

Case 1: **Smoke control by means of a ceiling vent or an opening in the upper layer.**
Case 2: **Smoke control by means of mechanically venting the smoke from the upper layer.**
Case 3: **Pressurization of the lower layer by mechanical ventilation.**

Before we introduce these cases we will discuss the mass and energy balances to be used.

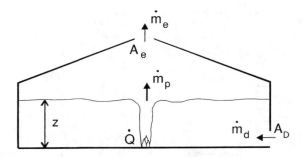

FIGURE 8.14   A room with a vent in the upper part.

Conservation of mass: Consider Figure 8.14 where a room has an inlet opening A_D at the lower part, with steady-state mass flow rate $\dot{m}_d$, and an outlet opening A_E at some height H_E (above the neutral layer) with a steady mass flow rate out of $\dot{m}_e$. The density and temperature of the upper layer are denoted ρ_g and T_g and of the lower layer ρ_a and T_a, respectively.

Rewriting Eq. (8.48), the conservation of mass for the upper layer can be written as

$$\frac{d}{dt}(\rho_g V_g) - \dot{m}_p + \dot{m}_e = 0 \tag{8.56}$$

Since we are considering the steady-state case we assume that the interface height is steady at some height z and therefore $dV_g/dt = 0$. We also assume that the temperature of the upper layer has attained a constant value and therefore that $d\rho_g/dt = 0$.

The mass balance for the upper layer can thus be written simply as

$$\dot{m}_p = \dot{m}_e$$

The same equations can be set up for the lower layer, resulting in

$$\dot{m}_d = \dot{m}_p$$

These equations are combined to give

$$\dot{m}_d = \dot{m}_p = \dot{m}_e \equiv \dot{m} \tag{8.57}$$

Note that in the above we have neglected any mass production in the room, i.e., we have neglected the burning rate of the fuel. We have shown, in Example 5.4, that even when the burning rate is very substantial, this has a very small effect on the mass flow rate through the openings.

In Chapter 5 we derived expressions for the mass flow into and out of openings and we will here use these, together with the expression for plume flow rates to calculate the steady-state layer interface height. However, we need to know the density of the upper layer as input to these equations and we must therefore set up the conservation of energy.

Conservation of energy: We have assumed that at steady state the temperature in the upper layer remains constant. Therefore, the energy produced in the upper layer, $\dot{Q}$, must equal the energy lost to the boundaries and the energy lost due to hot gas leaving the upper layer. The energy balance can thus be stated

$$\dot{Q} = \dot{m}_e c_p (T_g - T_a) + \dot{q}_{loss} \tag{8.58}$$

In Section 6.3.1 we set up the same energy balance, and Eq. (8.58) is identical to Eq. (6.2). As before, the term $\dot{q}_{loss}$ is the part of the energy produced that is lost to the enclosing surfaces of the compartment. This can be expressed as the net radiative and convective heat transfer from the upper layer to the surfaces that are in contact with the upper layer. This surface area is here denoted A_w and the heat will be conducted into these surfaces.

Note that in Section 6.3 we did not know the term A_w, since we did not calculate the height of the smoke layer. In Section 6.3, we therefore used the total boundary surface area, A_T, and used experimental correlations to express the gas temperature. Here, we calculate the smoke layer height and will therefore know A_w. Further, we assume that the plume flow rate at a certain height is the same as the flow rate of hot gases out through openings. We therefore know the terms in the energy balance expressed by Eq. (8.58), and the temperature can be calculated directly.

Heat losses to boundaries: For calculation of the heat lost to the boundaries we shall use an approach similar to that taken in Section 6.3.3, summarized here for completeness: The simplest way of taking into account the heat conducted into the wall is to neglect any cooling from the boundaries and assume that the boundary surfaces have a temperature $T_s = T_g$. Solving the general heat conduction equation for this case, assuming that the boundaries are semi-infinite in thickness, results in an expression for the heat flux per unit area into the wall given by

$$\dot{q}'' = \frac{1}{\sqrt{\pi}} \sqrt{\frac{k\rho c}{t}} (T_g - T_a) \tag{8.59}$$

where k, ρ, c are the conductivity, density, and specific heat of the compartment surface material, respectively. We can then write the last term in Eq. (8.58) as

$$\dot{q}_{loss} = hA_w (T_g - T_a) \tag{8.60}$$

where

$$h = \sqrt{\frac{k\rho c}{\pi t}} \tag{8.61}$$

for a semi-infinite solid. Since we are interested only in the smoke-filling period, which lasts only a number of minutes, the semi-infinite solid assumption will be sufficient. For very thin constructions the steady-state solution of the heat conduction equation gives $h = k/\delta$, where k is the conductivity of the solid material and d is its thickness.

The expression for h, the effective heat conduction coefficient given by Eq. (8.61), is similar to the corresponding term h_k, used by McCaffrey, Quintiere, and Harkleroad (see Chapter 6, Reference 1), and given by Eq. (6.15). The only difference is that in the latter case the constant $\frac{1}{\sqrt{\pi}}$ is omitted, since the McCaffrey et al. method for calculating gas temperatures is based on experiments, and this constant is taken up in the correlation coefficients.

Material properties for the boundaries, $k\rho c$, can be taken from Table 6.1 or from handbooks.

Expression for the gas temperature: We can now use the conservation of energy, Eq. (8.58), to express the upper-layer gas temperature as

$$T_g = T_a + \frac{\dot{Q}}{c_p \dot{m}_e + hA_w} \tag{8.62}$$

In the next three sections we use the above expressions when discussing the three cases mentioned earlier.

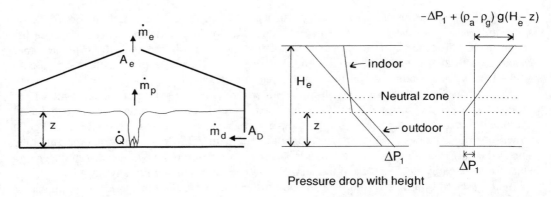

FIGURE 8.15 Schematic of a compartment with natural smoke ventilation from the upper layer and the resulting pressure differences.

8.6.3 CASE 1: NATURAL VENTILATION FROM UPPER LAYER

Here we set up a methodology where we guess the height of the layer interface, solve the mass balance and energy balance separately, and iterate forth a correct value for the steady-state interface height.

Consider Figure 8.15, where $\dot{m}_d$ and $\dot{m}_e$ are the steady-state mass flow rates into and out of the room and $\dot{m}_p$ is the plume mass flow rate into the upper layer. According to the mass balance (Eq. (8.57)) these are all equal and can be denoted $\dot{m}$.

In Section 5.5 we considered mass flows through openings where there is an upper hot layer of gases and a lower cold layer. We developed expressions for the pressure difference across the opening and the rate of mass flow through these, in terms of H_N, the height of the neutral layer. Here, we express the flows in terms of the pressure difference across the openings and we use the expressions given in Section 5.5 to do this.

Pressure difference across the lower opening, ΔP_l: Using the expressions discussed in Chapter 5 we know that the mass flow through the lower opening can be written

$$\dot{m}_d = C_d \rho_a v_d A_D \qquad (8.63)$$

where C_d is the flow coefficient (typically ≈ 0.6) and v_d is the velocity of gases through the opening. We are interested in expressing the pressure difference across the opening, ΔP_l, in terms of the mass flow, $\dot{m}_d$. From the Bernoulli equation we get $\Delta P_l = \frac{1}{2} \rho_a v_d^2$ so that $v_d = \sqrt{\frac{2\Delta P_l}{\rho_a}}$. Inserting this into the Eq. (8.63) and solving for ΔP_l gives

$$\Delta P_l = \frac{\dot{m}^2}{2\rho_a (C_d A_D)^2} \qquad (8.64)$$

Pressure difference across the upper opening, ΔP_u: For the upper opening we know that the mass flow rate can be expressed as

$$\dot{m}_e = C_d \rho_g v_e A_E \qquad (8.65)$$

Again, we seek an expression for the pressure difference across this opening, ΔP_u. This can be expressed as the hydrostatic pressure difference $\Delta P_u = (\rho_a - \rho_g) g (H_E - H_N)$, where $(H_E - H_N)$ is the height from the neutral layer to the opening (see Figure 8.15).

Since H_N, the neutral layer height, is not known, we can express this as the total pressure difference across both openings minus the pressure difference across the lower opening. The total pressure difference across both openings is expressed as $(r_a - r_g) g (H_E - z)$, and we can therefore write

$$\Delta P_u = (\rho_a - \rho_g)g(H_E - z) - \Delta P_l \qquad (8.66)$$

Mass flow rate through the upper opening: We can now use Eq. (8.65), the relationship $v_e = \sqrt{\dfrac{2\Delta P_u}{\rho_g}}$ and Eq. (8.66) to express the mass flow rate through the upper opening as

$$\dot{m}_e = C_d A_E \sqrt{2\rho_g\left(-\Delta P_l + \left(\rho_a - \rho_g\right)g(H_E - z)\right)} \qquad (8.67)$$

We can now proceed with our calculations.

Calculational procedure:

1. Give z a guess value.

2. Calculate $\dot{m} = \dot{m}_p = 0.21\left(\dfrac{\rho_a^2 g}{c_p T_a}\right)^{1/3} \dot{Q}^{1/3} z^{5/3}$. $\qquad (4.21)$

   For $T_a = 293K$ and $r_a = 1.2$ kg/m³, $\dot{m}_p = 0.076\ \dot{Q}^{1/3} z^{5/3}$.

3. Calculate $\Delta P_l = \dfrac{\dot{m}^2}{2\rho_a(C_d A_D)^2}$. $\qquad (8.64)$

4. Calculate $T_g = T_a + \dfrac{\dot{Q}}{c_p \dot{m}_e + h A_w}$. $\qquad (8.62)$

5. Calculate $r_g = 353/T_g$. $\qquad (5.9)$

6. Calculate $\dot{m}_e = C_d A_E \sqrt{2\rho_g\left(-\Delta P_l + \left(\rho_a - \rho_g\right)g(H_E - z)\right)}$. $\qquad (8.67)$

7. Check if $\dot{m} \approx \dot{m}_e$; if not, return to step 1.

For the special case where the lower opening A_D is very large, the pressure difference across this opening is very small. This simplifies Eq. (8.67) where DP_l can be taken to be ≈ 0. If the lower opening is very small, the pressure difference across it is very large. The expression under the root sign in Eq. (8.67) can then become negative, indicating that a steady-state layer height at the guessed value of z is not possible. A new, lower value of z must then be assumed and the calculations repeated.

The value of the effective heat conduction coefficient, h, must be approximated by some means, since time is involved in Eq. (8.61). The time at which steady state is achieved can typically be taken as roughly 10 minutes. For a concrete wall, the effective heat conduction coefficient at time 10 minutes can be estimated using Eq. (8.61) and Table 6.1 to be $h = \sqrt{\dfrac{2 \cdot 10^6}{\pi \cdot 600}} = 33$ W/m²K = 0.033 kW/m²K. For a more insulating material, fiber insulating board, the value is ten times less at 0.003 kW/m²K.

The value of A_w is given as the ceiling area plus the part of the walls that are in contact with the smoke layer. Also note that in Eq. (8.62), when using Q in [kW], h must be given in [kW/(m²K)] and c_p in [kJ/(kg K)].

EXAMPLE 8.7

Calculate the required ceiling vent area for an industrial building of floor area 30×40 m² and height 10 m if the smoke layer is not to sink lower than 6 m above the floor. An appropriate design fire for this case was found to be 1000 kW, and in case of fire, a door of area 4 m² is opened at the floor level. The boundaries can be assumed to be concrete. The ceiling vents are assumed to open a few minutes after the fire starts.

SUGGESTED SOLUTION

We follow the calculational procedure suggested above. The value of z is given as 6 m. Assuming $T_a = 20°C$, the plume mass flow rate is calculated as

$$\dot{m} = \dot{m}_p = 0.21\left(\frac{1.2^2 \cdot 9.81}{1.0 \cdot 293}\right)^{1/3} \cdot 1000^{1/3} \cdot 6^{5/3} = 15.14 \text{ kg/s}.$$

The pressure difference at the lower opening is $\Delta P_1 = \dfrac{15.14^2}{2 \cdot 1.2(0.6 \cdot 4)^2} = 16.6$ Pa. For concrete, the effective heat conduction coefficient can be taken to be ≈ 0.033 kW/(m² K) after 10 minutes. The boundaries in contact with the smoke $A_w = 30 \cdot 40 + 2 \cdot 30 \cdot (10 - 6) + 2 \cdot 40 \cdot (10 - 6)$

$= 1760$ m². The gas temperature is given as $T_g = 293 + \dfrac{1000}{1.0 \cdot 15.14 + 0.033 \cdot 1760} = 307$K $= 34°C$. The density of the upper layer is then $\rho_g = 353/307 = 1.15$ kg/m³. Equation (8.67) can now be solved for A_E as

$$A_E = \frac{15.14}{0.6\sqrt{2 \cdot 1.14\left[\underbrace{-16.6 + (1.2 - 1.14) \cdot 9.81 \cdot (10 - 6)}_{\Delta P_u \text{ is negative}}\right]}}$$

which does not give a solution since the pressure difference at the upper opening, ΔP_u, becomes negative. This can be amended either by allowing the smoke layer to descend further than to 6 m height or by increasing the area of the lower opening. The figure below shows the relationship between A_D and A_E for this case. When the lower opening is assumed to be infinitely large, $\Delta P_1 \approx 0$ and A_E is recalculated to give ≈ 11 m². Choosing $A_D = 16$ m² results in $\Delta P_1 = 1.04$ Pa and $A_E \approx 16$ m². The design is therefore based on the assumption that at least 16 m² vent area will be opened at floor level in case of fire.

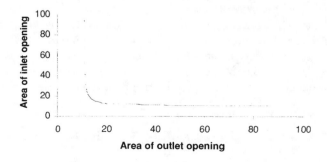

However, the figure above shows how important it is to provide ample area of inlet openings; a slight decrease in A_D may demand an extreme increase in A_E.

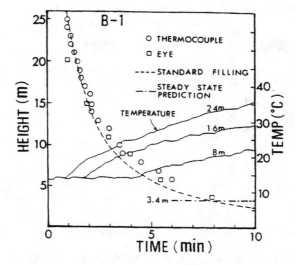

FIGURE 8.16 Smoke filling as a function of time. Steady-state layer interface height due to natural ventilation from the upper layer was calculated as 3.4 m. "Standard filling" curve shown for comparison. (From Yamana and Tanaka[6]. With permission.)

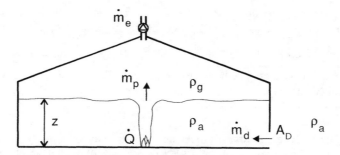

FIGURE 8.17 Schematic of a compartment with mechanical smoke ventilation.

For a large ceiling vent area there is a risk that not only smoke but also air from the lower layer will exit. A large ceiling vent area should therefore be divided into a number of vents; a rule of thumb requires each ceiling vent to be less than $2(H_E - z)^2$. In Example 8.7 this condition is fulfilled.

Comparison with experiments: The experiments by Yamana and Tanaka[6] were summarized in Section 8.6.1. These authors also carried out experiments for Case 1: Natural ventilation from upper layer. A 1300 kW fire was allowed to smoke-fill a compartment of 26.3 m height and 720 m^2 floor area. The lower ventilation opening was 3.23 m^2 and the ceiling vent was 4.46 m^2. Yamana and Tanaka calculated steady-state smoke layer height for this case to be 3.4 m. The results are shown in Figure 8.16, where the layer interface height was observed by thermocouple and by eye. The dotted curve is the calculated "standard filling" curve, presented for comparison, where it is assumed that no smoke exits from the upper layer (see also Figure 8.13).

8.6.4 CASE 2: MECHANICAL VENTILATION FROM UPPER LAYER

Consider Figure 8.17, where the smoke in the upper layer is vented out by mechanical ventilation. The volumetric flow of such fans, denoted $\dot{V}_e$, is usually given in [m^3/s] or [m^3/h]. The mass flow rate out by mechanical ventilation is then given by

$$\dot{m}_e = \dot{V}_e \rho_g \tag{8.68}$$

Calculational procedure:

1. Give z a guess value.

2. Calculate $\dot{m} = \dot{m}_p = 0.21\left(\dfrac{\rho_a^2 g}{c_p T_a}\right)^{1/3} \dot{Q}^{1/3} z^{5/3}$.   (4.21)

3. Calculate $T_g = T_a + \dfrac{\dot{Q}}{c_p \dot{m}_e + h A_w}$.   (8.62)

4. Calculate $\rho_g = 353/T_g$.   (5.9)

5. Calculate $\dot{m}_e = \dot{V}_e \rho_g$.   (8.68)

6. Check if $\dot{m} \approx \dot{m}_e$; if not, return to step 1.

EXAMPLE 8.8

Design the capacity of the mechanical ventilation system to keep the smoke layer above the height of 6 m for the building discussed in Example 8.7.

SUGGESTED SOLUTION

Example 8.7 gave the plume flow rate as 15.14 kg/s. The gas temperature and density of the upper layer were calculated to be 307K and 1.14 kg/m³, respectively. The volumetric flow rate of the fan must therefore be $\dot{V}_e = \dfrac{15.14}{1.14} = 13.3$ m³/s $= 48,000$ m³/h. This is a considerable volume flow rate. It may therefore be necessary to install a number of fans with a collective capacity of ≈ 50000 m³/h.

Comparison with experiments: Yamana and Tanaka carried out experiments for Case 2: Mechanical ventilation from upper layer.[6] A 1300 kW fire was allowed to smoke-fill a compartment of 26.3 m height and 720 m² floor area. The smoke was removed from the upper layer by a fan of a capacity ranging from 3.2 to 6.0 m³/s. Figure 8.18 shows the results for the case where the volumetric flow of the fan is 6.0 m³/s. The steady-state smoke layer height for this case was calculated to be 3.6 m. Figure 8.18 also shows the "standard filling" curve for comparison.

8.6.5 CASE 3: LOWER LAYER PRESSURIZATION BY MECHANICAL VENTILATION

By forced ventilation into the lower layer the whole compartment becomes pressurized. This smoke control method not only results in increased pressures across the smoke vents in the upper layer, but also prevents smoke originated outside the space to enter.

Consider Figure 8.19, showing a space that is pressurized by mechanical ventilation into the lower layer. The mass flow rate from the fan is denoted $\dot{m}_0$.

The mass balance for the lower layer is now written as

$$\dot{m}_0 = \dot{m}_p + \dot{m}_d \qquad (8.69)$$

and for the upper layer as

$$\dot{m}_p = \dot{m}_e \equiv \dot{m} \qquad (8.70)$$

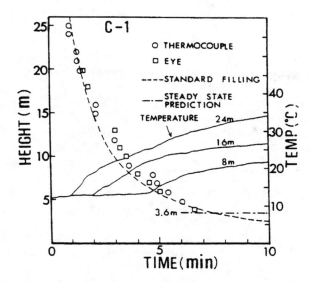

FIGURE 8.18 Smoke filling as a function of time. Steady-state layer interface height due to mechanical ventilation was calculated as 3.6 m. "Standard filling" curve shown for comparison. (From Yamana and Tanaka[6]. With permission.)

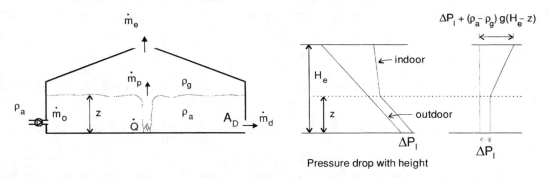

Pressure drop with height

FIGURE 8.19 Schematic of a compartment pressurized by mechanical ventilation into the lower layer.

The mass flow rate through the lower opening is given by Eq. (8.63) as

$$\dot{m}_d = C_d \rho_a v_d A_D \tag{8.63}$$

Since $\dot{m}_d = \dot{m}_0 - \dot{m}_p$ and $v_d = \sqrt{\dfrac{2\Delta P_l}{\rho_a}}$, we can solve for DP_l to give

$$\Delta P_l = \frac{\left(\dot{m}_0 - \dot{m}_p\right)^2}{2\rho_a (C_d A_D)^2} \tag{8.71}$$

To achieve positive pressure in the lower layer requires that $\dot{m}_0 > \dot{m}_p$.

As for Case 1, we can now give an expression for the mass flow rate out through the ceiling vent as

$$\dot{m}_e = C_d A_E \sqrt{2\rho_g \left(\Delta P_l + \left(\rho_a - \rho_g\right) g (H_E - z)\right)} \tag{8.72}$$

Note the similarity to Eq. (8.67).

Calculational procedure:

1. Give z a guess value.

2. Calculate $\dot{m} = \dot{m}_p = 0.21 \left(\dfrac{\rho_a^2 g}{c_p T_a} \right)^{1/3} \dot{Q}^{1/3} z^{5/3}.$ (4.21)

3. Calculate $T_g = T_a + \dfrac{\dot{Q}}{c_p \dot{m}_e + h A_w}.$ (8.62)

4. Calculate $\rho_g = 353 / T_g.$ (5.9)

5. Calculate $\Delta P_1 = \dfrac{(\dot{m}_0 - \dot{m}_p)^2}{2 \rho_a (C_d A_D)^2}.$ (8.71)

6. Calculate $\dot{m}_e = C_d A_E \sqrt{2 \rho_g \left(\Delta P_1 + (\rho_a - \rho_g) g (H_E - z) \right)}.$ (8.72)

7. Check if $\dot{m} \approx \dot{m}_e$; if not, return to step 1.

EXAMPLE 8.9

Use the concept of lower layer pressurization to design the capacity of the mechanical ventilation system used to keep the smoke layer above the height of 6 m for the building discussed in Example 8.7. Assume a ceiling vent area of $A_E = 6$ m² and an inlet opening $A_D = 4$ m².

SUGGESTED SOLUTION

Example 8.7 gave the plume flow rate as 15.14 kg/s. The gas temperature and density of the upper layer were calculated to be 307K and 1.14 kg/m³, respectively. Since $\dot{m}_0$ is not given, we must use Eq. (8.72) to solve for ΔP_1. This gives $\Delta P_1 = \dfrac{15.14^2}{(0.6 \cdot 6)^2 \cdot 2 \cdot 1.14} - (1.2 - 1.14) \cdot$ 9.81 $\cdot$ (10 − 6) = 5.4 Pa. We now use Eq. (8.71) to find $\dot{m}_0 = 15.14 + \sqrt{5.4 \cdot 2 \cdot 1.2} \cdot (0.6 \cdot 4)$ = 23.8 kg/s. A standard fan may supply a flow rate of, say, 5 m³/s = 6 kg/s. Four such fans would therefore be needed in this case. Alternatively, we could decrease the area of the inlet opening or increase the area of the outlet opening, but the flow supplied by the mechanical ventilation must always exceed the plume mass flow rate of ≈15 kg/s.

Comparison with experiments: Yamana and Tanaka carried out experiments for Case 3: Lower layer pressurization by mechanical ventilation.[6] Again, a 1300 kW fire was allowed to smoke-fill a compartment of 26.3 m height and 720 m² floor area. The lower layer was pressurized by a fan of a capacity ranging from 20 to 23.5 m³/s and the outlet opening ranged from 2.23 to 6.46 m². Figure 8.20 shows the results for the case where the volumetric flow of the fan was 20 m³/s and the outlet opening was 3.23 m². The steady-state smoke layer height for this case was calculated to be 7.5 m. Figure 8.20 also shows the "standard filling" curve, for comparison.

8.7 SUMMARY

In this chapter we discussed the enclosure fire at different stages of development. In order to summarize our findings we will use Figure 8.21 (reproduced from Figure 2.5) where we consider stages A, B, C, and D (see Section 2.3.2 and Section 5.3).

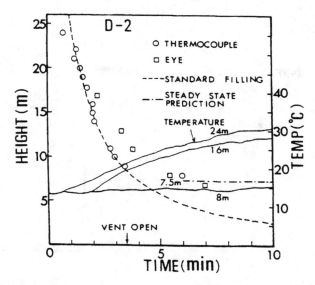

FIGURE 8.20 Smoke filling as a function of time. Steady-state layer interface height achieved by pressurization of the lower layer, calculated as 7.5 m. "Standard filling" curve shown for comparison. (From Yamana and Tanaka[6]. With permission.)

Stage A

Pressure profile: Pressure in enclosure always higher than outside. No neutral layer. Exiting flow either cold gases only (floor leak) or hot gases only (ceiling leak).

Calculation methods provided: Two cases: (a) constant $\dot{Q}$; (b) $\dot{Q} = \alpha t^2$.

(a) Hand calculations can give H_D (= z) and T_g (floor leak) as a function of time and steady state ΔP and $\dot{m}_e$. Equations used are (8.32), (8.36)–(8.38), (8.46), and Figure 8.9. Y_{O_2} (oxygen concentration) can also be calculated, considering mass flow rate out and oxygen used for combustion.

(b) Hand calculations can give H_D (= z) and T_g (floor leak) as a function of time. T_g, ρ_g, and z are calculated by simultaneously solving the coupled Eq. (8.48) and (8.53) (the mass and energy balance). In practice, these are solved using Eq. (8.50) and (8.52) and then checking ρ_g using Eq. (8.55).

Cannot be calculated by hand: Heat losses to boundaries, $\dot{q}_{loss}$ (adiabatic room).

Special notes: Valid for fire development until the hot layer reaches the top of the opening (when hot gases start flowing out). No heat losses are taken into account and the resulting T_g can be greatly overestimated at long times.

Stage B

Pressure profile: Pressure in enclosure always higher than outside, but exiting gas is both at temperature T_g and T_a.

Special notes: Stage B is valid if ΔP due to expansion of gases is greater than ΔP due to hydrostatic pressure difference. Stage B is usually of very short duration (few seconds if opening area is typical

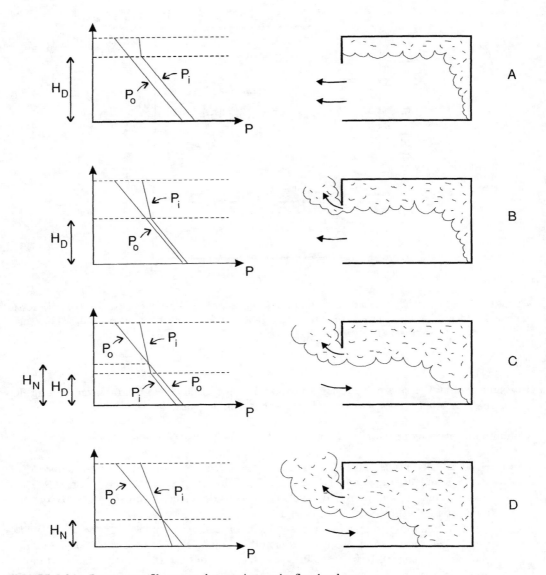

FIGURE 8.21 Pressure profile across the opening as the fire develops.

door or window), and quickly transcends to Stage C. No hand calculation methods are given here for Stage B, but many computer programs take this stage into account.

Stage C

Pressure profile: Positive pressure difference across upper part of opening, negative pressure difference across lower part (with respect to the enclosure). H_D and H_N define shape and magnitude.

Calculation methods provided: Two cases: (a) Constant $\dot{Q}$; (b) Variable $\dot{Q}$.

(a) Hand calculations can give *steady-state* values of H_D (= z) and T_g for a number of cases (natural-, mechanical-, and positive-pressure ventilation) using Eq. (4.21), (8.62), (8.64), and (8.67)–(8.72). The steady-state ceiling vent case is also given by Eq. (5.43)–(5.46).

Alternatively, T_g as a *function of time* can be calculated using McCaffrey et al., Eq. (6.11). $\dot{Q}_{FO}$ is calculated from Eq. (6.20), and $\dot{q}_{loss}$ as a function of time can be estimated from Eq. (6.15)–(6.19).

(b) No methods are provided for calculating H_D and H_N, one must use computer programs. T_g provided by the McCaffrey et al. correlation, Eq. (6.11), and related expressions.

Cannot be calculated by hand: H_D and H_N (but steady-state H_D can be calculated for constant $\dot{Q}$, as above).

Special notes: The opening is assumed so large that ΔP due to thermal expansion can be ignored. Valid for well-ventilated fire ($\dot{Q} < 1.5A_o\sqrt{H_o}$ (MW)) or until flashover occurs.

Stage D

Pressure profile: Simple pressure profile as shown in Figure 8.21. $H_D = 0$. H_N and T_g define magnitudes, but effect of T_g small for $T_g > 300°C$.

Calculation methods provided: H_N, $\dot{m}_g$, and $\dot{m}_a$ provided by Eq. (5.18)–(5.24). The terms in the energy balance can be calculated approximately using Eq. (6.28)–(6.30). Largest error in Eq. (6.29). Difficult to estimate how $\dot{Q}$ varies during the growth and decay stages. When $\dot{Q}$ is not known, Magnusson et al. provide T_g as a function of opening factor, fuel load density, and properties of bounding materials.

Cannot be calculated by hand: If $\dot{Q}$ is known, computer programs offer solutions for T_g, where conduction through walls is more sophisticated than given by Eq. (6.29).

Special notes: $\dot{Q}$ must be larger than $\dot{Q}_{FO}$. If $\dot{Q} >\approx 1.5A_o\sqrt{H_o}$ (MW), this results in underventilated fires and flames through openings.

REFERENCES

1. Klote, J.H. and Milke, J.A., *Design of Smoke Management Systems*, American Society of Heating, Refrigeration and Air-Conditioning Engineers, Atlanta, GA, 1992.
2. Welty, J.R., Wilson, R.E., and Wicks, C.E., *Fundamentals of Momentum Heat and Mass Transfer*, 2nd ed., John Wiley & Sons, London, 1976.
3. Zukoski, E.E., "Development of a Stratified Ceiling Layer in the Early Stages of a Closed Room Fire," *Fire and Materials*, Vol. 2, No. 2, 1978.
4. Hägglund, B., Jansson, R., and Nireus, K., "Smoke Filling Experiments in a 6×6×6 meter Enclosure," FOA Report C 20585-D6, National Defence Research Establishment, Sweden, 1985.
5. Yamana, T. and Tanaka, T., "Smoke Control in Large Scale Spaces, Part 1: Analytical Theories for Simple Smoke Control Problems," *Fire Science and Technology*, Vol. 5, No. 1, 1985.
6. Yamana, T. and Tanaka, T., "Smoke Control in Large Scale Spaces, Part 2: Smoke Control Experiments in a Large Scale Space," *Fire Science and Technology*, Vol. 5, No. 1, 1985.
7. Drysdale, D., "Heat Transfer and Aerodynamics," *An Introduction to Fire Dynamics*, Wiley-Interscience, New York, 1992.

PROBLEMS AND SUGGESTED ANSWERS

8.1 In a hermetically closed machine room on a ship, a pump causes 10 l of hot transformer oil to leak out. The oil is ignited by a spark. The room has a floor area of 6 m by 4 m and a height of 3 m. Assume that flaming combustion ceases when 40% of the oxygen content of the room has been used,

(a) How much is then left of the transformer oil?
(b) Calculate the pressure caused by the fire at this time.
(c) What is the significance of the spill area?

Suggested answer: (a) 3.2 kg oil have been used, assuming combustion efficiency to be 70%, fi 5.8 liters remain; (b) Noting that Q ◊ t = Q in Eq. (8.24), DP = 5.88 atm (not possible, no surrounding structure can withstand this pressure); (c) discuss.

8.2 The room described in Problem 8.1 has a 1-m-wide door with a 1-cm-high slit at the floor level. The oil spill has a rectangular area of sides 0.5 m by 0.5 m.

(a) Calculate the pressure increase, the air velocity, and the mass flow rate through the slit.
(b) The slit is now made 10 times higher. How will this influence the variables calculated in (a)?

Suggested answer: (a) DP = 1060 Pa, v_e = 40 m/s, $\dot{m}_e$ = 0.4 kg/s; (b) DP is 100 times smaller, v_e is 10 times smaller, $\dot{m}_e$ is the same.

8.3 A room has a floor area of 500 m² and a height of 4 m. The room is involved in a fire with a constant energy release rate of 1 MW. Calculate the time till the smoke layer is 1.5 m from the floor for

(a) The case where the leakage areas are at the floor level.
(b) The case where the leakage areas are at the ceiling level.

Suggested answer: (a) 250 s; (b) 520 s.

8.4 A room with a floor area of 200 m² and height 6 m goes through a smoke-filling process and is filled to the height of 2 m above the floor in 3 minutes. Calculate the energy release rate that causes this, assuming leakage areas at ceiling level.

Suggested answer: 1 MW.

8.5 A fire occurs with an energy release rate $\dot{Q}$ in a room of a certain geometry and a height H. The time until the smoke has reached the height $H/2$ was calculated to be 100 seconds. The same calculations are carried out in a room with the height $2H$, with all other parameters unchanged. How long time does it take for the smoke layer to reach the same absolute thickness (i.e., $H/2$)? The fire is assumed to be relatively weak and the leakage areas at the ceiling level.

Suggested answer: 21 s.

8.6 By making a number of assumptions, Eq. (8.30) expresses the first law of thermodynamics for the situation given in Figure 8.8:

$$\frac{c_v \cdot V}{R} \frac{dP}{dt} + \dot{m}_e c_p T_e = \dot{Q} \tag{8.30}$$

Some further assumptions allow the mass flow rate out through a floor leak to be given by Equation (8.34):

$$\dot{m}_e = \frac{\dot{Q}}{c_p T_a} \qquad\qquad (8.34)$$

(a) Explain the physical meaning of the terms in Eq. (8.30) and the assumptions made for arriving at Eq. (8.34).

(b) How is Eq. (8.34) used to derive an expression for the steady-state pressure rise?

(c) Derive a differential equation describing the smoke-filling process depicted in Figure 8.8 using Eq. (8.34), a plume equation, and the conservation of mass for the lower layer.

8.7 A room with a floor leakage has a height of 4 m and an area of 500 m². A fire with an energy release rate of 1 MW is ignited in the room. The time until the smoke layer is 1.5 m from the floor is 250 s. Give an equation for approximating the hot layer density at the time 250 s, when all heat losses are ignored. Calculate the corresponding gas temperature.

Suggested answer: 410°C.

8.8 In a nearly closed room a fire with a constant energy release rate of 500 kW is ignited. The room has a floor area of 100 m² and a height of 3 m. An opening near the floor is 2 m wide and 0.1 m high and one of the walls has a rectangular window with dimensions 0.9 m by 0.9 m. The flames are not in direct contact with the window. Assume a flow coefficient of 0.6.

(a) Calculate the time until the smoke layer has reached the floor.

(b) Will the fire go out due to oxygen deficiency in the smoke layer when the smoke layer reaches the floor? Assume the fire to go out when half the oxygen in the smoke layer has been used up.

(c) Will the window break due to the pressure build-up? Assume that the window can withstand 1 kPa pressure rise without cracking.

Suggested answer: (a) 150 s; (b) no; (c) no.

8.9 A fire releases 100 kW in a room with a floor area of 100 m² and a height of 4 m. The room has a door which is 1 m wide and 2 m high. At the moment when the smoke layer reaches the top of the opening, the gas temperature is found to be 300°C. Ignoring any heat losses to the boundaries, calculate how long the fire has been burning.

Suggested answer: (a) 340 s.

8.10 A hotel lobby has a floor area of 6 m by 8 m and is 6 m high. A large sofa consisting mostly of polyurethane is ignited. The flame spreads across the material, causing an energy release rate that increases linearly from 0 to 2400 kW in 4 minutes, after which the sofa has burnt out.

(a) Will the lobby be filled with smoke?

(b) How long will it take for the smoke layer to reach the height 1 m from the floor? What will the gas layer density be at this time (ignoring heat losses to boundaries)?

Suggested answer: (a) Yes, $z < 0.3$ m for $0 < r_g \leq 1.2$ (because Eq. (8.55) results in a negative value, it could not be used to check r_g in this case); (b) through iteration, $r_g \approx$ 0.79 kg/m³ and $t \approx 73$ s.

8.11 Do the same calculations as in Problem 8.10, using the Zukoski smoke-filling method and assuming the energy release rate to be constant at the average value of 1200 kW throughout.

Suggested answer: (a) Yes; (b) $\rho_g \approx 0.64$ kg/m^3 and t $\approx$ 34 s.

8.12 A storage room in an industry has a floor area of 30 m by 50 m and a height of 10 m. A garage door has a width of 3 m and a height of 3 m and the room has four windows that are 0.9 m by 0.9 m, situated at a height of 1 m from the floor. The fire load consists mainly of wooden pallets (1.2×1.2 m^2); the maximum amount stored is two pallet loads, side by side, which are 1.4 m high. The fire is assumed to attain maximum effect after 6 minutes and firefighters are assumed to arrive at the location 15 minutes after the fire starts. Design the size of smoke vents in the ceiling with the objective of avoiding flashover and only allowing the smoke layer to be 3 m thick. Assume a flow coefficient of 0.6 for the ceiling vents and assume the walls and ceiling to be of steel sheet, isolated with mineral wool with $k\rho c = 3.3 \cdot 10^3$ W^{2}s/m^{4}K^2.

Suggested answer: If $\dot{Q} = 7$ MW, then $\dot{Q}^* = 0.02 \Rightarrow$ relatively large fire. We must test both Thomas and Zukoski plumes. Thomas plume gives, for $z = 7$ m, $C_d = 0.6$, $A_D = 9$ m^2, $t = 6$ min $\Rightarrow A_v = 9.3$ m^2, $T_g = 310$°C. Using the Zukoski plume the problem cannot be solved and z must be allowed to be a lesser value. Therefore, the choice of plume equation and inlet area is very sensitive, and the design requirement is close to a criticality. Therefore, change the design requirement or solve in alternative ways.

8.13 An atrium has a floor area of 30 m by 20 m and is 20 m high. The ceiling ventilation is designed for a fuel area of 30 m^2 with an energy release rate of 0.5 MW/m^2. The inlet openings are very large. An effective heat conduction coefficient can be assumed to be 20 W/m^2°C.
   (a) Calculate the area of ceiling vents required for ensuring a maximum thickness of the smoke layer of 5 m.
   (b) A weak fire occurs in the atrium; a sofa burns for 3 minutes and releases energy at the constant rate of 500 kW. The ceiling vents are opened through signals from smoke detectors at the ceiling level, which activate at a temperature rise of 10°C. What will happen? Will the atrium be filled with smoke? How fast will this happen?

Suggested answer: (a) $A_E = 59$ m^2; (b) $T_g = 28$°C $\Rightarrow$ the smoke detectors may not activate. It will take over 10 minutes for the smoke layer to reach the height 2 m above the floor.

8.14 An industrial building is built mainly of lightweight concrete and has a floor area of 20 m by 30 m and a height of 10 m. Ceiling vents have been installed to remove smoke in case of fire. The inlet air vents consist of doors which are 6 m wide and 4 m high. The doors and the ceiling vents are assumed to open automatically in case of fire. A pool of flammable liquid on the floor has a diameter of 1.5 m and when ignited will have an energy release rate of 4600 kW. The energy release rate is assumed to increase linearly with the pool area, if the pool area increases in size. The time 5 minutes is chosen as a characteristic time.
   (a) Calculate the area of ceiling vents required if the smoke layer is to have a thickness of maximum 3 m.
   (b) Assume that the ceiling vents do not open but the inlet openings do. How long does it take for smoke to start flowing out through the inlet openings?

(c) Make the same assumption as in (b). After a while, steady-state conditions will arise with regard to the flow into and out of the inlet openings. Estimate the temperature of the gases flowing out through the opening.

(d) How large must the pool fire be (in m²) if flashover is to be achieved? Make the calculation for two cases: the construction is made entirely of lightweight concrete and the roof is made of lightweight concrete but the floor and walls are made of concrete.

Suggested answer: (a) 15 m²; (b) 106 s; (c) 180°C; (d) 9.5 m² and 19.1 m², respectively.

8.15 An industrial building has a floor area of 50 m by 30 m and is 8 m high. A 4 m² pool of heavy fuel oil occurs due to pipe leakage and this is ignited. An effective heat conduction coefficient can be assumed as 20 W/m² °C and the flow coefficient is assumed to be 0.6. The smoke layer is not to sink further than 5 m from the floor.

(a) Calculate the required area of the inlet openings if the ceiling vent is 10 m².

(b) Calculate the required capacity of a mechanical ventilation system, designed to remove the smoke.

(c) Positive pressure ventilation of the lower layer is to be installed to drive out the hot smoke gases. The ceiling ventilation area is limited to 8 m² and the inlet openings are 13 m². Calculate the required capacity of mechanical ventilation.

Suggested answer: (a) 13 m²; (b) 18 m³/s; (c) 23 kg/s or 19 m³/s.

9 Combustion Products

The ability to estimate the toxic hazards of combustion gases in a fire compartment is of great importance to the fire protection engineer. The species of interest to the fire engineer would most often be CO, CO_2, and O_2, but concentrations of other combustion products may also be of interest; for example, soot concentration can be directly linked to visibility through a gas mass. To allow an estimation of the hazard, the amount of each toxic product produced per unit fuel burned must be assessed, i.e., the species yield must be estimated. Once the production term is known (the yield), the concentration in the fire gases must be calculated. The products of combustion may be diluted by air entering the hot gas layer through the plume and gases may escape out through an opening, thus influencing the species concentrations in the hot gas layer. The concentration of species must therefore be calculated by considering a mass balance of the region of interest. For example, this region may be the hot gas layer in a room or a fire plume. The generation of combustion products is a very complex issue, and the engineer must rely on measurement and approximate methods for estimating the yield of a product. This chapter introduces some methods available to the engineer for estimating the yield of a species and discusses methods for calculating species concentrations.

CONTENTS

9.1 Terminology 227
9.2 Introduction 228
9.3 Fuel Chemistry 229
    9.3.1 Stoichiometry and Species 229
    9.3.2 Specific Yields 230
9.4 Conservation Equation for Species 238
    9.4.1 Control Volume Formulation 238
    9.4.2 Application to a Compartment 240
9.5 Using Experimental Data for Estimating Yields 242
    9.5.1 Data from Bench-Scale Tests 242
    9.5.2 Data from Hood Experiments 244
    9.5.3 Data from Compartment Fires 250
9.6 Predicting Species Concentrations in Compartment Fires 251
References 253
Problems and Suggested Answers 253

9.1 TERMINOLOGY

Complete combustion — Refers to the chemical reaction where all the product components are in their most stable state. For example, only CO_2 and H_2O for a general $C_xH_yO_z$ fuel.

Equivalence ratio — The ratio of the available fuel mass to the available oxygen (or air) mass, divided by r, the stoichiometric mass fuel to oxygen (or air) ratio.

Fuel mixture fraction — The mass of atoms that originally were fuel divided by the total mass of the gaseous mixture, given the symbol f. The magnitude of the fuel mixture fraction is an indicator of the degree of excess fuel in the gaseous mixture.

Mass fraction of a species — The mass of species i divided by the total mass of the gaseous mixture is defined as the mass fraction of species i and given the symbol Y_i.

Maximum yield of species — The maximum theoretical yield of species i, called $y_{i,max}$, based on stoichiometry that allows the maximum yield. For example, the yield of CO that requires all of the carbon to form CO would be $y_{CO,max}$.

Normalized yield of species — The normalized yield of species i is the yield divided by the unlimited air yield or by the maximum theoretical yield. We will thus be using two normalized yields: $y_i/y_{i,\infty}$ and $y_i/y_{i,max}$.

Stoichiometry — A balanced chemical equation defines the stoichiometry of a reaction; stoichiometry gives the exact proportions of the reactants for complete conversion to products, where no reactants are remaining. The stoichiometric ratio is the ideal reaction mass fuel to oxygen (or air) ratio and is given the symbol r.

Unlimited air yield of species — The yield of a species for combustion of any given fuel with an unlimited air supply, denoted $y_{i,\infty}$ for species i. It is assumed constant for a given burning condition. For our purposes, we use experimentally determined values that are measured under ample air supply. The measured values are often given the symbol $y_{i,Wv}$ where Wv stands for well ventilated. The two symbols $y_{i,\infty}$ and $y_{i,Wv}$ are taken to be identical in this text.

Yield of species — The mass of species i produced per mass supply rate of the gaseous fuel, denoted y_i.

9.2   INTRODUCTION

The production and disposal of combustion product by fire can affect people, property, equipment, and operations. It is surprising that relatively small fires can produce enough soot to damage items very far from the source, and the seemingly benign smoldering fire can produce enough carbon monoxide (CO) to kill people in a neighboring room. Other combustion products, e.g., hydrogen chloride (HCl), can cause corrosion leading to the failure of critical electronic components, and new materials will bring additional hazards from their combustion products.

Understanding the generation of combustion products involves a detailed knowledge of their chemistry. But this is very complex, and the fire protection engineer must rely on measurements, not fundamental theory, to make predictions.

For a given product, the nature of the combustion products will depend on the following:

- The model of combustion (flaming, smoldering, or thermal degradation, i.e., pyrolysis or evaporation);
- The availability of air; and
- The addition of chemical agents to retard.

All of these factors affect the pathway of oxidation to its complete state, or its interruption.

We will examine the nature of this fuel chemistry in fire, and how we can deal with its predictions in compartment fires. We are especially interested in predicting the concentrations of a number of species, such as CO, CO_2, and O_2, as well as concentrations of unburned hydrocarbons and soot.

We begin in Section 9.3 by defining the terms needed and give a general introduction to methods used for calculating the species yield, or amount of species produced per unit mass of fuel burned.

Once the production term is known (the yield), the concentration in the fire gases can be assessed. In order to do this one must calculate the concentration of species by considering a mass balance of the region of interest. Section 9.4 introduces how the law of the conservation of species can be applied to a control volume. This can then be applied to the region of interest to calculate the desired concentrations.

Most fuels are complex in nature, and the conditions they burn in are not often well defined. Such estimates must therefore be made using data from small- or large-scale experiments. In Section 9.5 we give an overview of the yield data available and summarize a number of engineering methods for calculating species produced per unit mass of fuel burned.

In Section 9.6 we discuss and give examples of how the information presented can be used in practical applications.

9.3   FUEL CHEMISTRY

In this section we introduce the terms and concepts necessary for estimating the yields of combustion products from fire. Later, in Section 9.5, we will show how experimental data from various sources can be used to estimate species yields in compartment fires, using these terms and concepts.

9.3.1   STOICHIOMETRY AND SPECIES

In general we would like to have detailed knowledge of the specific chemical equation for each fuel in a fire process. For example, the chemical equation for the "complete combustion" of methane (CH_4) in air is

$$CH_4 + \left(\frac{2}{0.21}\right)[0.21O_2 + 0.79N_2] \rightarrow CO_2 + 2H_2O + \frac{2(0.79)}{(0.21)}N_2$$

We see that atoms (and mass) are conserved, and the oxidation process is complete, yielding carbon dioxide (CO_2) and water (H_2O). In contrast, any interruption in this process would result in carbon monoxide (CO), hydrogen (H_2), soot (mostly C), and other hydrocarbons (HCs) due to recombination processes in the thermal degradation and oxidation. The appearance of these products becomes more likely as the incompleteness of combustion increases. It is interesting to note that while most of these products are gases (at least in the high-temperature reaction region), soot is a solid formed in the process.

For a particular fire process—flaming, smoldering, high-temperature oxidation—the chemical equations for a particular material can be very specific. Also, chemical retardants added to the material can alter the chemical equation. Initially, the availability of air can be a significant factor. The criterion for sufficient air is based on the air required for the complete chemical reactions. Sufficient air can be termed "over-ventilated" or "fuel-lean," and insufficient air "under-ventilated" or "fuel-rich." Here, we are discussing diffusion flames and related fire processes based on diffusion, not premixed flames. The perfect mixture of fuel and oxygen (or air) leading to complete combustion is termed *stoichiometry*. The terms "stoichiometry" and "stoichiometric coefficients" apply to any specific chemical equation, in general.

A complete chemical reaction is defined as follows for the fuel containing the elements indicated:

$$Fuel\ [C,\ H,\ O,\ N,\ Cl,\ F,\ Br] + Air \rightarrow Product[CO_2,\ H_2O,\ N_2, Cl_2, F_2, Br]$$

This is an ideal process that may represent real fire conditions for some of the products. However, under fire conditions the products are more likely to include the following:

$$CO_2,\ CO,\ H_2O,\ HCs,\ soot\ (C),\ N_2,\ HCN,\ HCl,\ HF,\ HBr$$

for the fuel considered. Under well-ventilated flaming conditions the products CO_2, H_2O, and N_2 will be nearly their ideal "complete" values. During under-ventilated, smoldering, and thermal

TABLE 9.1
Physicochemical Data for a Number of Fuels

Fuel	Chemical Structure	Molecular Weight	Stoich. fuel/ox Ratio, r
Propane	C_3H_8	44	0.276
Acetylene	C_2H_2	26	0.325
Ethanol (ethyl alcohol)	C_2H_5OH	46	0.480
Heptane	C_7H_{16}	100	0.284
Polystyrene	C_8H_8	104	0.325
Nylon	$C_6H_{11}NO$	43	0.428
Polyurethane (flexible) PU	$CH_{1.74}N_{0.07}O_{0.32}$	20	0.580
Polymethyl methacrylate	$C_5H_8O_2$	100	0.521
Wood (pine)	$C_{0.95}H_{2.4}O$	30	0.601
Polyvinyl chloride (PVC)	C_2H_3Cl	62.5	0.710

degradation conditions there will be increasing amounts of CO, soot, and hydrocarbons. Hydrogen cyanide (gas) and the halogen gases will only occur when N or the halogens are elements in the fuel. For these real fire processes, the particular chemical equation or stoichiometry for complex fuels must be formed from experiments. A considerable collection of such data is given in the appendices to the *SFPE Handbook of Fire Protection Engineering*. Table 9.1 gives a few examples.

9.3.2 SPECIFIC YIELDS

In order to make quantitative predictions of combustion product species, it has been found useful to express the resultant species mass emanating from the fire process in terms of the mass loss (or supply rate) of the fuel. This is distinct from the burning (reaction) rate of the fuel, since all of it may not react. In other words the system could be fuel-rich and have leftover fuel. It also should be realized that the chemical composition of the fuel may not be what burns. For pure liquid fuels, heat from the fire causes pure evaporation and there is no chemical change in going to the gaseous fuel. But for solid fuels, the gasification process is more complex. Charring (principally carbon) may or may not occur, but the gaseous fuel is now usually a mixture of more complex fuel species. Bound water or the release of retardants can also be part of the gaseous fuel mixture. (Figure 9.1.)

In the following we give the definitions of a number of terms that are necessary for our treatment.

Yield of species: Yield of species i is defined as

$$y_i = \frac{m_i}{m_f} \tag{9.1}$$

where m_i is the mass of species i produced (oxygen would have a negative yield since it is consumed), and m_f is the mass of the gaseous fuel supplied, or the mass lost in gasification.

Note that m_f could contain inert species and retardants. Examples of yields for various fuels are given in Table 9.2, taken from Tewarson.[1] Tewarson measured the yields and gives values for yields of species i under well-ventilated conditions (denoted $y_{i,Wv}$ and ventilation-controlled conditions (denoted $y_{i,vc}$). The former values are relatively constant for a given fuel and are usually slightly lower than the maximum possible yield (due to the fact that the combustion is nearly never 100% efficient). But under ventilation-controlled conditions, the combustion is inefficient to various degrees and the species yields depend very strongly on the availability of oxygen. Tewarson therefore only gives a few values of ventilation-controlled yields as examples.

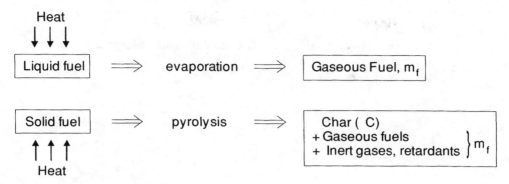

FIGURE 9.1   Fuel gasification in fire.

TABLE 9.2
Typical Combustion Product Yields for Various Fuels

	Well-ventilated (WV) fires ($\phi < 1$)							Ventilation-controlled (VC) fires ($\phi > 1$)	
Fuels	y_{CO_2} (g/g)	y_{CO} (g/g)	y_s (g/g)	$\Delta_{H_{eff}}$ (kJ/g)	Δ_{H_c} (kJ/g)	D_m (m²/g)	y_{HCl} (g/g)	y_{CO} (g/g)	y_{H_2} (g/g)
Propane	2.85	0.005	0.024	74	76.4	0.16	NA	0.23	0.011
Acetylene	2.6	0.042	0.096	37	48.2	0.32	NA	—	—
Ethanol (ethyl alcohol)	1.77	0.001	0.008	26	26.8	NA	NA	0.22	0.0098
Heptane	2.85	0.01	0.037	41	44.6	0.19	NA	—	—
Polystyrene	2.3	0.06	0.16	27	39.2	0.34	NA	—	—
Nylon	2.06	0.038	0.075	27	30.8	0.23	NA	—	—
Polyurethane (flexible) PU	1.5	0.031	0.23	19	27.2	0.33	NA	—	—
Polymethyl methacrylate	2.1	0.01	0.022	24	25.2	0.109	NA	0.19	0.032
Wood	1.33	0.005	0.015	12	17.7	0.037	NA	0.14	0.0024
Polyvinyl chloride (PVC)	0.46	0.063	0.14	5.4	16.4	0.40	0.5	0.4	—

Notes: NA = not applicable; — = not measured.

Source: From Tewarson, A., "Generation of Heat and Chemical Compounds in Fires," *SFPE Handbook of Fire Protection Engineering*, 2nd ed., National Fire Protection Association, 1995. With permission.

Free burn or unlimited air yield of species: This yield is denoted by $y_{i,\infty}$ and is to be taken as the value of $y_{i,WV}$, which is invariably measured under ample air supply. It depends on fuel and the burning configuration.

Maximum yield of species: This is the maximum theoretical yield of species i, denoted $y_{i,max}$, based on stoichiometry that allows the maximum. These values are very easily calculated from stoichiometry. Table 9.3 gives the values for the materials listed in Table 9.1.

Note that the maximum theoretical yields for CO_2 and H_2O are quite reasonable and are similar to the well-ventilated yield of these species. However, the maximum theoretical yield of CO gives an unrealistic number, since all of the carbon is assumed to result in CO, and no CO_2 is assumed to be produced. The maximum theoretical yield of CO is therefore nowhere near the well-ventilated yield of CO. The maximum theoretical yield is, however, a very useful number when calculating species concentrations, as is outlined in Section 9.5.2.

For example, when one mole of propane burns, the maximum yield of CO can be arrived at by assuming that all the carbon molecules produce CO and that no CO_2 is produced. From Tables 9.2

TABLE 9.3
Maximum Theoretical Yield of Species Based on Stoichiometry

Fuel	Stoich. fuel/ox Ratio, r	$y_{CO,max}$ (g/g)	$y_{CO_2,max}$ (g/g)	$y_{O_2,max}$ (g/g)
Propane	0.276	1.91	3.00	3.64
Acetylene	0.325	2.15	3.39	3.69
Ethanol (ethyl alcohol)	0.480	1.22	1.91	2.09
Heptane	0.284	1.96	3.08	3.52
Polystyrene	0.325	2.15	3.38	3.08
Nylon	0.428	1.48	2.32	2.61
Polyurethane (flexible) PU	0.580	1.41	2.21	2.05
Polymethyl methacrylate	0.521	1.4	2.2	1.92
Wood (pine)	0.601	0.89	1.40	1.13
Polyvinyl chloride (PVC)	0.710	0.903	1.42	1.42

and 9.3 we have for propane, $y_{CO_2,max} = 3.0$, $y_{CO_2,\infty} = 2.85$, quite similar numbers, but for CO we get $y_{CO,max} = 1.91$ and $y_{CO,\infty} = 0.005$.

The normalized yield of species: We will have occasion to use two normalized yields: $y_i/y_{i,\infty}$ (Tewarson[1]) and $y_i/y_{i,max}$ (Gottuk and Roby[7]).

Mass fraction of species: Another related variable is the mass fraction of species i, defined as

$$Y_i = \frac{m_i}{m} \qquad (9.2)$$

where m is the total mass of the gaseous mixture. This will be discussed further below; it is illustrated in Figure 9.2.

Heat of combustion: Also shown in the Table 9.2 are related yield quantities, namely the heat of combustion (see also Eq. (3.2)),

$$\Delta H_{eff} = \frac{Q}{m_f} \qquad (9.3)$$

where Q is the energy released in that fire process, in contrast to the maximum (complete) heat of combustion corresponding to the complete reaction for that fuel (ΔH_c). The fact that $\Delta H_{eff} < \Delta H_c$ indicates the degree of incomplete combustion and the degree to which char is not participating in flaming combustion.

Mass optical density: Another related yield is D_m, the "mass optical density." This property is defined in terms of the yield of soot (y_s) as

$$D_m = \alpha y_s \qquad (9.4)$$

where α is termed the aerosol particle optical density, which can roughly range from 1.9 m^2/g for nonflaming smoke to 3.3 m^2/g for flaming smoke, but α is generally not constant and can depend on y_s. The property D_m can be used to predict aspects of visibility, since the distance that lighted objects are discernable through smoke (visibility) is experimentally found to be nearly inversely proportional to D_m.

Let us focus on the combustion product species yields. An extensive compilation of yields has been developed by Tewarson.[1] Table 9.2 illustrates some of these data for a range of fuels. In some

FIGURE 9.2 Illustration of fuel and oxygen reaction.

gases (e.g., propane) the results can be taken as generic, but in others they represent the specific formulation (e.g., polyurethane). The results in Table 9.2 are for flaming fires. It is found that for over-ventilated fires, the yields are approximately constant. But for under-ventilated fires, the yields, in general, depend on the degree of excess fuel. Two alternative parameters that measure this degree are the equivalence ratio (ϕ) and the fuel mixture fraction (f). These are defined below.

The equivalence ratio:

$$\phi = \frac{m_f / m_{ox}}{r} \tag{9.5}$$

where m_f is the available fuel mass, m_{ox} is the available oxygen (or air) mass, and r is the *ideal* reaction stoichiometric mass fuel to oxygen (or air) ratio for *complete* combustion. Observe that this can also be calculated in terms of air (instead of oxygen) by replacing m_{ox} with m_{air} and by recalculating r to give the stoichiometric fuel/air ratio instead of the fuel/oxygen ratio.

The fuel mixture fraction:

$$f = \frac{m_{\text{fuel atoms}}}{m} \tag{9.6}$$

where $m_{\text{fuel atoms}}$ is the mass of atoms that were originally fuel, and m is the mass of the gaseous mixture.

Both of these parameters give a measure of the quantity of fuel relative to the oxygen. The fuel mixture fraction is the more difficult parameter to conceptually grasp. Imagine a steady-state process in which a fuel ($\dot{m}_f$) and oxygen ($\dot{m}_{ox}$) flow streams are mixed with a resulting combustion product flow stream ($\dot{m}$). This stream may have either excess fuel or oxygen (see Figure 9.2.) If $\dot{m}_f$ represents pure fuel (more general analyses could include inert species in the fuel stream such as N_2), then

$$f = \frac{\dot{m}_f}{\dot{m}}$$

since the fuel atoms are not destroyed, although they are rearranged in the product stream.

If the stoichiometry of the ideal complete reaction is

$$r \text{ g fuel} + 1 \text{ g oxygen} \rightarrow (r+1) \text{ g products} \tag{9.7}$$

then

$$\phi = \frac{\dot{m}_f / \dot{m}_{ox}}{r} \quad \text{as defined by Eq. (9.5).}$$

For the steady-state conditions given in Figure 9.2,

$$\dot{m} = \dot{m}_{ox} + \dot{m}_f$$

Therefore,

$$f = \frac{\dot{m}_f/\dot{m}_{ox}}{\dot{m}_f/\dot{m}_{ox} + 1}$$

or

$$f = \frac{\phi}{\phi + \dfrac{1}{r}} \tag{9.8}$$

Thus, we see how ϕ and f are related. But note that f ranges from 0 to 1, while ϕ ranges from 0 to infinity for pure oxygen to pure fuel, respectively.

Observe that in Figure 9.2 we used a mixture of pure oxygen and fuel. Again, using air instead of pure oxygen still results in Eq. (9.8) by replacing m_{ox} with m_{air} and by recalculating r to give the stoichiometric fuel/air ratio instead of the fuel/oxygen ratio.

In Section 9.5 we will examine data from experiments and introduce simple engineering methods for calculating yields. But to facilitate an understanding of the terms used above we first give some general examples of how the terms are used.

Practical application of the above equations: In order to calculate yields we must know the exact chemical formula for the reaction. The above equations can then be used to calculate the equivalence ratio, the fuel mixture fraction, and the yield of species i. However, the exact chemical formula for the reaction is hardly ever known in practical applications. It is only in cases where the products composition is directly measured in experiments that we can use the data and the above equations for calculations. We will discuss the experimental data in Section 9.5.

Alternatively, we must know the mass of fuel and the mass of air used in the reaction. This can often be estimated using methods for mass flow rates in plumes and mass flow rates through openings, as discussed in Chapters 4 and 5. The equivalence ratio can then be calculated and experimental data used (such as is given in Table 9.2) to estimate the yield of species i. This can give good results for over-ventilated fires ($\phi < 1$). It should be noted, however, that the yield data for under-ventilated fires in Table 9.2 is specific for the apparatus and conditions of the experiments and this must be kept in mind when applying the data to other scenarios.

EXAMPLE 9.1a

A propane burner releases gas at such a high rate that only half of it is combusted. Making the (relatively unreasonable) assumption that no CO is produced and that all of the hydrocarbons become CO_2 and H_2O, calculate the equivalence ratio, the fuel mixture fraction, and the CO_2 yield.

SUGGESTED SOLUTION

To calculate the stoichiometric ratio r we first set up the equation for complete combustion: $C_3H_8 + 5O_2 \rightarrow 3CO_2 + 4H_2O$. This gives the stoichiometric fuel/oxygen ratio as

$r = \dfrac{m_f}{m_{O_2}} = \dfrac{3 \cdot 12 + 8 \cdot 1}{5 \cdot 32} = 0.275$. When only half of the propane is combusted, assuming that

only CO_2 and H_2O are produced, we get the reaction formula $C_3H_8 + 2.5O_2 \rightarrow 1.5CO_2 +$

$2H_2O + 0.5C_3H_8$. The equivalence ratio becomes $\phi = \dfrac{44/80}{0.275} = 2$ and $f = \dfrac{44}{44 + 80} = 0.355$

(alternatively, Eq. (9.8) can be used with the same result). The yield of CO_2 is from Eq. (9.1)

$y_{CO_2} = \dfrac{1.5(12 + 32)}{44} = 1.5$ g/g.

Note that in reality considerable amounts of CO, soot, etc., will be produced in this under-ventilated case. However, the CO yield is an order of magnitude less than the CO_2 yield and the above estimate can therefore be considered to be reasonable. Note that the same result is obtained if the stoichiometric ratio and the equivalence ratio are calculated in terms of air instead of oxygen.

EXAMPLE 9.1b

The propane burner mentioned above is situated in a small combustion chamber with limited ventilation. The burner releases propane at a rate of 1.35 g/s and the measured air flow through the ventilation opening is 50 g/s. Assuming that all this air is used for combustion, calculate the equivalence ratio and the fuel mixture fraction. Give a rough estimate of the CO_2 and CO yields.

SUGGESTED SOLUTION

The stoichiometric ratio in terms of oxygen is as before $r = 0.275$. Since the mass fraction of oxygen in air is $\approx 23\%$, the mass flow rate of oxygen is 50 g/s $\cdot$ 0.23 = 11.5 g/s. Then

$\phi = \dfrac{1.35/11.5}{0.275} = 0.426$ and $f = \dfrac{1.35}{1.35 + 11.5} = 0.105$ Since $\phi < 1$ the fire is over-ventilated. We

can therefore use the data in Table 9.2 as a rough indicator of the yields. We thus find that $y_{CO_2} \approx 2.85$ g/g and $y_{CO} \approx 0.005$ g/g. We will return to this problem later and improve our estimates.

The relationship between Y_i, f or ϕ, and the yield, y_i: It can be shown that in diffusion flames, the species concentrations can be directly related to the fuel mixture fraction provided the exact stoichiometry is known. This says that as you mix fuel ($f = 1$) and oxygen ($f = 0$), the resultant state of the reacting mixture depends on the value of f. For deviations from ideal complete combustion, the stoichiometry is not necessarily known, but it still suggested that the species concentration of each component depends on f or ϕ. Such representations have had great utility in diffusion flame analysis and in compartment fire analysis. They are empirical, but have a theoretical basis for ideal conditions.

In the illustration given in Figure 9.2, we can calculate the mass fraction of species i from its definition (Eq. 9.1):

$$Y_i = \frac{m_i}{m} = \frac{\dot{m}_i}{\dot{m}} \tag{9.9}$$

assuming that the flow is steady.

By the stoichiometry given, *considering only one product* ($i = p$), we can show the relationship between Y_p, f or ϕ, and the yield, y_p. Stoichiometry has been represented in terms of r, and is based on the mass of fuel reacted, $\dot{m}_{f,\text{reacted}}$. Here, r is the stoichiometry of the particular reaction (it is not necessarily the ideal value of Eq. (9.7)). This is in contrast to the yield based on the mass of fuel lost or supplied, $\dot{m}_f$. If $\dot{m}_f > \dot{m}_{f,\text{reacted}}$, the condition is fuel-rich or under-ventilated. We now consider steady mass flow rates as illustrated by Figure 9.2. From the stoichiometry, and Eq. (9.7) and (9.9), we write

$$Y_p = \frac{\dot{m}_p}{\dot{m}} = \frac{\dfrac{(r+1)}{r}\dot{m}_{f,\text{reacted}}}{\dot{m}}$$

where $\dot{m}_{f,\text{reacted}} = \dot{m}_f$ (fuel-lean or over-ventilated) or $\dot{m}_{f,\text{reacted}} = r\,\dot{m}_{ox}$ for $\dot{m}_f > \dot{m}_{f,\text{reacted}}$ (fuel-rich or under-ventilated). From the definitions of f and ϕ,

$$\dot{m}_f = \dot{m}_{f,\text{reacted}} : \; Y_p = \frac{\left(\dfrac{r+1}{r}\right)\dot{m}_f}{\dot{m}_f + \dot{m}_{ox}} = \left(\frac{r+1}{r}\right)f = \left(\frac{r+1}{r}\right)\left(\frac{\phi}{\phi + \dfrac{1}{r}}\right) \tag{9.10a}$$

$$\dot{m}_f > \dot{m}_{f,\text{reacted}} : \; Y_p = \left(\frac{r+1}{r}\right)\left(\frac{r\dot{m}_{ox}}{\dot{m}_f + \dot{m}_{ox}}\right) = \left(\frac{r+1}{r}\right)\left(\frac{1}{\phi + \dfrac{1}{r}}\right) \tag{9.10b}$$

Note that the species mass fraction for a given fuel stoichiometry defined by r is only a function of ϕ (or f) and r.

Similarly, the yield of product (p) is given from Eq. (9.1) as

$$y_p = \frac{\dot{m}_p}{\dot{m}_f}$$

where as before $\dot{m}_f$ is the mass lost rate or supply rate of the fuel. Note the contrast between yields (based on mass loss of fuel) and stoichiometry (based on fuel reacted). It can be shown for over-ventilated conditions that the yield of product p, y_p, can be written

$$\dot{m}_f = \dot{m}_{f,\text{reacted}} : \; y_p = \frac{\left(\dfrac{r+1}{r}\right)\dot{m}_{f,\text{reacted}}}{\dot{m}_f} = \left(\frac{r+1}{r}\right) \tag{9.11a}$$

and under-ventilated conditions that

$$\dot{m}_f > \dot{m}_{f,\text{reacted}} : \; y_p = \frac{\left(\dfrac{r+1}{r}\right)r\dot{m}_{ox}}{\dot{m}_f} = \left(\frac{r+1}{r}\right)\Big/\phi \tag{9.11b}$$

Since

$$\phi = \frac{\dot{m}_f/\dot{m}_{ox}}{r} = \frac{\dot{m}_f/\dot{m}_{ox}}{\dot{m}_{f,\text{reacted}}/\dot{m}_{ox}} = \frac{\dot{m}_f}{\dot{m}_{f,\text{reacted}}}$$

we see that $0 \leq \phi \leq 1$ holds for the over-ventilated case, and $1 > \phi$ holds for the under-ventilated case. If r is the stoichiometric coefficient for the ideal complete reaction, then r is a constant (independent of ϕ) and Eq. (9.11) gives

$$\frac{y_p}{y_{p,\infty}} = \frac{1}{\phi} \qquad (9.12a)$$

where $y_{p,\infty}$ represents the over-ventilated value for an unlimited air supply, a constant value.

In general, we can empirically justify the following: For conditions of unlimited air supply for specific turbulent burning conditions it is found that $y_{i,\infty}$ is approximately constant. These are the over-ventilated yields given by Tewarson in Table 9.2.[1] As air becomes restricted, the stoichiometry can change. For a given fuel under turbulent burning conditions this can be expressed by r as a function of ϕ. Consequently, Eq. (9.12a) can be generalized as

$$\frac{y_p}{y_{p,\infty}} = f(\phi) \qquad (9.12b)$$

The actual functional dependence must be determined for each fuel. However, for major species such as CO_2 and H_2O we expect that the functional dependence is nearly fuel-independent. Also, it can be shown that the restriction of a single product does not apply to Eq. (9.12a) and (9.12b). The ratio of the oxygen yields (each negative) is also identical to Eq. (9.12a) and (9.12b).

EXAMPLE 9.1c

SUGGESTED SOLUTION

Use Eq. (9.11b) to calculate the yield of products from the reaction given in Example 9.1a.

SUGGESTED SOLUTION

Equation (9.11b) gives the yield of products as $((0.275 + 1)/0.275)/2 = 2.318$ g/g. How does this compare with the value obtained from Eq. (9.1)? The mass of fuel is 44 g and the mass of products is given from the reaction $C_3H_8 + 2.5O_2 \rightarrow 1.5CO_2 + 2H_2O + 0.5C_3H_8$ as being $1.5 (12 + 32) + 2(2 + 16) = 102$ g. Note that the product $0.5C_3H_8$ is not included in this mass, since Eq. (9.7) is based on fuel burnt and the unburned fuel is not assumed to enter the product stream. The yield of products calculated from Eq. (9.1) is therefore $102/44 = 2.318$ g/g.

EXAMPLE 9.1d

Use Eq. (9.12) to calculate the yield of CO_2 given in Example 9.1a.

SUGGESTED SOLUTION

The maximum possible yield of CO_2, $y_{CO_2,max}$ is arrived at from the stoichiometric reaction formula $C_3H_8 + 5O_2 \rightarrow 3CO_2 + 4H_2O$. Take this as the over-ventilated value, $y_{CO_2,\infty}$ (or alternatively use Table 9.3). Using Eq. (9.1) we get $y_{CO_2,\infty} = 3(12 + 32)/44 = 3$ g/g. The equivalence ratio was found to be $\phi = 2$ in Example 9.1a. The yield of CO_2 can then be found using Eq. (9.12) as being $y_{CO_2} = 3/2 = 1.5$ g/g, as was also found in Example 9.1a. Note, however, that Eq. (9.10) and (9.11) will not give these values, since they are valid for the case when the products are treated as one species.

The stoichiometry in a turbulent diffusion flame will depend on the nature of mixing. The mixing will affect the fuel-oxygen "contact" time relative to their "reaction" time. The functional dependence f(ϕ) in Eq. (9.12b) is an expression of this turbulent reaction process for a given fuel in a specific flame configuration. Limited data on turbulent buoyant flames indicative of natural fire conditions show a surprising universality in this dependence on f. However, results are not identical for species yield within flames, from flames and for changes in flame size and configuration.

We will return to this issue when we discuss experimental results in Section 9.5. But first, in Section 9.4, we examine how species yields can be used to calculate species concentrations in compartments. We examine the conservation equation for species.

9.4 CONSERVATION EQUATION FOR SPECIES

In previous sections of this chapter we defined and discussed many of the terms used when estimating species yields. In this section we discuss how species yields can be used to predict the resulting species concentration. We must therefore examine the conservation equation for species.

9.4.1 CONTROL VOLUME FORMULATION

We will consider macroscopic equations instead of point-wise partial differential governing equations. The former will apply to compartment fires, and can be applied to differential volume element to derive the latter. The equations will be represented for a control volume—a specified spatial region whose surface can be moving at the local velocity, $\overline{w}(x, y, z, t)$. This is distinct from the bulk fluid velocity, $\overline{v}$, and the diffusion velocity of each species, $\overline{V}_i$. The latter is due to molecular diffusions arising from a concentration difference in the species.

According to Fick's Law,

$$\rho_i \overline{V}_i = -\rho D \nabla Y_i \tag{9.13}$$

where D is the diffusion coefficient of species i in the mixture, ρ_i is the species density, ρY_i, and ρ is the mixture density.

The mass flux of species i across a boundary can be due to bulk flows as well as diffusion (Figure 9.3). The mass flow rate of species i flowing through surface dS is given by the species moving in the normal direction. This involves the projection of the relative velocity on to $\overline{n}$, or the vector dot product:

$$\dot{m}_{i,dS} = \rho_i (\overline{v} + \overline{V}_i - \overline{w}) \cdot \overline{n} dS \tag{9.14}$$

Now consider a finite control volume (CV). The conservation of species i states that

$$
\begin{bmatrix} \text{Rate of change} \\ \text{of species i} \\ \text{mass in the} \\ \text{CV} \end{bmatrix}
+
\begin{bmatrix} \text{Net rate of} \\ \text{mass flow} \\ \text{of i leaving} \\ \text{the CV} \end{bmatrix}
=
\begin{bmatrix} \text{Rate of production} \\ \text{of species i mass} \\ \text{due to chemical} \\ \text{reaction} \end{bmatrix}
-
\begin{bmatrix} \text{Mass loss rate} \\ \text{due to surface} \\ \text{deposition or settling} \\ \text{of particulates} \end{bmatrix}
\tag{9.15}
$$

These processes are illustrated for the CV in Figure 9.4.

Mathematically, Eq. (9.15) can be expressed as

$$\frac{d}{dt} \iiint_{CV} \rho Y_i dV + \iint_{CS} \rho Y (\overline{v} + \overline{V}_i - \overline{w}) \cdot \overline{n} dS = \iiint_{CV} y_i \dot{m}_f''' dV - \dot{m}_{i,loss} \tag{9.16}$$

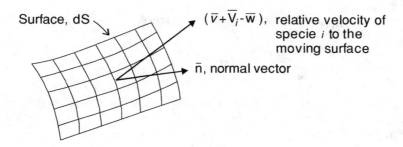

FIGURE 9.3   Mass flux of species i through a surface.

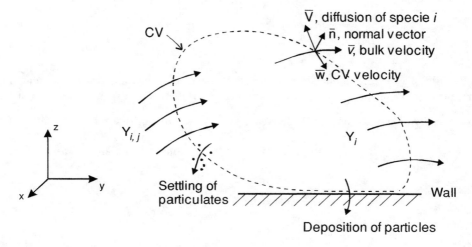

FIGURE 9.4   Conservation of species for CV.

The first term on the right is expressed in terms of yield and the fuel mass loss rate per unit volume, $\dot{m}_f'''$. It would more appropriately be expressed as the stoichiometry (s_i, mass of i produced per mass of fuel reacted) and the burning rate per unit volume, $\dot{m}_f'''$, reacted. The last term represents the losses due to surface deposition or settling of particulates.

From Fick's Law, Eq. (9.13), the diffusion term in Eq. (9.16) can be alternatively expressed as

$$\iint \rho Y_i \overline{V}_i \cdot \bar{n} dS = -\iint \rho D \frac{\partial Y_i}{\partial n} dS \qquad (9.17)$$

This term can be more practically determined by using macroscopic diffusion formulas of the form $\dot{m}_i'' \sim \Delta Y_i$, with the mass transfer coefficient as a proportionality constant.

For a CV with *uniform* properties, i.e., Y_i not dependent on x, y, and z, we can rewrite Eq. (9.16), since

$$m = \iiint \rho dV, \text{ mass in the CV}$$

and

$$\dot{m} = \iint \rho(\bar{v} - \overline{w}) \cdot \bar{n} dS, \text{ net mass flow rate out of the CV}$$

Equation (9.16) becomes

$$\frac{d}{dt}(mY_i) + \sum_{\substack{j=1 \\ \text{net out}}}^{N} \dot{m}_j Y_{i,j} = y_i \dot{m}_f - \dot{m}_{i,\text{loss}} \tag{9.18}$$

The last term now contains all of the diffusion and deposition loss terms. The second term on the left represents all of the mass flow rates of species i in the bulk flow leaving (+) or entering (−) the CV. $Y_{i,j}$ represents the ith species leaving or entering in the jth stream.

Conservation of the mass for the entire fluid, summing all the species would lead to the corresponding result:

$$\frac{dm}{dt} + \sum_{\substack{j=1 \\ \text{net out}}}^{N} \dot{m}_j = 0 \tag{9.19}$$

Thus the rate of mass accumulated in the CV must balance all of the rates of mass entering and leaving through all of the $j = 1, 2, \ldots, N$ streams. From this overall mass balance an alternative species conservation equation can be derived by using the product rule for derivatives $\frac{d}{dt}(mY_i) = Y_i \frac{dm}{dt} + m \frac{dY_i}{dt}$ and writing $\frac{dm}{dt}$ in terms of Eq. (9.19) and thus arriving at

$$m\frac{dY_i}{dt} + \sum_{\substack{j=1 \\ \text{net out}}}^{N} \dot{m}_j \cdot (Y_{i,j} - Y_i) = y_i \dot{m}_f - \dot{m}_{i,\text{loss}} \tag{9.20}$$

This form of the conservation equation for species can be useful for applications to compartments. Note that this is analogous to the governing point-wise equation,

$$\rho \frac{\partial Y_i}{\partial t} + \rho \bar{v} \cdot \nabla Y_i = \dot{m}_i''' + \nabla \cdot \rho D \nabla Y_i \tag{9.21}$$

We have shown in this section how the conservation equation for species i can be set up for a control volume and how this is analogous to the governing point-wise equation. We have also used the control volume approach to derive Eq. (9.20). This conservation equation for species i can be applied to a fire compartment, which we will detail in the next section.

9.4.2 APPLICATION TO A COMPARTMENT

Consider Figure 9.5 where a uniform property smoke layer (including the fire) and lower layer have been selected as control volumes, CV_1 and CV_2. The CVs are allowed to move at the smoke interface. This takes advantage of the observation that the interface is stably stratified with negligible bulk motion across it. Thermal and mass diffusion are present, but these are normally neglected. Also, wall soot deposition is not normally addressed, but these effects are presented in the equations and can therefore be included. If we assume no mixing at the interface, and a single opening as shown, Eq. (9.20) becomes for this case

$$m\frac{dY_i}{dt} + \dot{m}_o(Y_{i,o} - Y_i) - \dot{m}_p(Y_{i,p} - Y_i) - \dot{m}_f(Y_{i,f} - Y_i) = y_i \dot{m}_f \tag{9.22}$$

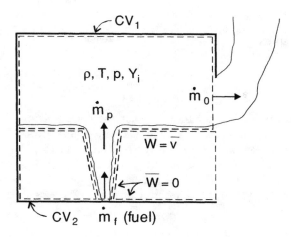

FIGURE 9.5   Control volumes selected in zone modeling.

where $\dot{m}_o$ is the rate of mass flowing out of the compartment and $\dot{m}_p$ is the mass flow rate entrained into the plume. We assume, quite reasonably, that the mass fraction of species i flowing out of the compartment, $Y_{i,o}$, is equal the mass fraction of species i in the upper layer, Y_i.

If there is no species i in the entrained stream and also in the fuel stream, then

$$m \frac{dY_i}{dt} + (\dot{m}_p + \dot{m}_f)Y_i = y_i \dot{m}_f \tag{9.23}$$

This gives us an illustrative equation, showing how species yield, y_i, can be used to calculate the mass fraction (and therefore mass concentration) of species i, Y_i, in a compartment smoke layer. The concentration is exponential, reaching a plateau depending on the rate of production and entrainment.

Yield data are also sometimes presented as a function of the fuel mixture fraction, f. An equation can easily be determined for the mixture fraction, since it can be considered a species—all of the original fuel atoms—that is not further created or destroyed. Therefore, from Eq. (9.23),

$$m \frac{df}{dt} + (\dot{m}_p + \dot{m}_f)f - \dot{m}_f = 0 \tag{9.24}$$

The drawback with directly applying the above equations is that the mass of the gas layer (or its volume and temperature) is often not known or is a strong function of time. These equations are therefore usually coupled to the differential equations for the conservation of energy and mass and solved simultaneously by computers. If the yield of species y_i is known, then the mass fraction (and therefore mass concentration) of species i, Y_i, can be calculated.

To illustrate the use of the above equations we now make the assumption that m, the mass of the upper layer, is constant and independent of time. In general, this may seem a rough assumption: even though the energy release rate may sometimes be constant and the volume of the upper layer may be constant, the temperature of the upper layer can seldom be assumed to be constant. However, when the gas temperature and layer volume are changing slowly, the assumption is quite reasonable.

We further assume that the plume and opening mass flow rates in Eq. (9.23) are steady and independent of time. The equation is then a linear first-order differential equation and can very easily be solved by integrating factor to give

$$Y_i(t) = y_i \frac{\dot{m}_f}{\dot{m}_f + \dot{m}_p} \left(1 - e^{\frac{\dot{m}_f + \dot{m}_p}{m} \cdot t} \right)$$

When time goes to infinity, the steady-state mass fraction of species i can then be calculated independent of the mass of the upper layer as

$$Y_i(t) = y_i \frac{\dot{m}_f}{\dot{m}_f + \dot{m}_p} = y_i \cdot f \qquad (9.25)$$

For a steady-state problem we can therefore calculate the mass fraction of species i in the upper layer (and therefore the mass concentration), if the species yield is known. Since y_i can be considered a function of ϕ, and f and ϕ are related for a given stoichiometry (Eq. 9.8), we can regard Y_i as a unique function of f or ϕ for a given fuel and fire condition.

In the next section we show how experimental data from various sources can be used to display these functional relationships for Y_i or y_i in terms of ϕ or f. We emphasize that these functional relationships are empirical.

9.5 USING EXPERIMENTAL DATA FOR ESTIMATING YIELDS

The generation of combustion products in a compartment fire is a complex issue and the engineer must rely on approximate methods for estimating the yield of a species. Once the yield is known, estimations on species concentration will have to be made. In this section we review a number ways in which experimental data can be used to estimate species yield. The types of experiments used for this purpose can be divided into three categories: data obtained from bench-scale tests, data obtained from hood experiments, and data obtained from compartment fire tests.

9.5.1 DATA FROM BENCH-SCALE TESTS

Tewarson collected a vast amount of data from experiments carried out in a bench-scale test apparatus called the Flammability Apparatus (Figure 9.6.)[1] The data are reported in the *SFPE Handbook of Fire Protection Engineering*.[1] A small example of the type of data reported is given in Table 9.2, where Tewarson reports species yields for CO_2, CO, and soot for well-ventilated fires ($\phi < 1$). He also reports some data on yields of CO, HCl, and H_2 for a number of fuels in the same table for under-ventilated conditions ($\phi > 1$). However, for nearly all fuels the yields are a strong function of the equivalence ratio for ($\phi > 1$) and a single yield value for under-ventilated fires is seldom seen.

As we examine Table 9.2, under-ventilated yields of CO are many times higher than their over-ventilated counter parts. But the yields of CO_2 are still nearly 10 times those of CO. Hence, the major species may follow Eq. (9.12a), and we shall be using this in Section 9.5.2.

Tewarson gives detailed results of the yield ratios for several fuels and for CO, soot, O_2, and CO_2.[1] These are sketched in Figure 9.7. For the fuels shown, only the CO yield ratio depends on the fuel, while the yield ratios of soot, O_2, and CO_2 are nearly the same. Both CO_2 and O_2 nearly follow Eq. (9.12a). The vertical axes represents the ratio of yield at under-ventilated conditions to the yield at over-ventilated conditions. The latter value is given in Table 9.2 and should be somewhat lower than the unlimited air yield, $y_{i,\infty}$.

Perhaps equally significant, at roughly $\phi = 4$, there is a transition from flaming to nonflaming conditions. These data represent relatively small turbulent fires under controlled air and fuel supply with the solids enhanced by external radiation.

In general, we can expect that a given fuel's products of combustion can be represented by a yield that depends on ϕ, alternatively, f. These relationships must be determined experimentally, since the stoichiometry will change markedly for under-ventilated conditions. Also the yield relation is likely to depend on the nature of mixing, e.g., a turbulent jet, plume, fire in a room, etc.

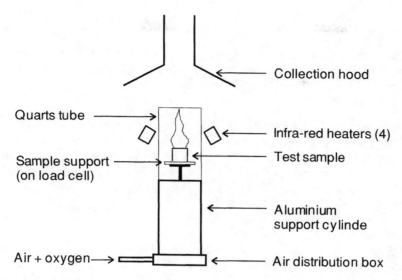

FIGURE 9.6 Flammability Apparatus designed by the Factory Mutual Research Corporation (FMRC). Sample configuration for ignition, pyrolysis, and combustion test.

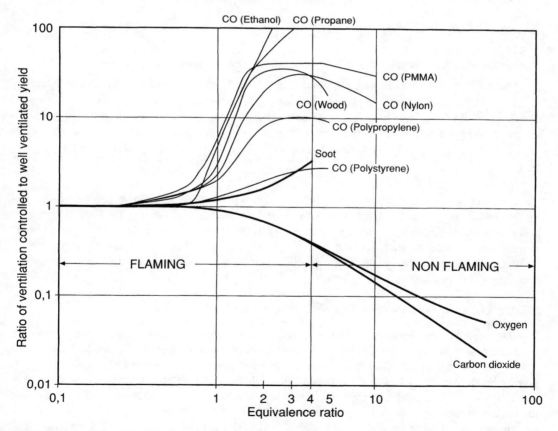

FIGURE 9.7 Effect of under-ventilation on yields of Oxygen, CO_2, and CO for many materials. (Adapted from Tewarson[1].)

EXAMPLE 9.2a

Use the information given in Example 9.1b to give an estimate of the yield of CO, CO_2, and soot, using the approximate data representation in Figure 9.7.

SUGGESTED SOLUTION

The equivalence ratio was found to be 0.426. From Figure 9.7 we see that the yields for CO, CO_2, and soot are very close to the over-ventilated value measured by Tewarson.[1] We therefore conclude that the values given in Table 9.2 can be used as a rough estimate and y_{CO_2} = 2.85 g/g, y_{CO} = 0.005 g/g, and y_s = 0.024 g/g. Observe that y_{CO_2} in this over-ventilated scenario is somewhat lower that the maximum possible CO_2 yield, $y_{CO_2,\bullet}$ = 3 g/g, as we found in Example 9.1d.

EXAMPLE 9.2b

Now assume that the airflow to the combustion chamber discussed in Example 9.1b is reduced from 50 g/s to 10 g/s. Estimate the yields of CO, CO_2, and soot.

SUGGESTED SOLUTION

The flow of oxygen is now 10 g/s · 0.23 = 2.3 g/s. The equivalence ratio is $f = \dfrac{1.35/2.3}{0.275} = 2.13$. From Figure 9.7 we find that for CO_2 the yield ratio at f = 2.13 is ≈0.5 and for soot the yield ratio is ≈1.5. Observe that these are roughly material-independent. The yield ratio of CO, however, is material-dependent and is found to be ≈60 for propane at f = 2.13. The under-ventilated yields can now be calculated using the over-ventilated values given in Table 9.2 so that y_{CO_2} = 0.5 · 2.85 g/g = 1.4 g/g, y_s = (1.5) · (0.024) g/g = 0.036 g/g, and y_{CO} = (60) · (0.005) g/g = 0.3 g/g.

It should be emphasized that this relationship for a given fuel should also depend on the fire mixing conditions. Table 9.2 and Figure 9.7 essentially were for data taken in the FMRC Flammability Apparatus (Figure 9.6) with a sample size of 100¥100 mm. We expect other fire conditions to behave in a similar manner, but quantitative accuracy limits have not been established. This must be kept in mind when applying the data in practice. Some general limitations in using experimentally derived yield data are discussed in Section 9.5.2.

9.5.2 DATA FROM HOOD EXPERIMENTS

The data presented by Tewarson in Figure 9.7 show some scatter, but the trends are clear.[1] Although his data are for the small-scale apparatus of Figure 9.7, others have produced similar data for larger scale.

Beyler was first to correlate his measurements of combustion gas concentrations in the upper layer with the so-called global equivalence ratio, GER.[2] He conducted his experiments under a hood. By extracting a known amount of gas from the hood to give a constant layer height and using the measured fuel flow rate and compositions of the upper layer, the global equivalence ratio could be calculated (assuming the layer to be "well stirred," allowing one "global" ratio to be calculated as opposed to many different "local" values within the layer). A range of values for the global equivalence ratio was achieved by varying the fuel supply rate and the height between the fuel bed and the layer interface. Several gaseous and liquid fuels were used.

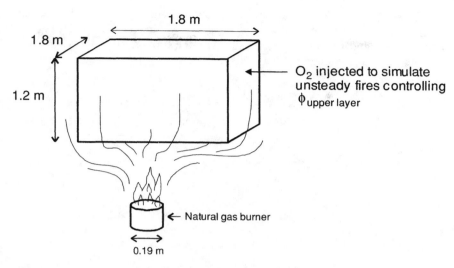

FIGURE 9.8   Schematic of the Zukoski et al.[3] hood experiments.

Several other experimenters also used hoods, and the concept of global equivalence ratio to correlate against the gas composition in the hot layer, such as Zukoski et al.,[3] Toner,[4] and Morehart.[5] Zukoski et al.[3] also injected oxygen into the hot layer to simulate unsteady fires (see Figure 9.8).[3]

For the hood experiments it is possible to define the global equivalence ratio in mainly two different ways.

The global plume equivalence ratio, ϕ_p, is defined as the fuel mass flow rate divided by the oxygen (air) mass entrainment into the plume normalized by the stoichiometric fuel to oxygen (air) ratio, or

$$\phi_p = \frac{\dot{m}_f}{r Y_{ox,p} \dot{m}_p} \tag{9.26a}$$

where the subscript "p" refers to the plume.

The global upper layer equivalence ratio, ϕ_{ul}, is defined as the ratio of the mass of gas in the upper layer originating from fuel sources to the mass of oxygen (air) in the air streams introduced into the upper layer, normalized by the fuel to oxygen (air) ratio, or

$$\phi_{ul} = \frac{m_f}{r m_{ox}} \tag{9.26b}$$

Note that from the definition in Eq. (9.5), r is the stoichiometric coefficient for complete combustion.

The latter ϕ is the instantaneous value, based on the total mass of fuel and oxygen available, associated with the unsteady case given by Eq. (9.23) and (9.24). The two equivalence ratios are not necessarily the same for unsteady cases. In an ideal two-layer fire, where all the air enters the upper layer through the plume, the two global equivalence ratios are the same for steady fires. Since we will not be dealing with unsteady cases for our engineering methods, we assume $\phi_p = \phi_{ul}$.

Equations for normalized yields can be developed as follows for the ideal case of a complete reaction. For this we assume that all of the H becomes H_2O and all C goes to CO_2 for $\phi < 1$, the well-ventilated condition. For $\phi > 1$, all of the excess fuel is considered to be total hydrocarbon (THC). Although the theoretical maximum yield of CO and H_2 are based on all of C and H forming these compounds only, for the ideal complete reaction no CO and H_2 are formed. These assumptions allow us to write our equations either in terms of unlimited air yield of species ($y_{i,\infty}$) or in terms of the maximum theoretical yield ($y_{i,max}$); we shall use the former. Since the yields of H_2O, CO_2, and O_2 are constant and maximum for $\phi < 1$, then

$$\frac{y_{CO_2}}{y_{CO_{2,\infty}}} = \frac{y_{H_2O}}{y_{H_2O,\infty}} = \frac{y_{O_2}}{y_{O_{2,\infty}}} = 1 \qquad (9.27a)$$

and

$$y_{CO} = y_{H_2} = y_{THC} = 0 \qquad (9.27b)$$

For $\phi > 1$, from Eq. (9.12a)

$$\frac{y_{CO_2}}{y_{CO_{2,\infty}}} = \frac{y_{H_2O}}{y_{H_2O,\infty}} = \frac{y_{O_2}}{y_{O_{2,\infty}}} = \frac{1}{\phi} \qquad (9.27c)$$

and

$$y_{CO} = y_{H_2} = 0 \qquad (9.27d)$$

The yield of the THC (or excess fuel) is by definition

$$y_{THC} = \frac{m_f - m_{f,reacted}}{m_f}$$

and by Eq. (9.5) $y_{THC} = 1 - \dfrac{rm_{ox}}{m_f}$ which leads to

$$y_{THC} = 1 - \frac{1}{\phi} \qquad \text{for } \phi > 1 \qquad (9.27e)$$

These are ideal results, assuming no production of CO and H_2. The corresponding yields are shown as the solid lines in Figure 9.9 (taken from Gottuk and Roby[7]), but here they are normalized by the theoretical maximum yield. The figure is based on data from Beyler, who conducted experiments for a wide range of gaseous and liquid fuels in an arrangement similar to Figure 9.8 (but no oxygen injected),[2] and later conducted studies with solid fuels.[6] His results for propane are plotted as normalized yields (based on the maximum possible yield of species i) vs. global equivalence ratio. The other fuels tested exhibited similar trends. The theory shows good agreement with the experimental results for the major species (H_2O, O_2, CO_2, and THC).

From fire investigations we know that at least two thirds of fire victims die from smoke inhalation and that CO is known to be the dominant toxicant in smoke; a relatively low concentration of CO can cause death within minutes. It is therefore of considerable importance to be able to predict the yield of CO, in order to be able to calculate concentrations using such Eq. as (9.23) and (9.24).

Pitts has given a review of models for estimating CO production in fires.[8] The global equivalence ratio concept seems to give good results for the main products of combustion. Figure 9.9 only

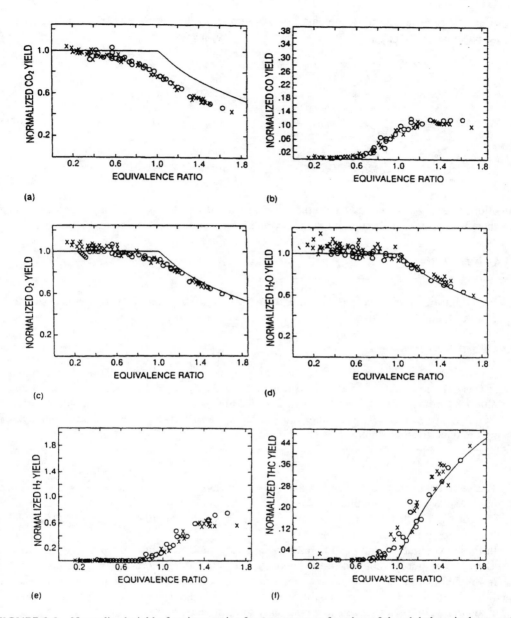

FIGURE 9.9 Normalized yield of major species for propane as a function of the global equivalence ratio. The drawn lines represent Eq. (9.27a, b, d, e). (Adapted from Gottuk and Roby[7]. With permission.)

shows results for propane, but other fuels show very similar results. However, when plotting normalized CO yields vs. equivalence ratio for many different fuels, the data are somewhat scattered.

Gottuk and Roby found, when plotting data for many different fuels, that a better agreement was obtained when plotting CO yields against equivalence ratio than when using normalized CO yields for correlations.[7] This is mostly due to the different mechanisms responsible for the formation of CO as opposed to the mechanisms for the formation of the main products of combustion. The results from Beyler's experiments are reproduced in Figure 9.10 for a number of different fuels,[2] but here the CO yields, instead of normalized CO yields, are plotted against plume equivalence ratio.

A simple representation for the data, from a number of different fuels, shown in Figure 9.10, would be to write

$$y_{CO} = 0 \qquad\qquad \text{for } \phi < 0.5 \qquad\qquad (9.28a)$$

$$y_{CO} = 0.3\phi - 0.15 \qquad\qquad \text{for } 0.5 < \phi < 1.2 \qquad\qquad (9.28b)$$

$$y_{CO} = 0.21 \qquad\qquad \text{for } \phi > 1.2 \qquad\qquad (9.28c)$$

Equations (9.28a, b, c) are drawn onto the data in Figure 9.10. Gottuk and Roby propose a number of other correlations for calculating the CO yields based on the data in Figure 9.10.[7]

EXAMPLE 9.2c

Use the information given in Example 9.2b and use Eq. (9.27) and (9.28) to give an estimate of the CO_2 and CO yields.

SUGGESTED SOLUTION

As before, f = 2.13. The unlimited air yield for CO_2 can be derived from the reaction formula (as was done in Example 9.1d) or read from Table 9.3 as $y_{CO_2,\bullet} = 3$ g/g (alternatively use the value 2.85 from Table 9.2). Using Eq. (9.27c) we find $y_{CO_2} = 3/2.13 = 1.41$ g/g and from Eq. (9.28c) $y_{CO} = 0.21$ g/g. These values are similar to those arrived at using Tewarson's data[1] and the data given in Figures 9.9 and 9.10.

The yield data can be correlated in a number of different ways, and we show some results as examples. Zukoski et al. essentially have shown that the upper smoke layer combustion products tend to correlate with f_{ul} for a 19 cm natural gas fire source ranging from 20 to 200 kW.[3] The results are shown in Figures 9.11 and 9.12.

In these experiments, the conditions were at steady state, but additional oxygen was injected into the layer of smoke captured in the hood, to add to the entrained air by the fire plume below the layer interface. This steady-state condition would correspond to unsteady compartment fire conditions where residual oxygen could be in the smoke layer. The correct correlating variable is f_{ul} (or alternatively, f), not f_p, which applies only to the steady case.

We end this section with a note on some limitations of applying the global equivalence ratio concept for estimating species yields and give some additional comments.

CO yield dependence on temperature: The global equivalence ratio concept has been found to be an excellent tool for determining major species concentrations. However, some discrepancies between the CO yield data of Morehart et al.[5] and Toner et al.[4] were found to be due to differences in the insulation of the hood used for the experiments. This resulted in gas temperatures being far higher for same values of equivalence ratio in one of the experimental series, and the resulting yields of CO reflected this. The CO yield has therefore been found to be not only a function of equivalence ratio but also a function of temperature. We have ignored this in the above equations, but they serve well for rough engineering estimates. Correlations of CO yield, where account is taken of temperatures, are given by Gottuk and Roby in the *SFPE Handbook*.[7]

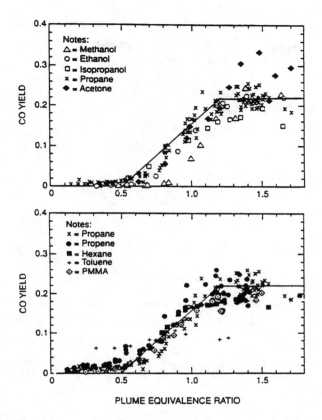

FIGURE 9.10 CO yields as a function of the plume equivalence ratio for various fuels studied by Beyler.[2] Equation (9.28a, b, c) is drawn onto the data. (Adapted from Gottuk and Roby[7]. With permission.)

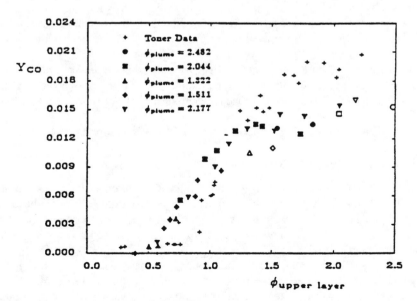

FIGURE 9.11 Mole fraction of carbon monoxide as a function of equivalence ratio for the upper layer. (From Zukoski et al.[3])

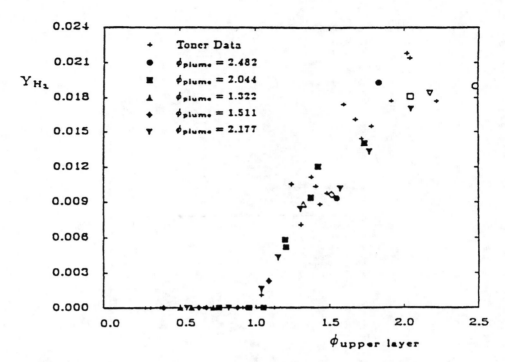

FIGURE 9.12 Mole fraction of hydrogen as a function of equivalence ratio. (From Zukoski et al.[3])

Some comments:

- We have in the above treatment assumed steady states, so that $\phi_p = \phi_{ul}$. When dealing with transient cases, account should be taken of this and the fact that the equivalence ratio and species yields then also become transient. The yields for the transient case follow the correlations in terms of f or ϕ_{ul} but not ϕ_p. Steady-state correlations apply as long as expressed and implemented in terms of f or ϕ_{ul}. The most expeditious approach is to compute f by Eq. (9.24).
- Pitts reports that when material undergoes pyrolysis within the upper layer (as in the case when the ceiling is lined with wood, for example), there is a very marked increase in the production of CO. So, if there is a second fuel burning in the layer, more complex results occur.

In the next section, we consider experiments carried out in compartments.

9.5.3   DATA FROM COMPARTMENT FIRES

The vast majority of fire victims die from smoke inhalation, often in locations remote from the fire, where CO is known to be the dominant toxicant in smoke. In the above we have summarized ways in which yields can be estimated for relatively pure fuels. Even for pure fuels, the chemistry of the reaction is very complex, and we have based our analysis on the ideal stoichiometric reaction of each fuel and experimental data.

Real fires are usually transient, and the fuel source is usually a combination of many different material types. Additionally, the smoke is often transported from the fire compartment, through corridors or other rooms, to a remote location where it may be a hazard to humans. Equation (9.23) can be used for calculating species concentrations within the fire compartment, but when the smoke exits the room, oxygen will be entrained at the door plume. This can drastically change the concentration of the gases entering the second room.

In the initial stage of the fire the burning is well ventilated with low CO yields as a result. As the fire reaches flashover, there is a jump in CO yields (as seen in the data presented above). During the post-flashover stage one might assume that some of the CO might burn as the door flame entrains additional air, resulting in CO yields at the door that are similar to the well-ventilated conditions.

Mulholland summarized some of the full-scale fire experiments that have been conducted at the National Institute of Standards and Technology (NIST) in Gaithersburg, Maryland.[9] He found that the CO concentration remained high remote from the fire even in flashover conditions and recommended that the high yields associated with flashover in the fire compartment be used for the burning in the doorway.

Mulholland and a panel of experts at NIST further recommended the following yield values to be used for hazard assessments:[9]

- For post-flashover conditions in an enclosure where the primary burning components are wood: Yield of CO = 0.3 g/g, yield of CO_2 = 1.1 g/g, depletion of O_2 = 0.9 g/g.
- For enclosure fires involving materials commonly found in offices, such as a computer console, TV monitor, electrical cables, and padded chair: Yield of CO = 0.2 g/g, yield of CO_2 = 1.5 g/g, depletion of O_2 = 1.8 g/g.

We see that the above values for CO and CO_2 are of the same order of magnitude as the values calculated in Examples 9.2a, b, and c for propane based on empirical correlations in terms of ϕ.

9.6   PREDICTING SPECIES CONCENTRATIONS IN COMPARTMENT FIRES

Although the generality of the experimentally derived yield data are not well established, they offer the best way to estimate species concentrations. This is especially true for the minor species such as CO, soot, and HCs during ventilation-limited conditions: $m_f > \dot{m}_{f,\text{reacted}}$ or $\phi > 1$. Once the yield has been established the conservation equations can be used to calculate concentrations.

We have discussed a number of methods available that allow the yield of species i to be estimated. In all cases the chemical composition of the fuel must be known; for complex fuels this can be taken from Table 9.1 or similar tables in the Appendix to the *SFPE Handbook*. The stoichiometric ratio can then be calculated (or taken from the above sources).

For an estimation of the equivalence ratio, the fuel mass flow rate must be given. Also, the mass flow rate of oxygen (or air) into the upper layer must be known. This can be roughly estimated using the plume equations discussed in Chapter 4 or by calculating the mass flow rate of air in through an opening, as discussed in Chapter 5.

Once the equivalence ratio is known, either Tewarson's data[1] or Beyler's data[2] can be used to estimate the species yields at that equivalence ratio.

- Using Tewarson's data requires that his value for well-ventilated yield, $y_{i,\text{Wv}}$, be taken from Table 9.2 or from the extensive tables in the *SFPE Handbook*. Figure 9.7 then provides the quotient between the well-ventilated and the ventilation-controlled case, $y_{i,\text{Wv}}/y_{i,\text{vc}}$. The yield value (at a certain equivalence ratio) for ventilation control, $y_{i,\text{vc}}$, can then be arrived at.
- Alternatively, Beyler's data can be used by first calculating (or reading from Table 9.3) the unlimited air yield, $y_{i,\infty}$, and using Eq. (9.27) and (9.28) (or Figures 9.9 and 9.10) to calculate the yield of species i. The maximum theoretical yield, $y_{i,\text{max}}$, is equivalent to $y_{i,\infty}$ in Eq. (9.27) since the production of CO and H_2 is neglected.

When gas concentrations are to be calculated for rooms remote from the fire room, rougher methods may be more appropriate, and Mulholland's yield values may be used for a post-flashover fire. For unsteady conditions correlations must be expressed in terms of f or ϕ_{ul} and not ϕ_p.

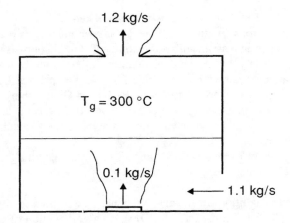

FIGURE 9.13   Steady-state fire in a compartment with ceiling ventilation.

Once the yield is known, Eq. (9.23) and (9.24) can be used to calculate the species concentrations. This usually requires that the conservation equations for energy and mass be solved simultaneously by computer to assess the mass of the upper layer, m. For the steady case, however, we can use Eq. (9.25).

EXAMPLE 9.3

A spill of heptane has ignited in a compartment (see Figure 9.13) and the radiation to the pool surface results in a fuel mass flow rate of 0.1 kg/s. The lower opening allows an incoming mass flow rate of air of 1.1 kg/s. The mass of smoke extracted through the ceiling vent is 1.2 kg/s. The compartment has a floor area of 25 m² and a height of 5 m. The smoke layer has stabilized at a height of 1 m from the floor and the upper layer gas temperature is stable at 300°C. Estimate the concentration of CO_2 and CO in the upper layer.

SUGGESTED SOLUTION

The stoichiometric fuel/oxygen ratio for heptane is 0.284 (from the ideal chemical reaction of C_7H_{16} or from Table 9.1). Since the mass fraction of oxygen in air is ≈23%, the mass flow rate of oxygen is 1.1 kg/s · 0.23 = 0.253 kg/s. Then $f = \dfrac{0.1/0.253}{0.284} = 1.4$. We find the unlimited air yield of CO_2 to be 3.08 (from Table 9.3 or by setting up the reaction formula). Equation (9.27c) gives the CO_2 yield as $y_{CO_2} = 3.08/1.4 = 2.2$ g/g and Equation (9.28c) gives the CO yield as $y_{CO} = 0.21$ g/g.

Since the problem is steady state, we can now use Eq. (9.25) to calculate the mass fractions of CO_2 and CO. The fuel mixture fraction is 0.1/(0.1 + 1.1) = 0.083. The mass fraction of CO_2 is $Y_{CO_2} = 2.2 \cdot 0.083 = 0.183$ g/g and the mass fraction of CO is $Y_{CO} = 0.21 \cdot 0.083 = 0.017$ g/g. The mass fractions in the upper layer of CO_2 and CO are therefore 18.3% and 1.7%, respectively.

We have dealt only with steady-state problems in this chapter. Real fires in compartments are transient; the fire grows through the well-ventilated stage to a stage where it becomes ventilation-controlled. The conservation equation for species must therefore be coupled to the conservation equations for mass and energy and the problem solved by computer. Several such computer programs are commercially available. Most of these allow only a single species yield to be specified for each species. The equivalence ratio and therefore the species yields, however, change with time, and it is important to carefully examine the technical reference guides to the computer programs before analyzing the results.

REFERENCES

1. Tewarson, A., "Generation of Heat and Chemical Compounds in Fires," *SFPE Handbook of Fire Protection Engineering*, 2nd ed., National Fire Protection Association, Quincy, MA, 1995.
2. Beyler, C.L., *Fire Safety Journal*, Vol. 10, No. 47, 1986.
3. Zukoski, E.E., Toner, S.J., Morehart, J.H., and Kubota, T., *Fire Safety Science—Proceedings of the Second International Symposium*, Hemisphere Publications, Washington, D.C., pp. 295–305, 1989.
4. Toner, S.J., Zukoski, E.E., and Kubota, T., Government Contractor's Report GCR-85-493, National Bureau of Standards, Washington, D.C., May 1985.
5. Morehart, J.H., Zukoski, E.E., and Kubota, T., Government Contractor's Report GCR-90-585, National Institute of Standards and Technology, Washington, D.C., December 1990.
6. Beyler, C.L., *Fire Safety Science—Proceedings of the First International Symposium*, Hemisphere Publications, Washington, D.C., pp. 431–440, 1986.
7. Gottuk, D.T. and Roby, R.J., "Effect of Combustion Conditions on Species Production." *SFPE Handbook of Fire Protection Engineering*, 2nd ed., National Fire Protection Association, Quincy, MA, 1995.
8. Pitts, W.M., "The Global Equivalence Ratio Concept and the Formation Mechanisms of Carbon Monoxide in Enclosure Fires," *Prog. Energy Combust. Sci.*, Vol. 21, pp. 197–237, 1995.
9. Mulholland, G.W., "Position Paper Regarding CO Yield," Appendix C in Nelson, H.E., *FPETOOL: Fire Protection Engineering Tools for Hazard Estimation*, National Institute of Standards and Technology Internal Report 4380, Gaithersburg, MD, pp. 93–100, 1990.

PROBLEMS AND SUGGESTED ANSWERS

9.1 An experiment is carried out in a large room with a 3.5 m-high opening. A 1.0 m-diameter container with heptane is ignited. After a short time the smoke layer height stabilizes at a height of 3 m from the floor.
   (a) Calculate the yield of CO_2.
   (b) Calculate the yield of CO.
   (c) Calculate the CO_2 concentration in the upper layer.
   (d) Calculate the CO concentration in the upper layer.

   Suggested answer: (a) Using equations and tables from Chapter 3, we find $\dot{m}_f = 0.053$ kg/s and $\dot{m}_p$ roughly 5.2 kg/s; $\dot{m}_{ox}$ is then $0.23 \cdot 5.2$. The equivalence ratio becomes f = 0.155 and $y_{CO_2} = 2.85$ g/g (Table 9.2). (b) $y_{CO} = 0.01$ g/g. (c) The fuel mixture fraction $f = 0.010$ and $Y_{CO_2} = 2.85\%$. (d) $Y_{CO} \approx 0\%$.

9.2 A 0.8 m-diameter pool of ethanol is ignited at floor level in a room with a normal door opening. After a while the smoke layer stabilizes at a height of 1.5 m from the floor. Calculate the CO_2 concentration in the hot layer.

   Suggested answer: Using equations and tables from Chapter 3, we find $\dot{m}_f = 0.0075$ kg/s, $\dot{m}_p$ roughly 0.8 kg/s, f = 0.086, $y_{CO_2} = 1.77$ g/g (Table 9.2), $f = 0.0094$, and $Y_{CO_2} = 1.7\%$.

9.3 A pool of ethanol is ignited in the scenario described in Problem 9.2, but now the door is closed. There is a well-specified leakage opening at floor level. The upper layer is now much hotter and the mass loss rate is measured to be $\dot{m}_f = 0.05$ kg/s. The mass flow rate in through the leakage opening is measured to be 0.3 kg/s. Estimate the concentration of CO_2 and O_2 in the hot layer.

Suggested answer: The equivalence ratio is found to be $\phi = 1.51$. Table 9.3 gives $y_{O_2,max}$ = 2.09 and $y_{CO_2,max}$ = 1.91. Therefore y_{O_2} = 2.09/1.51 = 1.38 g/g and y_{CO_2} = 1.91/1.51 = 1.26 g/g. Fuel mixture fraction is $f = 0.05/(0.05 + 0.3) = 0.143$, so $Y_{CO_2} = 1.26 \cdot 0.143$ = 18.0% and $Y_{O_2} = 1.38 \cdot 0.143 = 19.7\%$.

10 Computer Modeling of Enclosure Fires

The recent emergence of performance-based regulations for fire safety and the increased complexity of building design have led to a dramatic increase in the use of computer modeling of smoke and heat movement in buildings. To apply these programs correctly, the user must sufficiently understand the basis of the models to assess the accuracy and validity of the results. This chapter will provide a general discussion on the types of computer models used in fire safety engineering design. The main part of the chapter will deal with the so-called two-zone models, since this is the model type most frequently used in fire safety engineering design for smoke and heat movement. A more general discussion will then be given on the so-called CFD models, which allow much more detailed modeling of enclosure fires at the cost of complexity and economy. A vast number of computer programs for this purpose exist, and many such programs are available at little or no cost on the Internet; a final section of the chapter gives a list of Internet addresses where programs are available.

CONTENTS

10.1 Introduction 255
    10.1.1 Probabilistic Models for Building Fire Safety 256
    10.1.2 Deterministic Models..257
10.2 Zone Models 258
    10.2.1 Conservation Equations 258
    10.2.2 Source Term Submodels 262
    10.2.3 Mass Transport Submodels..263
    10.2.4 Heat Transport Submodels...266
    10.2.5 Embedded Submodels and Unresolved Phenomena 267
    10.2.6 Limitations with Respect to Building Geometry 268
10.3 Computational Fluid Dynamics Models...269
    10.3.1 General on CFD Models for Fire Applications.....................................269
    10.3.2 Turbulence Submodels...270
    10.3.3 Radiation Submodels 271
    10.3.4 Combustion Submodels 274
10.4 Computer Program Resources on the Internet 275
References 277

10.1 INTRODUCTION

The use of computer models for simulating fires in enclosures has increased dramatically in recent years. This is due to many factors, including the increased complexity of building design, the recent emergence of performance-based regulations, the rapid progress made in the understanding of fire phenomena, and the advances made in computer technology.

Using computer modeling in fire safety engineering design is, however, not simple or easy, no matter how "user friendly" the computer program may be. The user must have a fundamental understanding of the physics and chemistry of enclosure fires in order to assess the validity and accuracy of the simulation results.

The previous chapters have attempted to give such an understanding, and in this chapter we describe how the physics and chemistry of fire are applied in computer models for simulating enclosure-fire environment. We shall be concentrating on a special branch of such models: deterministic models for predicting the movement of smoke and heat in enclosures. Two different modeling techniques are commonly used: the zone modeling technique and the computational fluid dynamics (CFD) technique.

Section 10.2 gives a general description of the zone modeling approach, and Section 10.3 briefly discusses the CFD approach. Finally, Section 10.4 gives a list of some of the computer programs that are in use and World Wide Web addresses from which the programs can be downloaded.

Before going into detailed discussions on deterministic models for predicting the movement of smoke and heat, we give a brief overview of computer models used for building fire safety. In general, such models can be said to be either of a deterministic or a probabilistic nature. In this section we briefly discuss the probabilistic type of models and deterministic models for building fire safety.

10.1.1 PROBABILISTIC MODELS FOR BUILDING FIRE SAFETY

Purely probabilistic models do not make direct use of the physical and chemical principles involved in fires, but make statistical predictions about the transition from one stage of fire growth to another. The course of a fire is described as a series of discrete stages that summarize the nature of the fire. Time-dependent probabilities are ascribed to the possibility of the fire changing from one stage to another. These are determined from knowledge of extensive experimental data and fire incident statistics.

Probabilistic models can be combined in various ways with deterministic models to form **hybrid models**. Such procedures regard fires as being deterministic once the fire is fully defined, but the inputs are assumed to follow probabilistic models. Thus, the inputs to the deterministic models are treated as random variables. This methodology has, for example, been used for risk assessments and for analysis of uncertainties in deterministic models.

Three basic forms of purely probabilistic models are commonly used in fire safety engineering: network, statistical, and simulation models. Each of these deals with the uncertainties associated with fire growth processes.

A **network model** is a graphic representation of paths, or routes, by which objects, energy, information, or logic may flow, or move, from one point to another. Decision trees are one type of network model, where each event is associated with a fork (branch of the tree) that describes two or more possible outcomes following an event. Event trees or fault trees are a second type of network models. Relationships of causative events are shown by the use of two basic symbols of logic gates—the AND gate and the OR gate. Decision trees only represent OR gates, so the event tree type of a model provides greater flexibility in describing a process.

Statistical modeling involves the description of random phenomena by an appropriate probability distribution, while network models usually ascribe a certain, single-value probability to an event. Such a probability distribution can be thought of as a mathematical function that defines the probability of an event. Some of the more sophisticated statistical models use principles of probability theory to combine probability distribution of two or more random variables. The assigned probability distributions can be based on historical data, engineering evaluations, or both.

The term **simulation models** is used to describe computer simulations where different sets of conditions are tried out a great number of times to see how they affect the outcome. The most common simulation procedure is named *Monte Carlo simulation*. Fire development could, for example, be described as a progression from one state to another, determined by parameters, variables, or processes of the fire area such as ignition, flame spread, fire load, ventilation, etc. The computer randomly allows the fire to progress from one defined state to another and carries out a great number of such simulations. The output can, for example, be used to estimate the relative importance of each of the fire parameters, variables, and processes.

The model types mentioned above are often combined with each other to form hybrid type models. Further information on probabilistic fire models is given by Beyler,[1] Watts,[2] and Berlin.[3]

10.1.2 DETERMINISTIC MODELS

Generally, deterministic models used in fire safety engineering design of buildings can be divided into a number of categories depending on the type of problem to be addressed. Some of the main problem categories are smoke and heat transport in enclosures, detector/sprinkler activation, evacuation of humans, and temperature profiles in structural elements.

Deterministic computer models for simulating the **transport of smoke and heat in enclosures** are normally either of the zonal type or the CFD type. These will be further described in Sections 10.2 and 10.3. Section 10.4 shows where a number of such programs can be downloaded from the Internet.

Special models for simulating the **activation of detectors and suppression devices** have been developed. The results from a zone model will assume some average temperature for the upper part of the enclosure and does not reflect accurately the conditions for devices at ceiling level. This has led to the development of special programs for ceiling jet calculations to predict the time at which a fire will be detected or sprinkler will be activated. The DETACT model is one of these.[4] Such ceiling jet algorithms have also been introduced into zone models. CFD models can also be used for these calculations, since the fluid flow is modeled in great detail and no specific ceiling jet calculations are necessary.

Models for simulating **evacuation of humans** from buildings can be probabilistic, deterministic, or both. The deterministic type of model will typically include little or no physics and will to a considerable extent be based on information collected from statistics and evacuation experiments. They are still termed deterministic, since a single set of input values will always yield the same output. Some deterministic evacuation models will only attempt to describe the movement of humans; others will attempt to link movement with behavior. All of these represent the enclosure as a network, which can either be fine or coarse. A *coarse network* will be made up of nodes and arcs, where the occupants are located at the nodes and travel along the arcs. The models using *fine networks* will allow a more accurate description of the geometry and can take account of internal obstacles. Examples of the former are EVACNET,[5] ERM,[6] and EXITT;[7] examples of the latter models are SIMULEX[8] and EXODUS.[9] Section 10.4 shows where some of these programs can be downloaded from the Internet.

A great number of computer codes have been developed for calculating the **thermal and mechanical response of building elements** exposed to fire. These are based on the governing heat transfer and solid mechanics equations, which are usually solved using the finite element or finite difference methods. An example of such computer programs, where heat transfer analysis is combined with mechanical response analysis of concrete structures, is given by Anderberg.[10]

Additionally, a very large number of commercially available computer programs are available for the design and evaluation of automatic sprinkler systems. These are usually based on hydraulic analysis programs, which are commonplace in engineering offices.

The above list shows only a part of the very wide range of problems that can be encountered in fire safety engineering and indicates that there is a multitude of computer programs that may be of assistance in the design process. Our focus is, however, on enclosure fires, and the remainder of the chapter will therefore concentrate on deterministic computer models for simulating enclosure fires.

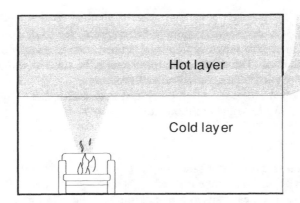

FIGURE 10.1   Two-zone modeling of a fire in an enclosure (reproduced from Chapter 1).

10.2   ZONE MODELS

The term "zone models" is used in fire safety engineering to identify a type of computer model for simulating the fire environment in an enclosure fire. Some such models consider only the fire room; others can apply the zone model technique in several rooms and thereby calculate the movement of smoke and heat through a building.

Most commonly, the zone model represents the system as two distinct compartment gas zones: an upper volume and a lower volume resulting from thermal stratification due to buoyancy. This type of model is therefore also termed a *two-zone model*. Conservation equations are applied to each zone and serve to embrace the various transport and combustion processes that apply. The fire is represented as a source of energy and mass, and manifests itself as a plume which acts as a "pump" of mass from the lower zone to the upper zone through a process called *entrainment*. Figure 10.1 (reproduced from Chapter 1) illustrates the basic two-zone model concept.

At this time numerous computer codes and software package exist based on the zone model approach. Friedman, in a recent survey, cited 21 zone models in use around the world.[11] This section will recap the basic conservation equations for the gas zones and list the various transport and combustion processes that make up the system. These processes would be referred to as the submodels of the system; they contribute subroutines to the computer codes for implementing a solution.

This section will therefore discuss

- conservation equations (referring to Chapters 8 and 9)
- source term submodels (referring to Chapters 3 and 9)
- mass transfer submodels (referring to Chapters 4 and 5)
- heat transfer submodels (referring to Chapter 7), and
- embedded submodels and unresolved phenomena

This section is mainly based on work previously written by one of the authors for the *SFPE Handbook* (Quintiere[12]).

10.2.1   CONSERVATION EQUATIONS

Control volume: The building block of the zone model is the conservation equations for the upper and lower gas zones. These are developed from the fundamental equations of energy and mass transport, either in control volume form as applied to the zones, or in the form of the differential equations representing the conservation laws, which are then integrated over the zones. Note that

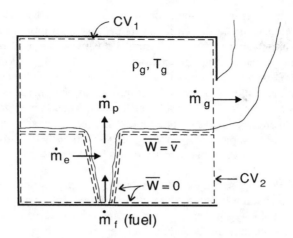

FIGURE 10.2 Control volumes selected in zone modeling. (Adapted from Quintiere[12].)

the conservation of momentum will not be explicitly applied, since information needed to compute velocities and pressures will come from assumptions and specific applications of momentum principles at vent boundaries of the compartment; these applications were discussed in Chapter 5.

Figure 10.2 illustrates a typical zone model representation of a compartment fire process. It shows a fire plume and a door vent. The hot combustion gases, which collect in the upper space of the room and spill out of the vent, constitute the "upper layer." A control volume (CV_1) is constructed to enclose the gas in this upper layer and the fire plume. The lower interface of the upper layer moves with the control volume such that no mass is transferred across this thermally stratified region. The velocity of the control volume along this interface, $\overline{w}$, is equal to the fluid velocity, $\overline{v}$.

The temperature of the upper layer is greater than that in the lower (zone) layer, which includes all the remaining gas in the room enclosed by a second control volume (CV_2). It has been assumed in zone modeling that the volume of the fire plume is small relative to the layer or gas zone volumes, and therefore its effect has been ignored. In general, multiple fire plumes can occur at any height in the room, and multiple vents or mass transport can take place between the zones (CV_1 and CV_2) and the surroundings. In each case these mass transports must be appropriately described in terms of the system variables; however, this may not always be easy or known.

Main assumptions: The properties of the upper and lower zones are assumed to be spatially uniform but can vary with time. Thus, temperature, T_g, and mass fraction of species i, Y_i, are properties associated with these ideal upper and lower homogeneous gas layers. Other assumptions in the application of the conservation laws to the zones are listed below:

1. The gas is treated as an ideal gas with a constant molecular weight and constant specific heats, c_p and c_v.
2. Exchange of mass at free boundaries is due to pressure differences or shear mixing effects. Generally these are caused by natural or forced convection, or by entrainment processes.
3. Combustion is treated as a source of mass and energy. No mechanism from first principles is included to resolve the extent of the combustion zone.
4. The plume instantly arrives at the ceiling. No attempt is made to account for the time required to transport mass vertically or horizontally in the compartment. Hence, transport times are not explicitly accounted for in zone modeling.
5. The mass or heat capacity of room contents is ignored compared to the enclosure structural wall, ceiling, and floor elements. Heat is considered lost to the structure, but

not to the contents. As contents shield boundary structural surfaces, some compensations can occur in the analysis, but for cluttered rooms this assumption may be poor.

6. The horizontal cross section of the enclosure is a constant area, A. In most all cases of zone modeling rectilinear compartments have been considered. However, this is not a necessary assumption and enclosures in which A varies with height can easily be handled.

7. The pressure in the enclosure is considered uniform in the energy equation, but hydrostatic variations account for pressure differences at free boundaries of the enclosure, i.e., $P \gg \rho g H$. In general the enclosure pressure, P, is much greater than the variations due to hydrostatics. For example, for $P = 1$ atm $= 101$ kPa (kN/m^2), the hydrostatic variation for a height, $H = 1$ m, gives a pressure difference of $\rho g H = 1.2$ kg/m$^3 \cdot 9.8$ m/s$^2 \cdot 1$m $= 10$ Pa (N/m^2).

8. Mass flow into the fire plume is due to turbulent entrainment. Entrainment is the process by which the surrounding gas flows into the fire plume as a result of buoyancy. Empirically, the inflow velocity linearly depends on the vertical velocity in the plume.

9. Fluid frictional effects at solid boundaries are ignored in the current models.

Conservation of mass: The conservation of mass for a control volume was earlier given by Eq. (8.1) as

$$\frac{dm}{dt} + \sum_{j=1}^{n} \dot{m}_j = 0 \qquad (8.1)$$

The equation states that the rate of change of mass in the volume plus the sum of the net mass flow rates out is zero for j flow streams. For the three flow streams shown in Figure 10.2 this would become

$$\sum_{j=1}^{3} \dot{m}_j = \dot{m}_g - \dot{m}_e - \dot{m}_f \qquad (10.1)$$

where $\dot{m}_g$ is the mass flow rate out of the door, $\dot{m}_e$ is the mass rate of entrainment into the fire plume, and $\dot{m}_f$ is the mass rate of gaseous fuel supplied. Note that $\dot{m}_e + \dot{m}_f = \dot{m}_p$ where $\dot{m}_p$ is the mass flow rate in the plume at the hot layer interface. The mass supply rate of fuel must in most cases be specified by the user of the zone model.

Conservation of species: The mass fraction of species i is given by Y_i. By using Eq. (8.1) and applying the conservation of mass for species i to a control volume, we arrive at Eq. (9.20), given as

$$m \frac{dY_i}{dt} + \underbrace{\sum_{j=1}^{n} \dot{m}_j \cdot \left(Y_{i,j} - Y_i \right)}_{\text{net out}} = y_i \dot{m}_f - \dot{m}_{i,\text{loss}} \qquad (9.20)$$

where m is the mass of the layer, $\dot{m}_{\text{react}}$ is the mass rate of gaseous fuel supplied, y_i is the mass yield of species i produced per mass rate of fuel supplied, and $\dot{m}_{i,\text{loss}}$ represents the losses due to surface deposition or settling of particulates.

The production term is written here as $y_i \dot{m}_f$, but the production is in reality linked to the mass rate of fuel reacted, $\dot{m}_{\text{react}}$. However, for most practical applications one must use experimental data to represent the production term and, experimentally, the yield of species i, y_i, is measured in terms of $\dot{m}_f$ and not $\dot{m}_{\text{react}}$.

The production term can, in principle, be described through a knowledge of the chemical equation of the reaction or its particular stoichiometry. Thus, stoichiometric coefficients can be used to represent the production of species and the consumption of oxygen in terms of the mass rate of fuel reacted.

Stoichiometry is, however, not easily determined, and the fuel gases as they emerge from the pyrolysis of solids can take many chemical forms that differ from the solid fuel's original molecular composition. This is why the mass production of species for fire is expressed in terms of the rate of mass loss for the pyrolyzing fuel, as was done in Chapter 9.

Hence, one must be careful to distinguish between the mass of fuel supplied and that reacted, and to relate available species yield data to the particular fire conditions of the application. That is, the yields or production rates may change with fire conditions, and therefore, in general, will not be consistent with data from small scale tests. For example, the production rate of CO changes markedly with air to fuel ratio.

Conservation of energy: We have earlier stated the conservation of energy through Eq. (8.21) as

$$\frac{d}{dt} \iiint_{CV} \rho u dV + \iint_{CS} \rho h v_n dS = \dot{Q} \tag{8.21}$$

By substituting $h - P/\rho$ for u in the first term, assuming quasi-steady state for the control volume (so that $dV/dt = 0$), and using the equation of state $P = \rho RT$, this can be rewritten as

$$V c_p \frac{dT_g}{dt} - V \frac{dP}{dt} + c_p \sum_{j=1}^{n} \dot{m}_j \left(T_j - T_g\right) = \dot{m}_{react} \Delta H_{eff} - \dot{q}_{loss} \tag{10.2}$$

where V is the volume of the control volume, P is the global pressure in the control volume, m_{react} is the rate at which fuel is reacted, ΔH_{eff} is the effective heat of combustion, and $\dot{q}_{loss}$ the rate of heat transfer lost at the boundary.

Usually in zone models it is assumed that all of the fuel supplied can react provided there is sufficient oxygen available. One assumption on the sufficiency of oxygen is to consider that all the fuel supplied is reacted as long as the oxygen concentration in that control volume is greater or equal to zero, i.e.,

$$\dot{m}_{react} = \dot{m}_f \qquad \text{if } Y_{ox} > 0. \tag{10.3}$$

Thereafter, an excess rate of fuel can exist to be transported into adjoining zones or control volumes where a decision must be made about whether it can continue to react. At the oxygen sufficiency condition, all of the net oxygen supplied to the control volume is reacted, so that as long as $Y_{ox} = 0$,

$$\dot{m}_{react} = r \cdot (\text{net mass rate of oxygen supplied}) \tag{10.4}$$

where r is the stoichiometric fuel to oxygen mass ratio.

This condition when $Y_{ox} = 0$ in compartment fires is termed the *ventilation-limited condition*. At this point, significant changes take place in the nature of the chemical reaction. Notably, incomplete combustion is more predominant for hydrocarbon fuels, leading to a significant increase in the yield of carbon monoxide and often soot. Thus, care must be used in interpreting the results of zone models once ventilation-limited conditions arise, particularly with respect to the prediction of species concentrations and the extent of burning. Material data used for well-ventilated conditions no longer apply. The whole issue of what constitutes a flammable mixture in a compartment gas layer and combustion in a vitiated layer has not yet been satisfactorily solved.

The first term on the left-hand side of Eq. (10.2) arises due to the change of internal energy with the control volume. If the temperature is not changing rapidly with time, this term can be small, and its elimination gives rise to a quasi-steady approximation for growing fires that allows a more simple analysis.

The second term arises from the rate of work done by pressure as the gas layer expands or contracts due to the motion of the thermal stratification interface. Having been rearranged, this term now is expressed as rate of pressure increase for the compartment that essentially can be caused by net heat or mass additions to the compartment gases. Except for rapid accumulation of mass or energy, or for compartments with small openings to the surroundings, this pressure rise is small and the pressure nominally remains at nearly the ambient pressure.

This was shown in Example 8.2, where an addition of 100 kW to a 72 m^3 gas volume in a room with a 0.1 m^2 vent area gave rise to roughly an increase of 10 Pa over normal ambient pressure of 101 kPa, which happens in less than a second. Any increase in pressure within the compartment could give rise to a flow of mass through a vent, and this term in Eq. (10.2) may be associated with "volumetric expansion." Conversely, a reduction in energy release rate will cause the pressure to drop relative to the ambient. This phenomenon, when cycling, explains the "breathing" effect for fires in closed buildings.

The third term of Eq. (10.2) accounts for the enthalpy flow rates and only applies to j flow streams that enter the control volume, since $T_j = T_g$ for all flow streams leaving, as long as the uniform temperature assumption still applies.

Summary: The zone model for the compartment fire system consists of two zones: the upper and lower gas layers. The solution process for the layer properties can be visualized by considering the conservation Eq. (8.1), (9.20), and (10.2) applied to each zone. The mass and energy equations comprise four equations that permit the determination of the two layer temperatures, one layer height (since the height of the other layer is directly found by difference from the total height of the compartment), and the compartment pressure (which was assumed uniform by assumption (7) given earlier). The densities are found from the ideal gas equation of state, in which approximately ρT is a constant.

To complete this solution process, each of the source or transport terms in the equations must be given in terms of the above layer properties or auxiliary relationships must be included for each new variable introduced. The source terms are associated with the $\dot{m}_{\text{react}}$ and y_i terms, and the transport terms include the j mass flow rates and the boundary heat transfer rates. The extent to which source and transport relationships are included reflect the sophistication and scope of the zone model. Some source and transport terms are essential to a basic zone model, others can be specified as approximations to reality, and others can be ignored when physically irrelevant. These source and transport relationships can be termed *submodels* and can comprise the subroutines of a zone model computer code. The nature of these submodels will now be discussed.

10.2.2 Source Term Submodels

The principal source term is the rate of fuel supplied, $\dot{m}_f$. In an experimental fire this can be known if the fire source is simulated by a gas burner. In the other extreme, the mass of fuel supply can be a result of a spreading fire over an array of different solid fuels. In general,

$$\dot{m}_f = \text{function of (fuel properties, heat transfer)} \tag{10.5}$$

in which the heat transfer to the fuel results from the flame configuration and the heated compartment. The fuel properties are still not completely defined or conventionally accepted for fire applications since no general theory exists for pyrolysis, and theories of flame spread and ignition are couched in terms of effective fire properties that are modeling parameters.

Nevertheless, data exist for fuel fire properties that can enable approximate models for $\dot{m}_f$ of reasonable accuracy. Chapter 3 describes how such approximations can be made from tabulated data or fuel properties for many solid and liquid fuels.

The rate of energy release, $\dot{m}_{react}\Delta H_{eff}$, required by Eq. (10.2) has already been discussed through Eq. (10.3) and (10.4). The point should be made that the heat of combustion employed must be with respect to the mass of fuel gases pyrolyzed, ΔH_{eff}, and is not the theoretical oxygen bomb value for the solid fuel, ΔH_c.

The production of species can be described in terms of species yields per mass loss rate of the fuel $\dot{m}_f$ instead of the mass rate of reacted fuel, $\dot{m}_{react}$. Such methods were described in Chapter 9, where y_i is defined as the mass of species i produced per mass supply of gaseous fuel.

For well-ventilated fires y_i may be reasonably constant for a given fuel, as given in Table 9.2. In general, it can vary with time and can significantly vary as ventilation-limited conditions are approached and achieved. In particular, y_i varies with the equivalence ratio, defined by Eq. (9.5) as

$$\phi = \frac{\text{mass of fuel available}}{\text{mass of oxygen available}}\Big/ r \tag{9.5}$$

where r is the stoichiometric value for complete combustion. Figures 9.7 and 9.9 show that y_i is relatively constant in the well-ventilated region ($\phi < 1$) but changes considerably in the under-ventilated region ($\phi > 1$).

The equivalence ratio may be computed in a zone (or upper layer) in which combustion has occurred by computing the mass concentrations of the available fuel and oxygen in the zone. This is done by Eq. (9.20) in which y_i is set equal to zero for both the fuel and oxygen, since this yields their availability, not their actual concentrations in the layer following combustion.

The generality of considering $y_i = y_i(\phi)$ for zone models is still under study, and its use must be considered as exploratory. Nevertheless, it currently offers the only practical zone model approach for estimating species, such as CO, under ventilation-limited conditions in compartment fires.

10.2.3 MASS TRANSPORT SUBMODELS

Entrainment: An essential feature of a zone model is the mass rate of entrainment relationship for the fire plume. This allows the principal mechanism for flow between the lower and upper stratified gas layers. Considerable work has been performed to develop entrainment relationships for pool fires or axisymmetric gas burner fires; much of this work was reported in Chapter 4.

Unfortunately, both the ideal theoretical plume models and correlations based on data vary widely, and no consensus exists among zone models in practice for the optimum pool fire entrainment model. Rockett has reported the variations in results he found using different fire entrainment models.[13] He found that the layer height, entrainment rate, and layer gas temperature varied by roughly a factor of two among the various models.

More useful data rather than ideal mathematical models are clearly needed to resolve this issue of accuracy for a simple pool fire. Yet even a perfect entrainment relationship for an axisymmetric pool fire would not necessarily be perfect in a zone model, because a plume in a compartment can be subject to nonsymmetric air flows that can bend the plume and thus affect its entrainment rate. Usually wind effects will increase the entrainment rate.

Rockett has shown that the effect of the entrainment model is crucial to predictions for the developing fire.[13] This suggests that the entrainment model must be representative of the actual object burning. However, no entrainment model exists for a wall, corner, or item of furniture; this dramatizes the lack of much-needed research in the area.

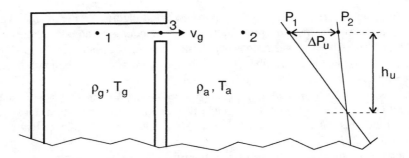

FIGURE 10.3   Flow through an opening (reproduced from Chapter 5).

Vent flows through openings in vertical partitions: The classical representation of fire in a room or building represents the structure with an opening such as a door or window to the ambient surroundings. Fire-induced flows through such openings have been discussed in detail in Chapter 5, and a widely accepted model exists to compute these flows based on the temperature distribution of the gases on either side of the opening.

The theoretical basis of the computation is orifice flow utilizing Bernoulli's equation along a streamline, as illustrated in Figure 10.3. The velocity at station 3, v_g, is given by Eq. (5.5):

$$v_g = \sqrt{\frac{2h_u(\rho_a - \rho_g)g}{\rho_g}} \tag{5.5}$$

where v_1 is assumed to be zero. The mass flow rate is computed by integration over a flow area of width W and height z using Eq. (5.14):

$$\dot{m} = C_d \int_0^z W\rho_g v(z)dz \tag{5.14}$$

where C_d is a flow coefficient. Emmons suggests that a value of 0.68 for C_d has an accuracy of +10% except at very low flow rates at the beginning at a fire.[14] In general, C_d will depend on the Reynolds number, as discussed in Chapter 5.

Figure 10.4 depicts two examples of vent flows through an opening in a vertical partition. In both cases the above equations apply, but the pressure distribution must be described appropriately. For example, in the pure natural convection case shown in Figure 10.4a, the pressure is determined by the static pressure with respect to the floor pressure, $P(0)$. Actually it is the floor pressure that applies in Eq. (10.2) and in the perfect gas equation of state.

The assumption is then that the flow velocities are small compared to the vent flow velocities to justify the static pressure computation. Thus, the vertical pressure distribution on either side of the opening is computed as McCaffrey and Rockett[15] illustrate the accuracy of the hydrostatic assumption in Figure 10.5. The sign of the pressure difference across the opening determines the flow direction.

Emmons presents the general equations that enable this computation to be included in a zone model.[14] It is by far the most accurate of the submodels, and provides the basis for linking rooms together in a zone model to enable smoke and fire growth computations for a large building.

Vent flows through openings in horizontal partitions: The flow through an opening in a horizontal partition can be considered in a manner similar to that for the vertical partition, provided the pressure difference is large enough. If there is only a single vent to the fire compartment through a horizontal partition, such as a ceilings, the flow must be oscillatory or bi-directional. The latter

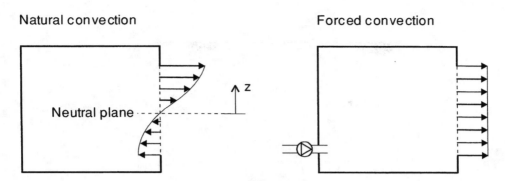

FIGURE 10.4 Typical vent flows. (Adapted from Quintiere[12].)

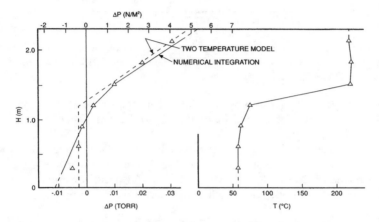

FIGURE 10.5 Vertical pressure difference across a room vertical partition compared to a computation based on room fire temperature distribution and a two-temperature zone model approximation using the hydrostatic pressure assumption. (From Quintiere[12].)

case implies a zero pressure difference with gravity solely determining the flow. Horizontal vent flows have not been treated in this textbook but a theory for this case has been developed by Epstein[16] and has been implemented by Cooper.[17]

Mixing between the layers: The primary exchange of fluid between the lower and upper gas layers is due to the buoyant effect of the fire plume. Secondary, but significant, mixing processes can occur due to the other effects. These are shown in Figure 10.6 and include

1. exchange due to a cold flow injected into the hot layer
2. exchange due to shear mixing associated with vent flows
3. exchange due to wall flows.

Phenomenon 1 is the inverse of the hot fire plume penetrating the upper layer. In both cases the fluid at the edge of the plume may not be buoyant enough to penetrate the respective layer. A related situation would be a cold forced jet introduced vertically into the lower layer. Depending on the relative temperatures, it may not escape the lower layer and penetrate into the upper layer. These are issues that can be resolved to some extent by work available in the literature on buoyant plumes and jets.

Phenomenon 2 has not been sufficiently studied, but data suggest that the mixed flow rate can be significant relative to the vent flow rate, especially for small vents.[18] A correlation for the existing rate has been developed from saltwater simulation experiments.[19]

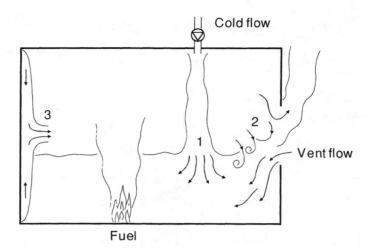

FIGURE 10.6 Secondary flows—mixing phenomena. (1) A cold plume descending from the upper layer into the lower layer; (2) shear mixing of an entering vent flow stream; (3) wall flows due to local buoyancy effects. (Adapted from Quintiere[12]. With permission.)

Phenomenon 3 has been discussed by Jaluria.[20] He presents relationships that would allow the estimation of the rate of transfer of cold fluid adjacent to the wall in the hot upper gas layer into the cold lower gas layer or vice-versa.

All of these flows tend to blur the sharp distinction between the upper and lower gas layers reducing their degree of stratification. Obviously, if sufficient mixing occurs, the layer may appear to become well-mixed or destratified. However, this should occur naturally in the context of the zone model and one should not have to switch to a well-mixed compartment model under these conditions.

Relationships for all of these secondary flows have not been developed with confidence or with full acceptance. Although they are important for improving the accuracy of a zone fire model, little work has been undertaken to establish their sound basis.

Forced flow effects: The effect of forced air flow on the fire conditions and smoke spread due to mechanical or natural wind forces has always been an issue in large building fires. Wind effects and the resultant pressure distribution around a tall building have become standard elements of design data for structural design, but have not been utilized for fire safety design. The motion of smoke through a building due to the mechanical ventilation system has been simulated by network models that treat the compartment volume as uniform in properties and the pressure losses due to vents and duct friction.

In order to link the mechanical ventilation system in a building to a two-zone model, one must include the full pressure-flow characteristics of the fires in both directions, to allow for the possibility of the back flow of smoke against the direction of air flow in the ducts. An attempt at this linkage has been put forth by Klote and Cooper,[21] who hypothesize a fan characteristic relationship. Ultimately, an experimental study will be needed to lay a foundation for this analysis.

10.2.4 HEAT TRANSPORT SUBMODELS

Convective heat transfer to surfaces: The $\dot{q}_{loss}$ term in Eq. (10.2) is composed of the convective and radiative heat loss to the boundary surfaces of the layer control volumes. This involves both heat transfer from the gas layers at their bulk temperatures and the heat transfer from the flame. Consistent treatment of the flame and layer gas heat transfer must be carried out for the zone model. If the flame becomes large and fills the upper layer, we cannot count the flame and gas heat transfer without being redundant.

Convective heat transfer to a ceiling by a fire plume has been widely studied at modest scales such that flame radiation may have been insignificant. Alpert specifically examined only convective heating,[22] in contrast to studies by You and Faeth[23] and Kokkala,[24] who included flame effects.

In general, convective effects will vary along the ceiling, walls, and floor, and depend on the nature and position of the fire. In some cases an "adiabatic wall temperature" has been appropriately introduced since, the driving force for convective heat transfer locally is not the bulk gas layer temperature but the local boundary layer temperature, which is not explicitly computed. Convective heat transfer data for the walls and floor of a fire compartment or for rooms beyond the fire compartment have not been developed. Hence, most zone models use estimates from natural convection correlations.

Radiative heat transfer: The theory of radiative heat transfer is adequate to develop the needed components for the zone model. However, the theory is not sufficiently developed to predict flame radiation from first principles without very sophisticated modeling of the soot and temperature distributions. Hence, flame radiation is relegated to empirical practices. Radiation from a smoke layer is easier to deal with within the context of a uniform property gas layer for the zone model. One unresolved difficulty is the availability of property data to determine the contribution of smoke particulates to the layer radiation properties. The discussion presented in Chapter 7 can be used to begin a development of the radiative equations needed by the zone model.

Conduction heat transfer: The radiative and convective heat transfer from the gas must be balanced by conduction heat transfer through the boundary surfaces. This requires a numerical solution to a partial differential equation in conjunction with the ordinary differential equations in time describing the conservation of energy and mass for the gas layers. Usually zone models have considered only one-dimensional conduction, which should be adequate for most applications. Most multiple compartment models do not consider communication by conduction into the next compartment, instead treating the structural elements as thermally thick. In principle, there is no difficulty with developing on accurate algorithm for conduction through the boundary elements for any conditions. For more information, the reader is referred to Rockett and Milke.[25]

10.2.5 EMBEDDED SUBMODELS AND UNRESOLVED PHENOMENA

The detailed physics that one can include in a zone model is limited only by our knowledge and imagination. The zone model can be very versatile in accommodating new physics even if it appears inconsistent with the uniform property layer assumption. By analogy to the relationship between inviscid flow and boundary layer flow in the analysis of aerodynamic bodies, the layer properties can be regarded as first-order approximation for higher-order analysis. Flame and boundary layer phenomena within the compartment can be computed by regarding the layer properties as infinite reservoirs. These phenomena can be computed after the primary layer properties are computed. Examples of embedded phenomena are shown in Figure 10.7.

Although the combustion region is assumed to be of negligible volume at the zone model for mutation, the flame height can be computed along with the velocity and temperature distributions in an axisymmetric fire plume (see Chapter 4). Other embedded phenomena can include the ceiling jet, the computation of temperature distributions over the ceiling, the deposition of soot and other products of combustion to surfaces, and the heating degradation of structural elements.

Some significant phenomena are absent from consideration by the zone modeling approach for fire. These include (1) vent flames; (2) transient flow in corridors; and (3) shaft flows (see Figure 10.8).

These phenomena require more research, and new strategies to enable them to be included into a zone model. Vent flames are significant for fire growth into the next space and usually follow flashover. Information about their heat transfer and extent need to be computed. Transient corridor flows are important in the analysis of smoke transport along long corridors. The current zone model methodology would yield an instantaneous layer that would descend, but the actual process produces a transient ceiling jet. Flows up vertical shafts involve the interaction of plumes with walls, pressure-driven effects, and turbulent mixing.

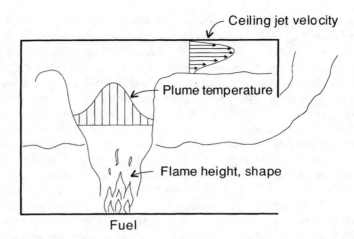

FIGURE 10.7   Examples of embedded phenomena. (Adapted from Quintiere[12]. With permission.)

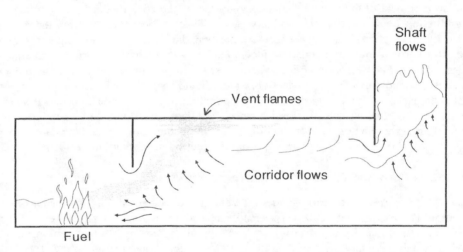

FIGURE 10.8   Examples of significant phenomena absent from zone models. (Adapted from Quintiere[12]. With permission.)

10.2.6   LIMITATIONS WITH RESPECT TO BUILDING GEOMETRY

The two-zone models have been constructed with the purpose of treating a fire in a single enclosure or a series of connected enclosures whose sizes are representative of domestic rooms, offices, or small industrial units. Simulations show a good agreement with experiments carried out in such enclosures, for the fire compartment and the nearest rooms.

The zone modeling technique may not be suitable for some other geometries, such as smoke spread in rooms with a large length-to-width ratio or rooms where the horizontal length to vertical length ratio is very large or very small. The models are, however, often applied to such geometries, and the user must be acutely aware of the modeling limitations. A brief discussion on these effects for a number of typical geometries is given below.

A very weak fire in a large space will not necessarily result in a two-zone situation. A weak plume may not be able to drive the fire gases to the ceiling, and a layer may be formed at mid-height of the building and not under the ceiling. The ceiling ventilation may therefore not be effective in venting out the smoke. The plume in the two-zone model will, however, instantaneously send the smoke to the ceiling, where it will instantaneously spread across the ceiling. The model will take into account the ceiling ventilation and thereby predict a less hazardous situation than might occur for this scenario in a real fire situation. In order to use the simulation results for design purposes, the user must be aware of this limitation and be able to justify the design.

Similarly, a relatively large fire in a small enclosure will not necessarily lead to a two-zone situation. The powerful turbulence will disturb the zones, and a "well-mixed" situation may result.

A very large fire in an enclosure with a relatively low ceiling can result in a situation where the flame reaches the ceiling and stretches horizontally under it. The plume model predictions will to a considerable extent be applicable to this scenario, since the entrainment occurs below the hot gas layer. However, the zone model will assume that the smoke layer forms instantaneously under the entire ceiling and that cooling of the smoke due to contact with the ceiling will begin at that time. This may result in a slightly too cold and thin layer of smoke, whereas in reality the smoke will form a "pillow" at ceiling level above the fire, and may spread across the entire ceiling at a relatively slow rate.

If a sprinkler is activated, the zone model is no longer valid. The sprinkler flow will cool the smoke and mix it so that the two-zone analogy is no longer valid.

The above are only a few examples of geometries where the zone model limitations must be seriously considered. In most design situations the user must simplify a relatively complex building geometry to allow input to the zone models. Knowledge of how this should be done can be achieved only by testing the model for many scenarios, bearing in mind the assumptions discussed earlier in this section.

Due to the limitations, comparative runs must be made for each scenario, and the results must be examined carefully. The models can only be seen as one of many engineering tools for examining different design options. The design must, as always, be based on engineering estimates, practical experience, and common sense.

10.3 COMPUTATIONAL FLUID DYNAMICS MODELS

The most sophisticated deterministic models for simulating enclosure fires are termed "field models" or "CFD models" (computational fluid dynamics models). In this section we give a general description of CFD models and discuss the sub-models that are especially important for application to fire problems, i.e., submodels for turbulence, heat transfer, and combustion.

10.3.1 GENERAL ON CFD MODELS FOR FIRE APPLICATIONS

The CFD modeling technique is used in a wide range of engineering disciplines and is based on a complete, time-dependent, three-dimensional solution of the fundamental conservation laws. The volume under consideration is therefore divided into a very large number of subvolumes and the basic laws of mass, momentum, and energy conservation are applied to each of these. Figure 10.9 (from Chapter 1) shows a schematic of how this may be done for a fire in an enclosure.

The governing conservation equations for mass, energy, and momentum contain as further unknowns the viscous stress components in the fluid flow. Substitution of these into the momentum equation yields the so-called Navier–Stokes equations, and the solution of these is central to any CFD code.

Several CFD codes, developed for use in a wide range of engineering disciplines, are in existence, and many are available commercially. A CFD code consists of

- A preprocessor. Here the geometry of the region of interest is defined, the grid is generated, the physical and chemical phenomena that need to be modeled are selected, fluid properties are defined, and boundary conditions are specified.
- A solver. Here the unknown flow variables for a new time step are approximated. The approximations are discretized by substitution into the governing flow equations and the algebraic equations are solved.

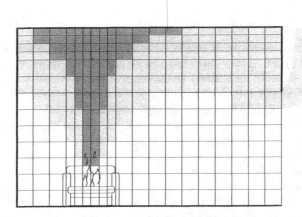

FIGURE 10.9 Computational fluid dynamics models divide the enclosure into a large number of subvolumes (reproduced from Chapter 1).

- A postprocessor. This allows a display of both input and output data in various forms (grid display, vector plots, contour plots, 2D and 3D surface plots, particle tracking, etc.). A considerable number of commercially available software packages are commonly used for this purpose.

Engineering applications of CFD involve not only fluid flow and heat transfer, but can also involve combustion, phase change, multiphase flow, and chemical reactions, to name a few processes. Examples of such complex flow systems are furnaces, internal combustion engines, heat exchangers, etc. Adequate physical models that are appropriate to the problem under consideration must be incorporated into a CFD code if it is to be used successfully.

The very wide range of engineering problems that can be addressed by CFD models is such that no single CFD code can incorporate all of the physical and chemical processes of importance. There exist, therefore, only a handful of CFD codes that can be used for problems involving fire. These, in turn, use a number of different approaches to the subprocesses that need to be modeled. Some of the most important of these subprocesses are

- Turbulence modeling
- Radiation and soot modeling
- Combustion modeling

CFD models for fire are currently used mainly in fire research; their use in fire safety engineering design is limited due to the expert knowledge required for the processes listed above. To set up the problem, execute it, and extract the output can also be very time consuming, but these problems are currently being addressed by the development of pre- and postprocessors for a number of CFD codes for fire.

As a result, the core curriculum in fire safety engineering would usually not include an in-depth study of the physics and chemistry used in CFD modeling of fires in enclosures. Nevertheless, CFD modeling may in some cases be the only way to proceed with certain design problems.

The following sections will therefore not explicitly state the equations used for the subprocesses listed above, but we shall give a brief description of these. The interested reader is referred to Cox[26] and Stroup[27] for a somewhat more detailed description.

10.3.2 TURBULENCE SUBMODELS

All flows encountered in engineering practice become unstable above a certain Reynolds number and are said to be turbulent. The velocity fluctuations associated with turbulence give rise to additional stresses on the fluid, the so-called *Reynolds stresses*, described by the Reynolds equations. Furthermore, visualizations of turbulent flows reveal rotational flow structures, the so-called *turbulent eddies*, with a wide range of length scales. High Reynolds number turbulent flows may contain eddies with length scales down to 10^{-6} m. Additionally, the fluctuations can occur very fast and can have a frequency in the order of 10 kHz.

A direct solution of the time-dependent Navier–Stokes equations of fully turbulent flows at high Reynolds numbers therefore requires extremely fine geometric grids and extremely small time steps. The computing requirements for the direct solution are thus truly phenomenal and must await major developments in computer hardware.

Certain assumptions must therefore be made to avoid the need to predict the effects of each and every eddy in the flow. Several such turbulence modeling approaches have been used and depend mostly on the type of engineering problem to be solved.

The **k-ε model** is based on the time-averaged Reynolds equations. Two transport equations (partial differential equations or PDEs) are solved, one for the turbulent kinetic energy, k, and another for the rate of dissipation of turbulent kinetic energy, ε. The equations used contain a number of empirical constants, determined from experimental results. A number of variations of the k-ε model exist; the so-called standard k-ε model is widely used in CFD codes. One of the main drawbacks of this model is that the eddy viscosity is assumed to be identical for all the Reynolds stresses, so that the turbulence has no preferred direction.

The mass of air entrained into a fire plume controls to a considerable degree the process of smoke filling, the concentrations and temperatures in the hot layer, and the combustion in the flame. Since gravitational forces apply only in the vertical direction, the standard k-ε model does not model plume entrainment correctly. This has been partly amended by using a k-ε model with buoyancy modifications. Tuovinen reported that even this type of k-ε model predicts a plume that does not widen with height, as it does in reality.[28] By further changing of the constants in the Reynolds stress equation, the plume properties can be made to fit experimental data better, but this has adverse effects in other regions, such as in the ceiling jet.

Further work is clearly needed on the turbulence models used in CFD codes for fire applications. Any progress in such modeling must be based on relevant experimental data, for a wide range of flow situations.

Another common way of modeling turbulence is termed **large eddy simulation,** where the time-dependent flow equations are solved, not only for the mean flow but also for the largest eddies, and some effects of the smaller eddies are taken account of. This technique has been used to model outdoor plumes with no combustion, but simulations of fire phenomena in a compartment are scarce.

The implementation of a large eddy simulation turbulence model into a CFD code is a cumbersome and costly operation, since a number of other subprocesses must be completely remodeled and new solvers may be required. Large eddy simulations are very costly with regard to computing time and are presently not considered economical for practical simulations in fire safety engineering. But this may, in the future, be the best alternative to k-ε modeling for fire applications.

10.3.3 Radiation Submodels

The radiative transfer equation is an integro-differential equation, and its solution even for a one-dimensional, planar, gray medium is quite difficult. In fires, the multidimensional combustion system consists of highly nonisothermal and nonhomogeneous medium where spectral variation of the radiative properties of the medium must be accounted for. It is therefore necessary to introduce some simplifying assumptions and strike a compromise between accuracy and computational effort.

The problem is usually divided into two parts: first, an appropriate solution method must be chosen for the integro-differential equation, and second, assumptions must be made on the radiative properties of the medium (combustion gases and particles).

Solution methods: The methods for solving the integro-differential radiative heat transfer equation can be divided into the following categories:

- exact models
- statistical methods
- zonal methods
- flux methods
- hybrid methods

The most desirable solution of any equation is its **exact** closed-form solution. The exact solution of the radiative transfer equation can, however, only be obtained after some simplifying assumptions, such as uniform radiative properties of the medium and homogeneous boundary conditions. Considering that in most engineering systems the medium is nonhomogeneous and radiative properties are spectral, it can be concluded that the exact solutions are not practical for engineering applications. Nevertheless, exact solutions for simple geometries and systems are needed, as they can serve as benchmarks against which the accuracy of other approximate solutions can be checked.

A **statistical method,** such as the Monte Carlo method, consists of simulating a finite number of photon histories through the use of a random number generator. For each photon, random numbers are generated and used to sample appropriate probability distributions for scattering angles and pathlengths between collisions. The Monte Carlo calculations yield answers that fluctuate around the "real" answer and, as the number of photons initiated from each surface and volume element increases, this method converges to the exact solution. For practical, time-dependent problems, however, this method requires immense computational power.

In the **zonal method** the volume of the enclosure is divided into a number of zones, each assumed to have a uniform distribution of temperature and properties. Then the direct exchange areas between the surface and volume elements are evaluated and the total exchange areas are determined using matrix inversion techniques. There is some difficulty in adopting the zonal method for problems having complicated geometries, since numerous exchange factors between the zones must be evaluated and stored in memory. This method is therefore very expensive in terms of storage capacity and CPU time. Otherwise, this method is considered to be attractive for practical engineering calculations.

The **flux methods** simplify the problem by assuming that the radiation intensity is uniform on given intervals of the solid angle. Then, the integro-differential equation can be reduced to a series of coupled linear differential equations in terms of average radiation intensities, or fluxes. By changing the number of solid angles over which radiative intensity is assumed constant, one can obtain different flux methods, such as two-flux, four-flux, or six-flux methods. It is also possible to use nonuniform solid angle divisions, which results in the so-called discrete ordinates approximation to the radiative transport equation. Cox states that the flux methods are acceptable when determining, for example, the radiant transfer from a hot ceiling layer to the floor of an enclosure, while these methods are unsuitable in the vicinity of the source.[26]

The **hybrid methods,** as the name suggests, consist of a combination of the methods described above. Most of the above methods have their flaws. In order to take advantage of the desirable features of different models, various combinations of the models have been proposed and used. The Monte Carlo method suffers from the extensive computational time required for the calculations. If the direction of each ray is given deterministically rather then statistically and if all the directions constitute an orthogonal set, then the solution would be less time-consuming and the accuracy would increase with the increase in the number of directions. With this in mind, Lockwood and Shah proposed a **discrete transfer model** that combines the virtues of the zonal, Monte Carlo, and discrete ordinates methods.[29] They showed that very accurate results could be obtained with this method by increasing the number of ray directions.

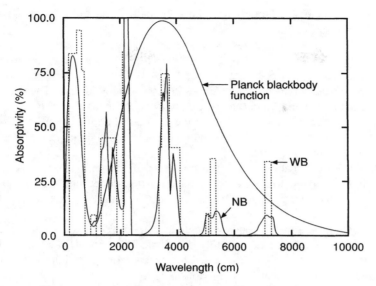

FIGURE 10.10 Spectral absorptivities of an H_2O–CO_2 air mixture at a temperature of 1000K, pressure 1 atm, and path length 1 m. NB stands for narrow band, WB for wide band. The normalized Planck blackbody function is plotted to show the total radiation absorbed by the medium.

The discrete transfer model has been the most commonly used method for solving the integro-differential radiative transfer equation when dealing with fire problems. The method seems to combine the advantages of accuracy and computational economy required for the consideration of engineering problems involving fires in compartments.

Radiative properties: Radiative properties of combustion systems are a complicated function of wavelength, temperature, pressure, composition, and path length. The products of combustion usually consist of combustion gases such as H_2O, CO_2, CO, etc., and particles such as soot. The radiative properties of the gases and particles are often treated separately.

The combustion gases are strong absorbers and emitters of radiant energy, but these radiative properties are a strong function of wavelength. Consequently, the variation of the radiative properties with the electromagnetic spectrum must be accounted for. Spectral calculations are performed by dividing the entire wavelength (or frequency) spectrum into several bands and assuming that the absorption/emission characteristics of each species remain either uniform or change smoothly over these bands. As one might expect, the accuracy of the predictions increases as the width of these bands become narrower.

A number of approaches to solve this problem have been suggested, among them

- total absorptivity–emissivity models
- wide-band models
- narrow-band models

As an example, Figure 10.10 shows the spectral absorptivities of a H_2O–CO_2 air mixture. The two curves NB and WB show the difference in applying the narrow-band and wide-band models, respectively. A third curve shows the normalized Planck blackbody function, to show the relative contribution of each gas band to the total radiation absorbed by the medium.

The crudest, but most economical, of the above-mentioned model types are the **total absorptivity–emissivity models,** and these are currently the most commonly used in CFD models for fire applications. Here, the spectral, or band, absorptivities for each species are first integrated over the entire spectrum for a given temperature and pressure to obtain total absorptivity and emissivity curves. Then, appropriate polynomials are curve-fitted to these families of curves at different temperatures and pressures using regression techniques.

Sometimes, these curve-fitted expressions can be arranged so that they present the sum of total emissivity or absorptivity of the clear or gray gases. These are known as the "weighted sum-of-gray-gases" models.

Other models give the curve-fitted expressions in terms of polynomials. The one most frequently used for fire applications is the Modak model. One of the advantages of the Modak model is that it accounts for the soot contribution along with the gas contribution. The model assumes the total pressure of 1 atm and assumes that the gas radiation occurs along a homogeneous path, i.e., uniform temperature and/or uniform pressure.

Meanwhile, in fire applications, the steep gradients of temperature and composition concentrations cause significant nonhomogeneity of the radiation properties. The applicability of the Modak model in nonisothermal, and nonhomogeneous combustion system can therefore been questioned.

As a result, the **wide-band models** have been developed, taking advantage of the fact that the infrared radiation from each species is mainly contributed by several wide-band spectral regions, as shown in Figure 10.10. These can be applied to nonisothermal, nonhomogeneous medium, but the more detailed spectral information is not considered.

Another alternative is the **narrow-band model,** which is based on spectrum integration and expected to be the most applicable engineering method to calculate radiation for nonisothermal, nonhomogeneous radiating paths. Molecular models and tabulated spectral data are used to calculate the spectral absorption coefficients of the combustion products. This relatively more detailed spectral calculation offers good accuracy and generality (see Figure 10.10) but at a high computational cost. This method is therefore not currently used for engineering applications in fire, but Yan and Holmstedt have developed and implemented a relatively fast narrow band model in a CFD code for research purposes.[30] Developments in computer technology may allow wider use of this technique in the future.

For an excellent review of radiation heat transfer in combustion systems, the interested reader is referred to Viskanta and Menguc.[31]

10.3.4 COMBUSTION SUBMODELS

The mechanism by which species are formed and destroyed in fire is extremely complex and involves chemical and physical processes on a molecular and a macroscopic level. Ignition, combustion, and extinction occur, all at the same time, within the microstructure of the turbulent flame. These events occur at high frequencies with spatial separation of only a few millimeters. The mixture of gases can be diluted by complete or incomplete products of combustion at a given location. Thousands of different states can thus exist at different points within the flame, at a given time.

In order to avoid these complications, one can give the heat release rate in certain control volumes as user input, and therefore not deal with combustion at all. But for fire applications it is important to allow the process of fuel and air mixing, so that the heat release rate and the position of the flame be determined by the actual flow conditions. This also allows the prediction of species concentration and estimation of soot concentrations, which has important significance for the radiation calculations.

In combustion studies, the eddy dissipation concept (EDC) model and the laminar flamelet model are the two most widely used combustion models.

CFD codes for fire applications most often use the **eddy dissipation concept** model, developed Magnussen and Hjertager,[32] due to its simplicity and reasonably good correlation with measured data. The model assumes a single, one-step reaction and infinitely fast reactions.

The effects of finite rate chemical kinetics must, however, be taken into account for a realistic prediction of partially oxidized species such as CO and soot. One of the most promising approaches for predicting the effects of finite rate kinetics in fires is the **laminar flamelet** model (Peters[33]). In studies of combustion in internal combustion engines, jet engines, etc., this model is often used. This approach assumes that the combustion occurs locally in thin laminar flamelets embedded within the turbulent flow field. For simple fuels, such as methane or propane, for which the chemistry is sufficiently well known, the relationships between the instantaneous species composition and mixture fraction can be computed directly.

This requires that laminar flamelet libraries be established from experiments, where the state relationships of species concentrations, temperature, enthalpy, viscosity, density, and soot concentration are stored as a function of mixture fraction. Such libraries exist only for a number of simple fuels, but the number of such libraries are steadily growing.

The interested reader is referred to the excellent discussion by Cox,[26] where combustion submodels for CFD are summarized.

10.4   COMPUTER PROGRAM RESOURCES ON THE INTERNET

This section gives some examples of computer programs that are available on the Internet. Most of the programs listed are available at no cost and can be downloaded directly. Some models require registration and a small handling fee; the program is then sent by post. A few programs are commercially available at a relatively high price.

Since Internet web sites have a tendency to change URL addresses, this list is by no means complete or up-to-date. Some of the addresses quoted below are, however, kept updated on http://www.brand.lth.se. This is the Web site of the Department of Fire Safety Engineering at Lund University, and links to many of the sites quoted below are given there.

We divide the programs into three types: CFD models, two-zone models, and special-purpose models. Finally, we point to locations where some evacuation models can be found.

CFD models

Program name: SOFIE (Simulation of Fires in Enclosures)
Description: A CFD program developed within, and accessible to, the fire science community. Contains a multitude of submodels specially developed for fire applications. Currently not very user-friendly, requires extensive training.
Institution: University of Cranfield
Availability: A small license fee is required before the program can be downloaded; only available for noncommercial purposes.
URL address: http://www.cranfield.ac.uk/sme/sofie/

Program name: SMARTFIRE
Description: A CFD model developed by the Fire Safety Engineering group at the University of Greenwich with U.K. Home Office collaboration. The program is user-friendly; several add-on packages are available for input, output, and other purposes. Some general knowledge on the CFD technique is required, and courses for this purpose are given at the University of Greenwich.
Institution: Fire Safety Engineering group, University of Greenwich
Availability: At a relatively high cost for commercial purposes, not directly downloadable.
URL address: http://fseg.gre.ac.uk/fire/lsmrt.html

Program name: JASMINE
Description: A CFD program developed especially for fire applications. Therefore includes many submodels that are important for fire simulations. User-friendly preprocessor is available.
Institution: Fire Research Station
Availability: At a relatively high cost, not directly downloadable.
URL address: http://www.bre.co.uk/frs/frs2_1.html

Two-zone models

Program: HAZARD I
Description: The HAZARD I program package contains several computer programs, some of which are also available as stand-alone programs. This multicomponent fire-hazard calculation method was developed by the National Institute of Standards and Technology's Building and Fire Research Laboratory in the USA. HAZARD I calculates the development of fire effects (thermal, toxic gases, smoke) as a function of time and the reactions and movements of occupants. It also calculates detector activation times. Among the programs in this package are the two-zone model CFAST, the input program CEdit, and the output program CPlot (see below for download information on these stand-alone programs).
Institution: NFPA, the National Fire Protection Association
Availability: A small fee is required; the program is mailed by post on CD.
URL address: http://www.nfpa.org/datamod.html, orders through OSDS@NFPA.org.

Program: CFAST
Description: CFAST is probably the most popular two-zone model in use. It can simulate smoke movement in up to 18 connected rooms, and includes submodels for many enclosure fire phenomena.
Institution: Building and Fire Research Laboratory, NIST
Availability: Free, can be directly downloaded.
URL address: http://www.bfrl.nist.gov/864/hazard/cfasthom.html

Program: FASTlite
Description: A user-friendly software package that builds on the core routines of an earlier produced program package (FPEtool) and the computer model CFAST. The two-zone model is limited to treating three rooms, but very many other calculational tools are available in this package.
Institution: Building and Fire Research Laboratory, NIST
Availability: Free, can be directly downloaded.
URL address: http://flame.cfr.nist.gov/fire/fastlite.html

Special-purpose programs

The Building and Fire Research Laboratory at the National Institute of Standards and Technology (NIST) has made a large number of programs freely available. Some of the programs are for special purpose (for example prediction of glass breakage) and other programs are a collection of several useful calculation routines.

The programs are made for a DOS environment, but can be run under the DOS prompt in Windows 95. Some of these programs are listed below. The programs can be directly downloaded from http://www.bfrl.nist.gov/864/fmabbs.html. Further information is available at that URL address.

ALOFT-FTTM - A Large Outdoor Fire plume Trajectory model—Flat Terrain
ASCOS - Analysis of Smoke Control Systems
ASET-B - Available Safe Egress Time—BASIC
ASMET - Atria Smoke Management Engineering Tools
BREAK1 - Berkeley Algorithm for Breaking Window Glass in a Compartment Fire
CCFM - Consolidated Compartment Fire Model version VENTS
CFAST - Consolidated Fire and Smoke Transport Model
DETACT-QS - Detector Actuation—Quasi Steady
DETACT-T2 - Detector Actuation—Time squared
ELVAC - Elevator Evacuation
FIRDEMND - Handheld Hosestream Suppression Model
FIRST - FIRe Simulation Technique

FPETool - Fire Protection Engineering Tools (equations and fire scenarios)
LAVENT - Response of sprinkles in enclosure fires with curtains and ceiling vents

Evacuation models

Program: SIMULEX
Description: A Windows program for modeling evacuation. The program can simulate evacuation of geometrically complex buildings, with many occupants. It allows input of CAD drawings for a plan of the building. The program is user-friendly compared to other evacuation models.
Availability: Free, can be directly downloaded.
URL address: http://www.brand.lth.se/dator/utr.htm

Program: EXODUS
Description: A sophisticated evacuation model, taking account of people–people, people–fire and people–structure interactions. It allows evacuation simulations of complex buildings with many occupants and is user-friendly.
Availability: At a relatively high cost for commercial purposes, not directly downloadable.
URL address: http://fseg.gre.ac.uk/exodus/

Program: EVACNET+
Description: A classical "network" type of model. The user defines a system of nodes and arcs, where the occupants are positioned at the nodes and move along the arcs towards the exit. Creating the input file is relatively time-consuming.
Availability: Free, can be directly downloaded.
URL address: http://www.brand.lth.se/dator/utr.htm

Program: ERM
Description: The Escape and Rescue (ERM) model is based on the same node and arc method as EVACNET+, but is developed especially for hospitals and healthcare facilities. The program allows the user to specify occupants of various degrees of mobility.
Availability: Free, can be directly downloaded.
URL address: http://www.brand.lth.se/dator/utr.htm

REFERENCES

1. Beyler, C., "Introduction to Fire Modeling," *Fire Protection Handbook*, National Fire Protection Association, Quincy, MA, 1991.
2. Watts, J.M., "Probabilistic Fire Models," *Fire Protection Handbook*, National Fire Protection Association, Quincy, MA, 1991.
3. Berlin, G.N., "Probability Models in Fire Protection Engineering," *SFPE Handbook of Fire Protection Engineering*, SFPE, Quincy, MA, 1988.
4. Evans, D.D. and Stroup, D.W., "Methods to Calculate the Response Time of Heat and Smoke Detectors Installed below Large Unobstructed Ceilings," *Fire Tech.*, Vol. 22, No. 1, 1986.
5. Kisko, T.M. and Francis, R.L., "EVACNET+: A Computer Program to Determine Optimal Building Evacuation Plans," *Fire Safety Journal*, Vol. 9, 1985.
6. Alvord, D.M., "Status Report on the Escape and Rescue Model and the Fire Emergency Evacuation Simulation for Multifamily Buildings," Report NBS-GCR-85-496, National Bureau of Standards, Gaithersburg, MD, 1995.
7. Levin, B., "EXITT—An Evacuation Model of Occupant Decisions and Actions in Residential Fires," *Proc. of the Second International Symposium on Fire Safety Science*, Hemisphere Publishing, Washington, D.C., 1989.
8. Thompson, P. and Marchant E., "A Computer Model for the Evacuation of Large Building Populations," *Fire Safety Journal*, Vol. 24, No. 2, 1995.

9. Owen, M., Galea, E.R., and Lawrence, P., "Advanced Occupant Behavioural Features of the Building—EXODUS Evacuation Model," *Fire Safety Science—Proc. of the Fifth International Symposium*, International Association for Fire Safety Science, 1997.

10. Anderberg, Y., "User's Manual for TCD 3.0 with SUPER-TEMPCALC," Fire Safety Design, Lund, Sweden, 1990.

11. Friedman, R., "Survey of Computer Models for Fire and Smoke," 2nd ed., Factory Mutual Research Corp., Norwood, MA, 1991.

12. Quintiere, J.G., "Compartment Fire Modeling," *SFPE Handbook of Fire Protection Engineering*, 2nd ed., National Fire Protection Association, Quincy, MA, 1995.

13. Rockett, J.A., "Using the Harvard/NIST Mark VI Fire Simulation," NISTIR 4464, National Institute of Standards and Technology, November 1990.

14. Emmons, H.W., "Vent Flows," *SFPE Handbook of Fire Protection Engineering*, 2nd ed., National Fire Protection Association, Quincy, MA, 1995.

15. McCaffrey, B.J. and Rockett, J.A., "Static Pressure Measurements of Enclosure Fires," *J. of Research, Nat. Bur. Stand.*, June 1977.

16. Epstein, M., "Buoyant Driven Exchange Flow through Small Openings in Horizontal Partitions," *Jour. of Heat Transfer*, Vol. 110, ASME, 1988.

17. Cooper, L.Y., "An Algorithm and Associated Computer Subroutine for Calculating Flow through a Horizontal Ceiling Flow Vent in a Zone-Type Compartment Fire Model," NISTIR 4402, National Institute of Standards and Technology, October 1990.

18. McCaffery, B.J. and Quintiere, J.G., "Buoyancy Driven Countercurrent Flows Generated by a Fire Source," *Heat Transfer and Turbulent Buoyant Convection*, Vol. 2, D.B. Spalding and N. Afgan (Eds.), Hemisphere Publishing, Washington, D.C., pp. 457–472, 1977.

19. Lim, C.S, Zukoski, E.E., and Kubota, T. "Mixing in Doorway Flows and Entrainment in Fire Flames," Cal. Inst. of Technology, NBS Grant No. NB82NADA3033, June 1984.

20. Jaluria, Y., "Natural Convection Wall Flows," *SFPE Handbook of Fire Protection Engineering*, P.J. DiNenno (Ed.), Chap.1–7, SFPE/NFPA, Quincy, MA, 1988.

21. Klote, J.H. and Cooper, L.Y., "Model of a Simple Fan-Resistance Ventilation System and Its Application to Fire Modeling," NISTIR 89-4141, National Institute of Standards and Technology, September 1989.

22. Alpert, R.L., "Convective Heat Transfer in the Impingement Region of a Buoyant Plume," *Jour. of Heat Transfer*, Vol. 109, ASME, February 1987.

23. You, H.Z. and Faeth, G.M., "Ceiling Heat Transfer during Fire Plume and Fire Impingement," *Fire and Materials*, Vol. 3, No. 3, 1979.

24. Kokkala, M.A., "Experimental Study of Heat Transfer to Ceiling from an Impinging Diffusion Flame," *Fire Safety Science—Proc. of 3rd Inter. Symp.*, G. Cox and B. Langford (Eds.), Elsevier Applied Science, London, 1991.

25. Rockett, J.A. and Milke, J.A., "Conduction of Heat in Solids," *SFPE Handbook of Fire Protection Engineering*, 2nd ed., National Fire Protection Association, Quincy, MA, 1995.

26. Cox, G., "Compartment Fire Modelling," Chapter 6 in *Combustion Fundamentals of Fire*, Cox, G. (Ed.), Academic Press, London, 1995.

27. Stroup, D.W., "Using Field Modeling to Simulate Enclosure Fires," The SFPE Handbook of Fire Protection Engineering, Second Edition, The National Fire Protection Association, Quincy, MA, 1995.

28. Tuovinen, H., "Simulation of Combustion and Fire-Induced Flows in Enclosures," Report LUTVDG/(TVBB-1010), Department of Fire Safety Engineering, Lund University, Lund, Sweden, 1995.

29. Lockwood, F.C. and Shah, N.G., *18th Symposium (International) on Combustion*, Combustion Institute, Pittsburgh, PA, pp. 1405–1414, 1981.

30. Yan, Z. and Holmstedt, G., "Fast, Narrow-Band Computer Model for Radiation Calculations," Numerical Heat Transfer, Part B, Vol. 31, pp. 61–71, 1997.

31. Viskanta, R. and Menguc, M.P., "Radiation Heat Transfer in Combustion Systems," *Prog. Energy Combust. Sci.*, Vol. 13, pp. 97–160, 1987.

32. Magnussen, B.F. and Hjertager, B.H., "On Mathematical Modelling of Turbulent Combustion with Special Emphasis on Soot Formation and Combustion," *16th Symp. (Int.) Combust.*, Combustion Institute, Pittsburgh, PA, 1976.

33. Peters, N., "Laminar Flamelet Concepts in Turbulent Combustion," *21st Symp. (Int.) Combust.*, Combustion Institute, Pittsburgh, PA, 1986.

Appendix A

Fire Safety Engineering Resources on the Internet

The material in this appendix was collected and summarized by Johan Lundin, Department of Fire Safety Engineering, Lund University, Sweden. The appendix contains a list of Internet addresses with material of interest for fire safety engineering. Such addresses have a tendency to change from time to time, but an up-to-date document is accessible at *http://www.brand.lth.se/english/*.

CONTENTS

Organizations Dealing with Fire Safety Issues 280
Tools 282
Laboratories...283
International Organizations...284
Education at University Level 285
Magazines...286
Fire Modeling...287
Evacuation Modeling 288

ORGANIZATIONS DEALING WITH FIRE SAFETY ISSUES

NATIONAL INSTITUTE OF STANDARDS AND TECHNOLOGY (NIST), BUILDING AND FIRE RESEARCH LAB
http://www.bfrl.nist.gov/

NIST is a national laboratory dedicated to enhancing the competitiveness of U.S. industry and public safety performance prediction methods, measurement technologies, and technical advances needed to assure the life cycle quality and economy of constructed facilities. Its products are used by those who own, design, construct, supply, and provide for the safety or environmental quality of constructed facilities.

- Publications online (download)
- Fire modeling computer programs (download)
- International conferences
- Search their library FIREDOC
- Project summaries
- Fire Test Data
- Research

NATIONAL FIRE PROTECTION ASSOCIATION (NFPA)
http://www.nfpa.org/

The mission of this international nonprofit organization is to reduce the burden of fire on the quality of life by advocating scientifically based consensus codes and standards, research, and education for fire and related safety issues.

- National Fire Codes
- Fire Investigations (download)
- Education
- Research
- Certification
- Engineering Advisory Service
- Statistical data service
- Library
- Publications
- Seminars, workshops, and conferences
- Periodicals

SOCIETY OF FIRE PROTECTION ENGINEERS (SFPE)
http://www.sfpe.org/

SFPF is a professional society representing those practicing in the field of fire protection engineering. The purpose of the society is to advance the science and practice of fire protection engineering and its allied fields, to maintain a high ethical standard among its members, and to foster fire protection engineering education.

- Publications
- Fire modeling computer programs (download)
- International conferences
- Certification

INTERNATIONAL ASSOCIATION FOR FIRE SAFETY SCIENCE (IAFSS)
http://www.iafss.org/

IAFSS was founded with the primary objective of encouraging research into the science of preventing and mitigating the adverse effects of fires and of providing a forum for presenting the results of such research. Every third year the association organizes an international symposium on fire safety science.

- Bulletin board
- Links to various resources on the Internet
- International symposium
- IAFSS—Educational Subcommittee
  http://www.brand.lth.se/iafss-es/iafss-es.html

UNITED STATES FIRE ADMINISTRATION (USFA)
http://www.usfa.fema.gov/

USFA provides national leadership in fire training, data collection, technology, and public education and awareness, supporting the efforts of local communities to save lives and reduce injuries and property loss due to fire.

- National Fire Academy
- National Fire Programs
- Education
- Publications (download)

TOOLS

NORTH AMERICAN EMERGENCY RESPONSE GUIDEBOOK

http://www.tc.gc.ca/canutec/english/guide/toc/toc_e.htm

This is primarily a guide to aid first responders in quickly identifying the specific or generic hazards of the material(s) involved in an incident, and protecting themselves and the general public during the initial response phase of an incident.

- ID number index
- Name of material index
- List of dangerous water-reactive-materials
- Actions
  - Safety precautions
  - Protective actions
  - Protective action decision factors to consider
  - Who to call for assistance
  - Fire and spill control

THE IAFSS MAIL LIST

http://www.wpi.edu/Academics/Depts/Fire/new-iafss/mailList.html

The International Association for Fire Safety Science operates a mailing list devoted to discussion and exchange of information related to research into fire safety and fire prevention. To subscribe to the list, please send an e-mail message, **subscribe iafss firstname lastname**, to mailserv@cc.newcastle.edu.au (replace firstname and lastname with your real names). No subject is needed.

PAPERS AND PUBLICATIONS

- Lund University
  http://www.brand.lth.se/english/bibl/public.htm
  http://www.brand.lth.se/english/utbild/senior.htm

- NIST
  http://flame.cfr.nist.gov/bfrlpubs/

- NFPA
  http://www.nfpa.org/investigations.html

- USFA
  http://www.usfa.fema.gov/pub/online.htm

LABORATORIES

FIRE LABORATORIES

- Fire Research Station (FRS), U.K.
  http://www.bre.co.uk/frs/frs1.html

- VTT Building Technology, Finland
  http://www.vtt.fi/rte/firetech/

- Swedish National Testing and Research Institute (SP)
  http://www.sp.se/

- Norwegian Fire Research Laboratory (SINTEF)
  http://www.sintef.no/units/civil/nbl/

- Western Fire Center, U.S.
  http://www.westernfire.com/

- Factory Mutual, U.S.
  http://www.factorymutual.com/

WHAT KIND OF INFORMATION?

- certified products
- facilities
- research and development
- testing procedures
- research results
- projects
- publications
- specialties
- other information

INTERNATIONAL ORGANIZATIONS

THE INTERNATIONAL ORGANIZATION FOR STANDARDIZATION (ISO), TECHNICAL COMMITTEE 92
http://www.bre.co.uk/iso/

The mission of ISO is to promote the development of standardization and related activities in the world with a view toward facilitating the international exchange of goods and services, and to developing cooperation in the spheres of intellectual, scientific, technological, and economic activity.

SCOPE FOR TECHNICAL COMMITTEE 92
Standardization of the methods of assessing

- fire hazards and fire risk to life and to property
- the contribution to fire safety of design, materials, building materials, products, and components

and methods of mitigating the fire hazards and fire risks by determining the performance and behavior of these materials, products, and components, as well as of buildings and structures.

INTERNATIONAL COUNCIL FOR BUILDING RESEARCH (CIB), WORK-GROUP 14
http://www.vtt.fi/rte/firetech/cibw14/

CIB is a worldwide network of over 5000 experts from about 500 organizations, who actively cooperate and exchange information in over 50 commissions covering all fields in building and construction-related research and development.

The purpose of W14 is to

- provide a strategic overview of fire safety technology needs over the next 10 years;
- provide an ongoing research focus for the development of a sound technical basis for fire safety engineering methods;
- promote the acceptance of fire safety engineering methods and their relationship with performance based codes;
- provide fire safety technology input to other CIB working commissions as appropriate; and
- transfer fire safety engineering outputs internationally, including the standards community.

EDUCATION AT UNIVERSITY LEVEL

Universities developing a model curriculum in FSE

- Lund University, Sweden
  http://www.brand.lth.se/english/

- Worcester Polytechnic Institute, U.S.
  http://www.wpi.edu/Academics/Depts/Fire/

- University of Maryland, U.S.
  http://www.enfp.umd.edu/

- University of British Columbia, Canada
  http://www.ubc.ca/

- University of Edinburgh, U.K.
  http://oats.civ.ed.ac.uk/research/fire/index2.htm

- University of Ulster, U.K.
  http://www.engj.ulst.ac.uk/SCOBE/FIRE/fire.html

- University of Canterbury, New Zealand
  http://www.civil.canterbury.ac.nz//fire.html

- Victoria University, Australia
  http://kirk.vut.edu.au/cesare/www.html

- Oklahoma State University, U.S.
  http://www.fireprograms.okstate.edu/index.ssi

What kind of information?

- programs
- courses
- projects
- research
- resources
- senior project publications
- other information

MAGAZINES

There are a number of journals presented on the Web. It is possible to download some of them; others offer a list of contents and order forms.

- IAFSS Newsletter
  http://www.wpi.edu/Academics/Depts/Fire/new-iafss/newsletter.html

- Combustion and Flame
  http://www.elsevier.nl/inca/publications/store/5/0/5/7/3/6/

- Fire Technology
  http://www.nfpa.org/firetech.html

- Fire and Materials
  http://journals.wiley.com/wilcat-bin/ops/ID1/0308-0501/prod

- Journal of Applied Fire Science
  http://literary.com/baywood/pages/AF/index.html

- Journal of Fire Science
  http://www.techpub.com/tech/default.asp

- Journal of Hazardous Materials
  http://www.elsevier.nl/inca/publications/store/5/0/2/6/9/1/

- American Institute of Chemical Engineers Journal
  http://198.6.4.175/docs/publication/journal/index.htm

- Risk Analysis
  http://198.6.4.175/docs/publication/journal/index.htm

- Risk Management Quarterly Newsletter
  http://www.dne.bnl.gov/rmq.html

FIRE MODELING

HAZARD I
http://cfast.nist.gov/hazardi.html

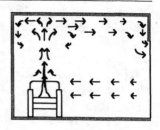

HAZARD I involves an interdisciplinary consideration of physics, chemistry, fluid mechanics, heat transfer, biology, toxicology, and human behavior. As an implementation of the hazard assessment method, the HAZARD I software consists of a collection of data, procedures, and computer programs which are used to simulate the important time-dependent phenomena involved in residential fires. The following models are included:

- Smoke transport model (CFAST)
- Evacuation (Exit)
- Detector and sprinkler activation (Detact)
- Toxicity (Tenab)

FASTLite
http://flame.cfr.nist.gov/fire/fastlite.html

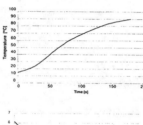

FASTLite is a user-friendly software package that builds on the core routines of FPEtool and the computer model CFAST to provide calculations of fire phenomena for use by the building designer, code official, fire protection engineer, and fire-safety related practitioner. FASTLite includes a number of tools of use to the fire safety practitioner:

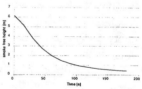

- three-room fire model
- heat and smoke detector activation
- suppression by sprinklers
- lateral flame spread
- mass flow through a vent
- atrium smoke temperature
- ceiling jet temperature plume
- filling rate
- radiant ignition of a nearby fuel

MORE FIRE MODELING SOFTWARE (AT NO COST)

- NIST
  http://www.bfrl.nist.gov/864/fmabbs.html

- SFPE
  http://www.wpi.edu/Academics/Depts/Fire/SFPE/software.html

EVACUATION MODELING

SIMULEX

http://www.ies4d.com/page13.html

Simulex is a computer package for PCs that is able to simulate the escape movement of many people from large, geometrically complex building structures.

- Fast, easy-to-use importing of CAD DXF files, and the automatic assessment of routes and travel distances through the whole building.
- Precise modeling of each person's position, orientation, walking speed, side-stepping, overtaking movement, and route-assessment at every 1/10th of a second.
- Full hard-disk recording of the evacuation, which can be played back at any time for repeated viewing and analysis.
- Final output file describing the building geometry, the number of people in different parts of the building, and a breakdown of the number of people leaving the building over regular time-steps to enable flow rate analysis

EXODUS

http://fseg.gre.ac.uk/exodus/

EXODUS—an evacuation tool for the safety industry—has been developed to meet the challenging demands of performance-based safety codes. Based on a highly sophisticated set of submodels, it shatters the mold of traditional engineering analysis to produce realistic people–people, people–fire, and people–structure interactions. As a result, the safety engineer can test more designs in less time to reach the optimal solution, free of the high cost and potential danger associated with human evacuation trials.

Appendix B

Suggestions for Experiments and Computer Labs

It is of great importance to provide the student some knowledge of experimental measurements and a sense of how accurate (or inaccurate) such measurements can be. It is equally important to enhance the student's awareness of the limitations and applicability of two-zone modeling, since the two-zone assumption is used in very many fire safety engineering applications (in both computer models and hand-calculation equations). This is best done by combining experimental activity with computer simulations using two-zone models and hand-calculation methods. This appendix outlines an example of such an activity, which starts with a brief introduction to a well-known two-zone model, then gives examples of 1/3-scale and full-scale compartment fire experiments and, finally, suggests how the experiments can be simulated and results compared.

CONTENTS

B1 Introduction to HAZARD I (or CEdit, CFAST, and CPlot)..289
    The HAZARD Shell 290
    The CEdit Program 290
    The CFAST Program 292
    The CPlot Program 292
    Adding a Second Compartment and Plotting Results...292
B2 Experiments in a 1/3-Scale Room...293
B3 Large-Scale Experiments 296
B4 Simulation of Experiments Using HAZARD I 297
    The 1/3-Scale Experiments..298
    Large-Scale Experiments 298

B1   INTRODUCTION TO HAZARD I (OR CEdit, CFAST, AND CPlot)

Goal: The student is to (1) get acquainted with the computer package HAZARD I (or the programs CEdit, CFAST, and CPlot, which are parts of the HAZARD I package); (2) know how to specify the room geometry, openings, and the energy release rate; (3) know how to view the results and present these in graphs and tables.

Reference: Peacock, R.D., Jones, W.W, Forney, G.P., Portier, R.W., Reneke, P.A., Bukowski, R.W., Klote, J.H., "An Update Guide for HAZARD I Version 1.2", NISTIR 5410, National Institute of Standards and Technology, Gaithersburg, MD, 1994.

Overview of HAZARD I, CEdit, CFAST, and CPlot: The HAZARD I package contains a number of programs, which are summarized below. Note that the programs CEdit, CFAST, and CPlot can be downloaded directly from the NIST Website (see Appendix A).

CEdit	This program asks questions on room geometry, thermal properties of the building materials, energy release rate of the fire, etc., and generates an input file to the two-zone model CFAST.
CFAST	A two-zone model for calculating gas temperatures, smoke layer height pressure, and much more. The input file created by CEdit is used here.
MLTFUEL	A program that allows the user to specify a number of different burning objects. The program uses this information to create a single energy release rate curve. This program will not be discussed further here.
FIREDATA	A database containing information on material reaction-to-fire properties. This program will not be discussed further here.
CPlot	A program that allows the user to view output from the two-zone model CFAST.
EXITT	This program uses information on building geometry, building population, and the results from CFAST to make estimations on evacuation. This program will not be discussed further here.
DETACT	A simulation program for calculating detection time for heat detectors. This program will not be discussed further here.
TENAB	This program estimates tenability for humans exposed to radiation and toxic gases. This program will not be discussed further here.

Getting acquainted with the whole program package will take considerable time. Here we discuss only CEdit, CFAST, and CPlot.

Note that this text assumes that one is using the full HAZARD I package. Some of the instructions will not apply if the stand-alone programs CEdit, CFAST, and CPlot are used. If this is the case, the user simply starts CEdit to create input, CFAST to run the simulation, and CPlot to inspect the results.

THE HAZARD SHELL

The HAZARD shell is a program that brings together all the ingredients of the HAZARD I package and controls the flow between the different programs mentioned above. Start the HAZARD shell by running the "Hazardi.exe" file. The screen will show four menus. Investigate these by using the "arrow" keys. We will only be using three of these menus: "Create Input Data," "Run Fire Model," and "Analyze Results."

THE CEDIT PROGRAM

We will start by creating an input file for CFAST. Go to the "Create Input Data" menu. Choose "Interactive Data Input" and push the Enter key. This starts up the CEdit program (or simply run the "Cedit.exe" program if the HAZARD I package is not being used).

First, the "Initialization" window will appear. The program assumes that we wish to load an existing input file. The "Function" keys at the bottom of the screen allow the user navigate through different parts of the program (keys F1 to F10). By choosing the F2 key (FILE), a list of all existing input files appears. One can choose any of these, but we will start by choosing the file "DATA.DAT" and push the Enter key.

This leads us to the "Overview" window. The name of each window is given at the top of the screen. Furthest down on the screen, a list of "Function" keys appear, showing which options can be chosen. We will start by investigating these. Push the F2 key (or the "go" key). There are

11 windows in the CEdit program, which we describe briefly below. Push the "esc" key (escape key) to get back to the "Overview" window.

1. The "Overview" window: How many rooms have been specified? How many openings? The numbers in gray can not be changed in this window, only the white numbers. Change the Simulation time to 5 minutes (300 seconds), History interval to 5 seconds, and Display interval to 5 seconds (The History command will cause results to be recorded every 5 seconds, the Display command causes the graphs that are drawn while CFAST is running to be updated every 5 seconds).

2. Push F2 and go to the "Ambient conditions" window: Investigate the data but change nothing.

3. Go to the "Geometry" window: Change the room size to 3×4 m^2 and the height to 3 m.

4. Go to the "Vents (doors)" window. "Sill" indicates the height from the floor to the top of the threshold or the height from the floor to the window sill. "Soffit" indicates the height from the floor to the top of the door or window. Change the door so that it becomes 1 m wide and 1.8 m high. We have specified only one room, so the outdoors will defined as room number 2.

5. Go to the "Vents (ceiling)" window. This is where one specifies ceiling openings, or stairways to an upper floor. Investigate this window, but change nothing.

6. Go to the "Fans, Ducts..." window. Investigate this window but do not specify any mechanical ventilation at this time.

7. Go to the "Thermal properties" window: Specify the ceiling, walls, and floor as being made of concrete. Start with the ceiling. Push the F6 key ("DATA") and choose CON-CRETE. Pick material properties for concrete by pushing the F6 key ("PICK"). Do the same for the walls and the floor.

8. Go to the "Fire Specification" window: Change the Heat of Combustion value to 20 MJ/kg. "Lim O2" (limiting oxygen index) indicates that combustion will become limited when the oxygen level has dropped under a certain value. Change this to 10%.
   "Rel Hum" indicates the initial relative humidity in the room (for most cases not a very important parameter).
   "GMW" has to do with the molecular weight of the fuel and is a relatively unimportant parameter in most cases.
   "Pos" indicates where in the room the plume is located (x, y, z).
   "Type" indicates the type of combustion one wishes to simulate. 1 is chosen when no account is to be taken of oxygen starvation and 2 when the oxygen concentration will be taken into account when considering energy release rate. Most often one would choose 2.
   "Pyrolysis" states the fuel pyrolysis in kg/s. It is not necessary to specify this if the user specifies "Heat_release" (which is more commonly done).
   "Heat_release" is the energy release rate in W (or J/s).
   "Height" is the height from the floor to the fuel base.
   "Area" is the fuel area.
   The other parameters have to do with the concentration of gases; we will not discuss this at this time.
   Change the energy release rate so that it is 0 kW at time 0 and 1 MW at around 150 seconds. Change the fuel height so that the fuel base is at floor level for all times. Observe that one can change the time periods as well, but this will be discussed later.

9. The "Objects" window allows the user to place combustible objects in the room. These can have certain ignition criteria. At a given surface temperature or incident radiation the object ignites and a new fire source is created. The objects can be picked from a database. Investigate this window but do not make changes. In the current version of HAZARD I (Ver. 1.2) this feature is not fully functional.

10. The "Files,..." window: Push the F6 key. This causes graphs to appear on the screen once the CFAST simulation starts. Hit the F7 key and specify the name of the output file (history file) as "TEST1.HIS" and save the input file as "TEST1.DAT." Note that a * may appear in front of "Main data file . . . ," which indicates that some changes made to the input file have not been saved (for example, the name of the output file). If this is the case, save the input file again. Push the "esc" key to get back to the "Files, . . ." window.

11. We will not be using the "Version and Settings" window. Hit F10 ("quit") to exit the input program CEdit.

THE CFAST PROGRAM

We have now arrived back at the HAZARD shell. Go to the "Run Fire Model" menu. Our input file will automatically be "TEST1.DAT," since this is the last file saved by CEdit. Now we will run the CFAST program and see a graphical representation of the results as the simulation proceeds. Choose "Begin Simulation." This will execute the CFAST program (if not using the HAZARD I package, simply run the "Cfast.exe" program, and the latest input file will be chosen).

Once the simulation is done, the graphs on the screen disappear. You can quit the simulation at any time by hitting the "esc" key. The results (up to that time) will be saved in the output file "TEST1.HIS."

THE CPLOT PROGRAM

We now wish to view the results of the simulation and will use the CPlot program for that purpose. Go to the menu "Analyze Results." Choose "Examine data," which will start the CPlot program (if not using the HAZARD I package, simply run the Cplot.exe program).

First use the command help (user input is given in **bold**):

Command>**help**
A list of commands that can be used in this program will appear. First a file with simulation results must be read into the program, for example, the results file we saved earlier, "TEST1.HIS."

Command>**file test1.his**
This file now contains a considerable amount of data from the simulation. We must indicate which data we are interested in. This is done by using the command "add."
Now we shall retrieve the hot gas temperature using the command "add":

Command>**add temperature**
What compartment number (1 ... 2) for Temperature **1**

And which layer **U**
The "U" stands for "Upper Layer." Now use the command "variable" to get a list of the output variables that are available. Then retrieve data on smoke layer height, energy release rate, and wall temperature.

Use the command "plot" to get curves on the screen. Also try to get the data as a list of numbers on the screen. IMPROVISE! Leave CPlot using the command "end."

ADDING A SECOND COMPARTMENT AND PLOTTING RESULTS

CEdit: Now we add a second compartment. Start CEdit ("Interactive Data Input" in the HAZARD shell). Read in the earlier input file "TEST1.DAT," go into the "Geometry" window, push the "Add" key. Add a room with dimensions 3×3×3 m³. Go to the "Vents (doors…)" window. Add an opening 0.5 m wide and 1.5 m high between the first and the second room by using the command "add."

Go into the "Files…" window, push F6 for graphical output during the simulation, save your input and output files using the names "TEST2.DAT" and "TEST2.HIS," and exit CEdit.

CFAST: Go into the "Run Fire Model" in the HAZARD I shell (or start the program "Cfast.exe") and start the simulation. Investigate the results using CPlot.

Changing the energy release rate: Now we wish to change the energy release rate. We assume that there is a 2 m² pool of gasoline on the floor of room 1. We can calculate the maximum energy release rate by using equations for pool fires. We know from Table 3.3 that $\dot{m}''_\infty = 0.055$ kg/(s m²) and $\Delta H_c = 43.7$ MJ/kg. χ can be assumed as 0.7. Now calculate the maximum energy release rate using Eq. (3.5) and (3.6).

We do not want the fire to reach the maximum energy release rate instantly. We therefore assume that the fire growth in the initial stage is "ultra fast," i.e., that the fire grows as

$$\dot{Q} = \alpha t^2$$

How long does it take to reach the maximum release rate $t_{Q=max}$? Calculate four points on the energy release rate curve from $t = 0$ to $t_{Q=max}$ and draw the curve roughly on a piece of paper.

Start CEdit, go into the "Fire specification" window and enter the energy release curve. Modify the time-points by pushing the F6 key ("MOD"), which stands for "modify." Save your files as "TEST3.DAT" and "TEST3.HIS" and run the simulation using CFAST. IMPROVISE! Open and close the windows and doors. Simulate. Compare results.

Comparing variables in different rooms: Start Cplot and read the output file "TEST3.HIS" using the "file" command. Create a graph where the upper layer temperature in room 1 and room 2 are shown in the same graph by using the "plot" command and specifying variables to be plotted in parenthesis (where (1,2) means that variables 1 and 2 will be plotted on the same graph).

Comparing results from different simulations: We assume that the Cplot program has read some variables from the "TEST3.HIS" output file as outlined above. Now read the file "TEST2.HIS" into CPlot. Select a number of variables using the "add" command. Compare variables from the two different runs, for example temperatures in room 1, by using the "plot" command, and by selecting variables in parentheses as above.

Preparing a graph of simulations for your report: We will now save our data in column format so that the data can be imported to an Excel spreadsheet. Importing the data to Excel allows us to manipulate the graph and lay it out as we wish. Later, we can even import experimental data to Excel and compare calculated and experimental data.

In CPlot, choose the command "def" (default) to make certain that the values you have added will be saved as columns in ASCII-format (not RAPID save). Now use the command "save," choose variables you wish to save, and choose a name for your results file, for example "TEST.RES." This file will then be saved in the HAZARDI directory (not HAZARDI/DATA). Start Excel, open the file "TEST.RES" (the file will be converted to Excel format), and draw your graphs.

Observe that in some national versions of the Excel program (for example, the Swedish version) the default decimal point is a comma (,) and not a period (.). The period must then be changed to comma, using the "seek/replace" command in Excel, before a graph can be created.

B2 EXPERIMENTS IN A 1/3-SCALE ROOM

This section gives an example of an experiment that can be carried out in a typical university department lab. The example given here requires a 1/3-scale room, a hood for collecting combustion gases, oxygen concentration measurements, and some other experimental apparatus. Further, the fuel consists of particle board mounted on walls and ceiling in order to simulate a growing fire resulting in flashover.

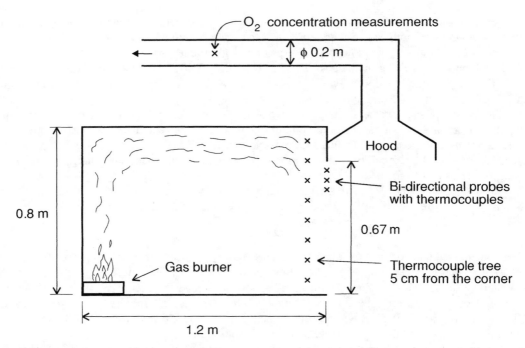

FIGURE B1   Schematic of the experimental set-up.

The experiment can be simplified considerably if these resources are not available. Using a simple gas burner and a tube of gas allows the energy release rate to be estimated by weighing the gas tube. Temperatures can be measured using thermocouples and hand-held thermometers.

Goal: The purpose of the experiment is to give the student

- Some knowledge of experimental measurements and a sense of how accurate (or inaccurate) such measurements can be
- A sense of how the theoretical two-zone assumption compares to controlled experiments
- Experimental data that can be used for comparing results from computer simulations and hand calculations

General: In this experiment we simulate fire growth in a small room with an opening and with combustible material mounted on walls and ceiling. We compare the results to calculations using expressions for pressure differences and gas flow velocities in the opening as well as calculations of temperature in the room. We see how the fire will grow and become fuel controlled and check this with equations for mass flow rates through the opening for a fully developed fire.

A small room, 1.2 m long, 0.8 m wide, and 0.8 m high, has particle board mounted on three of the walls and on the ceiling. The experiment is carried out as a corner test, i.e., the ignition source is a small gas burner located in one of the corners furthest from the opening. The gas burner energy release rate is roughly 15 kW, allowing the flame to reach the ceiling.

The room is placed under a hood such that the combustion gases are collected and the oxygen concentration can be measured in the duct leading from the hood (see Figure B1).

Measuring instruments: Three bi-directional probes are placed in the opening together with three thermocouples, as shown in Figure B1. This allows the pressure differences across the opening to be measured and the mass flow rate through the opening to be calculated. A thermocouple tree with eight thermocouples, evenly distributed by height, is placed in one of the front corners in the room, roughly 5 cm from the corner (see Figure B1).

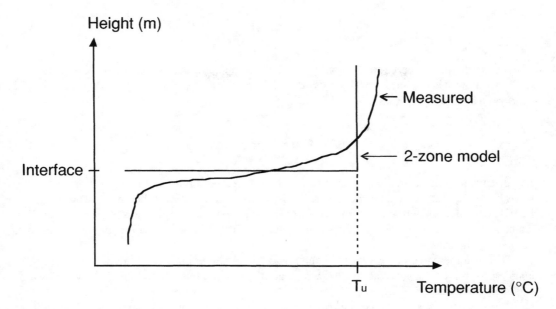

FIGURE B2   A schematic of the two-zone concept vs. measured temperatures.

Oxygen concentration is measured in the duct leading from the hood, which allows the energy release rate to be estimated.

Visual observations: Visual observations are made of the hot gas layer height as a function of time, the flashover process, and time to flashover.

Calculations: We are interested in comparing measured temperatures with calculated temperature results using a two-zone model. The two-zone model assumes a single temperature in the upper gas layer and a single temperature in the lower gas layer. Figure B2 shows a schematic of measured and calculated temperatures at a given time as a function of the room height.

We also wish to compare calculated and experimentally derived mass flow rates through the opening. We cannot directly measure the height of the neutral layer in the experiments. An alternative is to measure the pressure difference at many different heights across the opening and thereby obtain a pressure profile and an estimation of the neutral layer height. However, the pressure differences are relatively small (except at the top of the opening) and are difficult to measure. One can instead use a single measurement of pressure difference and a number of temperature measurements in order to determine the height of the neutral layer.

The background of this method is as follows: Assume that the neutral layer height is z_N and the pressure at this height is $P(z_N)$. We are interested in the pressure at height z. The temperature and density of air outside the room is ρ_a and T_a. The pressure outside the room can then be written

$$P_u(z) = P(z_N) + \rho_a \cdot g \cdot (z_N - z)$$

Since $\rho_a = 353/T_a$ and $g = 9.81$ m/s^2, this can be written as

$$P_u(z) = P(z_N) + \frac{3461}{T_a} \cdot (z_N - z) \tag{B1}$$

Inside the room the temperature is not constant with height. The pressure inside the room, P_i, can then be written as a function of height (where the height z_N is assumed to be 0) and temperature

$$P_i(z) = P(z_N) + \int_z^0 \frac{3461}{T_i(z')} \cdot dz' \tag{B2}$$

where z' is the height between the points where the temperature is measured.

If we now wish to measure the pressure difference across the opening we subtract Eq. (B2) from Eq. (B1) and get

$$\Delta P(z) = 3461 \int_z^0 \left[\frac{1}{T_i(z')} - \frac{1}{T_a} \right] \cdot dz' \tag{B3}$$

Say that we have measured the temperature at four different heights in the hot gas layer. We then know the temperature in three height intervals; each interval has the height z'. The integral in Eq. (B3) can then be written as sums. If we have measured the pressure difference at the height z, we can use Eq. (B3) to calculate the height z since $\Delta P(z)$ is known. Use this method to estimate the neutral layer height.

Then use the equations given in Chapter 5 to calculate the mass flow rate into and out of the opening for the two stages: the fire growth stage (the stratified case), and the fully developed fire (the well-mixed case).

Observe that this process is not simple or trivial. Many assumptions and some engineering judgment must be used to arrive at what we call "experimental data." Measuring small pressure differences is not easy, and the experiment may not necessarily confirm your ideas of the two-zone model.

Security: The experiments described here will result in flashover and a fully developed fire, with toxic gases and flames out through the opening as a result. It is therefore important to take precautions as the fire approaches flashover. There is also some risk associated with the unburnt gases that collect in the hood and duct when the fire is very underventilated. The fire should therefore be extinguished shortly after flashover.

Report: Give a short description of the experimental set-up, room geometry, openings, surface materials, and energy release rate. Show graphs of experimental data and how the final data is arrived at from the raw data. Choose two time-points: one before and one after flashover. Show how the pressure differences at these times are used to estimate the height of the neutral layer and velocity and mass flow rate in the opening. Observe that this is not a trivial process and that engineering judgment must be used to arrive at the results. Show comparisons with calculated mass flow rates using equations from Chapter 5. Describe the growth to flashover and the flashover period.

Simulate this fire using HAZARD I (or CEdit, CFAST, and CPlot). Section B4 specifies this task further. You must therefore collect information that is needed as input for the simulation, such as geometry, openings, surface materials, and measured energy release rate.

B3   LARGE-SCALE EXPERIMENTS

This section gives an example of an experiment that can be carried out at a fire department or a rescue school practice field. The example given here uses an apartment and stairway structure located at the National Rescue Service School at Revinge, Sweden. The experiment can be carried out in any reasonable scale structure, since only a few relatively simple measuring devices are used.

Building geometry: The building used at Revinge consists of two apartments on two floors, each around 40 m², connected by a four-floor stairway (see Figure B3). The supporting structure is made of prefabricated concrete beams and columns. The floors and ceilings are made of concrete slabs that have been specially insulated against high temperatures. The walls are made of light weight concrete blocks.

On the inside the apartment surfaces are covered with special mortar that can withstand high temperatures and quick cooling. A number of window and door openings are provided to the outside.

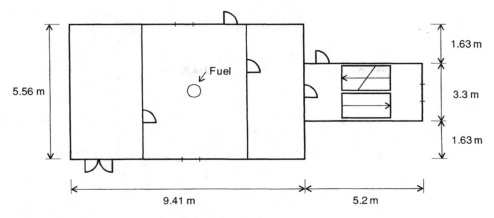

FIGURE B3 Plan view of the experimental facility at Revinge.

Two small openings are provided at floor level (0.6 m wide and 0.2 m high) to allow extinguishing water to exit.

The stairway supporting structure is made of prefabricated concrete, stairs made of a gitter of galvanized iron. The stairway ceiling is fitted with an opening for ventilating smoke.

Measuring instruments and procedure: One thermocouple is placed near the ceiling (10 cm from the ceiling) in the apartment on the second floor. Another thermocouple is placed near the stairway ceiling; these are connected to hand-held thermometers.

- All doors are kept shut throughout the experiment, but a window is kept open in the second-floor apartment.
- To allow visual observation of the smoke layer height, a scale is drawn on a length of paper fastened to the wall, from floor to ceiling. The observations are made through the small openings at floor level.
- A small vessel, 0.8 m diameter with 0.15 m rim, is placed in the apartment, on a weighing scale (Figure B4). The scale is insulated by mineral wool and aluminum foil. The reading instrument for the scale is placed outside one of the openings at floor level and the cable from the scale to the instrument is protected by mineral wool and aluminum foil.
- The vessel is filled with heptane and the liquid ignited.
- The temperature in the apartment upper layer is measured for the first 5 minutes. Then the door from the apartment to the stairway is opened and the temperature measured in both compartments. The smoke fills both compartments very rapidly, but rough visual observations are made of smoke layer height.
- The door from the stairway to the outside is kept open throughout the experiment.

Report: To allow computer simulation of the experiment, all relevant geometrical information must be noted (building geometry, opening geometry, and position), the internal surface material properties must be known and the mass loss rate of the fuel (to allow estimation of the energy release rate). The experimentally measured temperatures and smoke layer heights are then compared to results from computer simulations (see Section B4).

B4 SIMULATION OF EXPERIMENTS USING HAZARD I

Goal: The main goal of this computer lab is to enhance understanding of the limitations and applicability of two-zone models by attempting to simulate the experimental scenarios described in Sections B2 and B3.

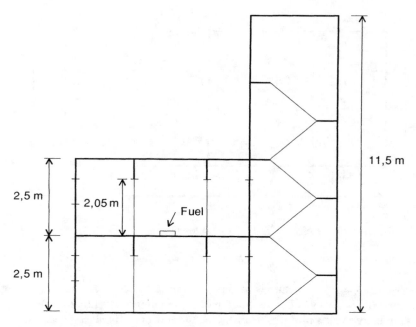

FIGURE B4   A section of the experimental facility at Revinge.

THE 1/3-SCALE EXPERIMENTS

Simulation: Retrieve the measured energy release rate from the experiment, and simplify this for use as input to HAZARD I. Retrieve the experimentally measured temperatures, the derived mass flows, and the visually observed smoke layer height. Organize this in an Excel spreadsheet.

Simulate the experimental scenario and compare results. Observe the engineering assumptions made, for example the fact that as the corner flame travels upward on the combustible lining material, the axisymmetrical plume assumption is no longer valid. The fuel base height may become a sensitive parameter due to limited oxygen entrainment to the plume. Note, however, that engineers in all disciplines make engineering judgments. The important issue is to be acutely aware of the assumptions made, and thereby be able to draw logical conclusions.

Report: Give a description of the experimental scenario and a description of the input file. Give a comparison of experimentally measured or derived temperatures, mass flows, and smoke layer height with calculated values. Discuss the results.

LARGE-SCALE EXPERIMENTS

Simulation: Retrieve the measured fuel mass loss rate from the experiments. Use this as input to HAZARD I. Input an effective heat of combustion, ΔH_{eff}, so that the model can estimate a realistic energy release rate. Input the geometry of the building and indicate the material properties of the interior surface materials.

Start by specifying the whole apartment as a single room, where the floor level has a certain height above ground level. Then specify the stairway as a second room, where the door between the two rooms is closed for the first 5 minutes.

Simulate the experimental scenario and compare calculated and experimentally measured temperatures and smoke layer heights.

Report: Give a description of the experimental scenario and a description of the input file. Give a comparison of experimentally measured temperatures and smoke layer height. Discuss the results.

Appendix C

A Simple User's Guide to CEdit

One of the more widely used computer programs for simulating fires in enclosures is the two-zone model CFAST. The CEdit program is commonly used as a preprocessor or input program to CFAST, even though other such programs have recently been developed and are available on the Internet. These programs pose questions to the user in order to ease the onerous task of inputting the relatively large number of parameters that need to be specified before the CFAST simulation can be run. Even though the preprocessor eases the task considerably, the user must frequently seek further information on very many of the parameters. The intention of this appendix is to give the user a handy and simple user's guide to CEdit where the parameters are very briefly discussed. A much more detailed discussion is given in the main reference: Peacock, R.D., Jones, W.W, Forney, G.P., Portier, R.W., Reneke, P.A., Bukowski, R.W., Klote, J.H., "An Update Guide for HAZARD I Version 1.2," NISTIR 5410, National Institute of Standards and Technology, Gaithersburg, MD, 1994. This appendix is based on a document written in Swedish by Jonas Nylén, a Fire Safety Engineer at Helsingborg Fire Department in Sweden.

CONTENTS

C1 Overview 300
C2 Ambient Conditions 300
C3 Geometry 301
C4 Vents, Doors 301
C5 Vents, Ceiling. 302
C6 Fans, Ducts 302
C7 Thermal Properties 303
C8 Fire Specification 304
C9 Objects 305
C10 Files 305

C1   OVERVIEW

Having started CEdit and chosen an input file, the Overview window appears. The Overview window presents a summary of the input file. The only parameters that can be changed are the title and the time periods for the simulation. To do this, the cursor is placed on the actual parameter and a new value is typed in. The allowed range of the value is given at the bottom of the window. The newly typed value is displayed at the lower right-hand corner of the window. Pushing the "enter" key activates the newly typed value. The default SI units are given in the tables below.

Command	Comments	Unit	Can be changed?
File	Shows which file was retrieved from the data directory.		
Title	Allows the user to write comments to distinguish different input files.		Yes
Compartments	Specifies how many rooms are defined in the input file.		No
Doors	Specifies how many horizontal openings there are between the different rooms.		No
Ceiling vents	Specifies how many vertical openings there are between the different rooms, like ceiling vents, staircase openings, etc.		No
MV connects	The number of connections for mechanical ventilation		No
Simulation	Specifies the length of the simulation. Maximum length is 86,400 s (24 hours).	s	Yes
Print	Specifies the time interval between outputs to the printer file.	s	Yes
History	Specifies the interval between outputs to the output file.	s	Yes
Display	Specifies the interval between outputs to the screen.	s	Yes
Restart	Allows a time to be specified if the simulation is to be restarted at a given time.	s	No
Temperature	Initial temperature inside and outside the room.	°C	No
Pressure	Initial pressure inside and outside the room.	Pa	No
Station elv	Reference height for specifying initial pressure and temperature.	m	No
Wind speed	Wind speed.	m/s	No
Ref height	Reference height for wind speed.	m	No
Power law	A power law allows the wind speed to be calculated as a function of height. The default value of the power used in the equation is 0.16.	—	No
Species tracked	Specifies the chemical species that are included in the simulation.	—	No
Ceiling jet	Specifies whether a ceiling jet is taken into account.	—	No

C2   AMBIENT CONDITIONS

The Ambient conditions window defines the initial conditions inside and outside the room.

Command	Comments	Unit	Can be changed?
Temperature	Initial temperature inside and outside the room.	°C	Yes
Pressure	Initial pressure inside and outside the room.	Pa	Yes
Station elv	Reference height for specifying elevation, initial pressure, and temperature.	m	Yes
Wind speed	Wind speed	m/s	Yes
Ref height	Reference height for wind speed.	m	Yes
Power law	A power law allows the wind speed to be calculated as a function of height The default value of the power used in the equation is 0.16.	—	Yes

There are a number of parameters in this window that cannot be changed and are therefore not commented upon here.

C3   GEOMETRY

The Geometry window specifies the room geometry such as length, width, and height. If rooms are to be added, the F4 key ("add") is pressed and the program will ask for input. For other changes place the cursor on the parameter, enter a new value, and press the "enter" key.

Command	Comments	Unit	Can be changed?
Compartment no.	Gives the number that has been allocated to the room.	—	No
Width	Width of the room.	m	Yes
Depth	Depth of the room.	m	Yes
Height	Height of the room.	m	Yes
Floor elevation	Floor height above the station elevation.	m	Yes

C4   VENTS, DOORS

The Vents, doors window specifies the horizontal openings between the rooms and out to the open (horizontal, since the flow through these is assumed to be horizontal). If ventilation openings are to be added, the F4 key ("add") is used and the user is prompted for input.

Command	Comments	Unit	Can be changed?
Width	Width of opening.	m	Yes
Sill	Height from floor to top of threshold or to window sill.	m	Yes
Soffit	Height from floor to top of opening.	m	Yes
Wind	The cosine of the angle between the wind vector and the opening (only for openings connecting to the outside). Default value is 0.	m/s	Yes
A-Sill	Window sill height with respect to station elevation (see Figure C1).	m	No
A-Soffit	Height to the top of the opening with respect to the station elevation (see Figure C1).	m	No
Vent (1→4)	If more than one opening is defined between two rooms, the user must number the openings. The maximum number of openings from each room is 4.	—	Yes

To open and close the openings as a function of time, see Section C10.

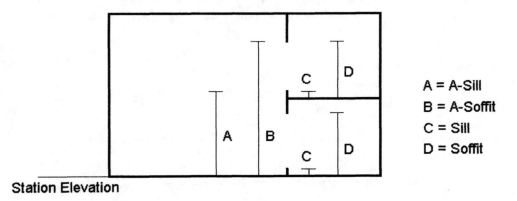

FIGURE C1   A schematic showing how opening heights are defined.

C5   VENTS, CEILING

The Vents, ceiling window specifies the vertical openings between rooms or to the outside (vertical indicating that the flow is vertical through the opening). The openings can, for example, be between floor levels or be ceiling vents to the outside. These openings cannot be opened or closed as a function of time, such as the horizontal openings. To add vertical openings use the F4 key ("add").

Command	Comments	Unit	Can be changed?
Top	Specifies the number of the top compartment.	—	Yes
Bottom	Specifies the number of the bottom compartment.	—	Yes
Shape	The shape of the opening can be specified to be circular (C) or square (S).	—	Yes
Area	Specifies the opening area.	m^2	Yes
Relative height	Specifies the opening height in relation to the bottom compartment.	m	No
Absolute height	Specifies the opening height in relation to the station elevation.	m	No

C6   FANS, DUCTS

The Fans, ducts window specifies the mechanical ventilation system, together with the windows that appear when using the F6 and F7 keys. This window specifies how the duct system is connected to the compartments. The F6 key allows specification of the nodes and the geometry of the duct system and the F7 key allows the properties of the fan to be entered.

Command	Comments	Unit	Can be changed?
Compartment	Number of a compartment where mechanical ventilation inlet or outlet is provided.	—	Yes
Node	Number of the node in the actual compartment.	—	Yes
Orientation	Indicates the orientation of the ventilation inlet or outlet. V = vertical orientation, H = horizontal orientation.	—	Yes
Height	Height from the station elevation to the center of the ventilation opening.	m	Yes
Area	Area of the inlet or outlet duct	m^2	Yes

The window appearing when hitting the F6 key (shown below) allows the duct system to be specified. The nodes are those that have been given in the previous window (see Figure C2).

Command	Comments	Unit	Can be changed?
Node 1 and 2	Node numbers.	—	Yes
Len	Length of the duct.	m	Yes
Diam	Diameter of the duct.	m	Yes
Abs rough	Roughness for the inside of the duct.	m	Yes
Flow coeff. node 1 and 2	Flow coefficient to allow for an expansion or contraction at the end of the duct. Default = 0 (straight through connection).	—	Yes
Expanded joint 1 and 2	Area of the expanded joint.	m^2	Yes
Absolute node hght 1 and 2	Height above station elevation for the actual node.	m	Yes

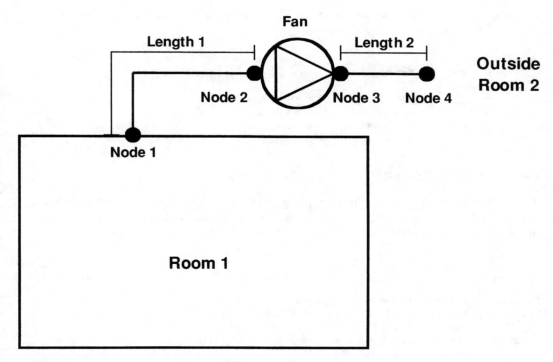

FIGURE C2 Definitions of parameters when specifying fans and ducts.

To specify the fan capacity, a fan curve can be specified in the window appearing when the F7 key is pressed (shown below). The fan is assumed to be positioned between the nodes, which are specified as nodes 1 and 2.

Command	Comments	Unit	Can be changed?
Fan	Fan identification number.	—	Yes
Node 1 and 2	The fan is placed between these node numbers.	—	Yes
Pmin and flow	Pressure and flow in the right-hand side of the fan curve.	Pa and m³/s	Yes
Pmax and flow	Pressure and flow in the left-hand side of the fan curve.	Pa and m³/s	Yes
No Coeff	Number of coefficients necessary to specify the assumed fan curve.	—	Yes

C7 THERMAL PROPERTIES

The Thermal properties window allows the user to specify thermal properties of the ceiling, wall, and floor material in the different compartments. This can be done by typing in the material name, if the material is listed in the thermal database. The default thermal database is contained in a file called "THERMAL.DF." To check which materials exist in the database, the F6 key is used. To return to the main window the escape key ("esc") is used. Another way of selecting materials is to view the thermal database using the F6 key ("data"), placing the cursor on the preferred material and again hitting the F6 key ("pick"). The material will then be chosen. Repeat until walls, floor, and ceiling have been assigned a material. If a material is missing the user can manually enter a new material using the F4 key ("add"), or simply by reading the default database file "THER-MAL.DF" into any editor and adding a material of choice.

Note that the thickness of the material is also specified in the database; the thickness cannot be specified separately in the input program.

Command	Comments	Unit	Can be changed?
Compartment No	Compartment number.	—	No
Ceiling properties	Ceiling material.	—	Yes
Floor properties	Floor material.	—	Yes
Wall properties	Wall material.	—	Yes

C8   FIRE SPECIFICATION

The Fire specification window allows specification of the energy release rate, fuel mass loss rate, heat of combustion, and other parameters to do with combustion. The user can also specify where in the room the fire is positioned and how ventilation control is taken into account.

Command	Comments	Unit	Can be changed?
Heat of C	Effective heat of combustion of the fuel.	J/kg	Yes
Lim O_2	This gives the limiting oxygen index and specifies the oxygen concentration below which a flame will not burn. The default value is 10%.	%	Yes
Rel Hum	The initial relative humidity in the system. This is not often taken into account, so the default value is 0.	%	Yes
GMW	Molecular weight of the fuel vapor. Used for converting to ppm, has no influence on the calculations themselves. Default value 16.0.	g	Yes
Pos	The position of the fuel base in the x, y, and z direction.	—	Yes
Room	This specifies which room the fire is in.	—	Yes
Type	This is to specify how the energy is released. 1 = the release rate is independent of oxygen concentration, 2 = the release rate is dependent on the oxygen concentration, once this is below a given value.	—	Yes
Cjet	This allows specific calculations to be performed for the ceiling jet, which can be useful when investigating detector response. For other applications, the ceiling jet calculations are normally turned off. OFF = No ceiling jet, CEILING = Ceiling jet included, WALL = influence of walls, ALL = Both taken into account. Choose OFF to diminish calculation time.	—	Yes
Pyrolysis	Specifies the fuel mass burning rate. If this is specified the energy release rate is automatically calculated.	kg/s	Yes
Heat release	The energy release rate. If this is specified the mass burning rate is automatically calculated. See Eq. (3.5).	J/s = W	Yes
Height	The height of the fuel base from the floor. This value is used in the plume calculations.	m	Yes
Area	Area of the fuel base as a function of time.	m^2	Yes

Species concentrations can be calculated but this input is needed only if tenability limits are to be estimated. For the parameters listed below, the required input demands that the fuel is well known. If this is not the case, default values can be used, but the output must be viewed with this in mind.

Command	Comments	Unit	Can be changed?
H/C	Fuel mass ratio of hydrogen to carbon.	—	Yes
CO/CO2, C/CO2	For calculation of CO, the mass ratio of CO to CO_2 must be given. For calculation of optical density the mass ratio of C to CO_2 must be given.	—	Yes
HCN, HLC, CT	For calculation of species such as HCN, the yield (kg of species produced per kg of fuel pyrolized) must be specified.	kg/kg	Yes

While the energy release rate or the pyrolysis rate is being specified, the time intervals can be changed. The F6 key ("mod" for modify) is used for this purpose.

C9   OBJECTS

The CFAST program contains a model that estimates the spread of fire from one object to another. The database OBJECTS.DF contains different objects with information on ignition temperatures and time to ignition at a given radiant heat flux (see window below). The model is under development and should be used with caution.

Command	Comments	Unit	Can be changed?
Name	Name of the object.	—	Yes
Compartment	Compartment number where the object is placed.	—	Yes
Start time	The earliest time during simulation that CFAST should check whether the ignition criteria for the object are met.	s	Yes
First element	Indicates a position on the object surface where burning starts. This parameter is not currently used by CFAST but must be set for future compatibility. Default value is 1.	—	Yes
Position x, y, z	The position of the object in the room.	m	Yes

C10   FILES

This is the window that allows files to be saved or specified and is normally the last window used in CEdit. Different choices with regard to presentation of the results are also given.

Command	Comments	Key
Quick estimates	This allows a quick estimate to be made as a control of the input file. If the results are unreasonable the input file should be adjusted and the simulation repeated.	F5
Run time graphics	By hitting the F6 key the comment on the screen is changed to "append display file," which allows the results to be displayed on the screen during the simulation.	F6
Save data file	This is where the input (.DAT) and output files (.HIS). are named. Observe that if a star (*) is displayed on the input file line, then the latest changes to the input file have not been saved. The cursor must then be moved on to the input file name and the "enter" key hit.	F7
Write to log file	A log file can be activated so that the progress during the simulation can be logged. If there seems to be a problem simulating a run, the log file can give information on when exactly the problems arise.	F8

Index

A

Absorptivity
    definition, 141
    and emissivity, models for, 273
    incident energy absorbed, 143
    spectral, in radiative heat transfer, 163–164
Alcohols
    combustion efficiency, 31–32
    mass loss rate, 33
ALOFT-FITM (A Large Outdoor Fire plume Trajectory model), 276
American Institute of Chemical Engineers Journal, 286
ASCOS (Analysis of Smoke Control Systems), 276
ASET-B (Available Safe Egress Time - BASIC), 276
ASMET (Atria Smoke Management Engineering Tools), 276
Atmospheric pressure, 83

B

Babrauskas gas temperature calculations, 135–136
Backdraft, 11, 16
Bench-scale test, 242–244
Bernoulli equation, 84, 86
Bounded plumes, 71–72
Bounding surfaces
    combustible wall lining materials, 126
    effect of properties of, 23–24
    heat loss to, 211
Boussinesq approximation, 55, 58
BREAK1 (Berkeley Algorithm for Breaking Window Glass in a Compartment Fire), 276
Building safety probabilistic models, 3, 256–257
Buoyancy
    axisymmetric plume zones, 52
    described, 48–49
    equation for, 57–58
    force, 56
Burning rate
    described, 25
    energy release rate assessment, 30
    formula for, 27
    well-mixed case, 101–102

C

Candles, 12–14
Carbon monoxide (CO)
    estimating production in fires, 246–247
    yield as function of equivalence ratio and temperature, 248
    yields in compartment fires, 251

CCFM (Consolidated Compartment Fire Model version VENTS), 276
CEdit program
    fire specifications, 293
    second compartment, 292–293
    single component, 290–292
    user's guide
        ambient conditions, 300
        fans, ducts, 302–303
        fire specifications, 304–305
        geometry, 301
        objects and fire spread, 305
        overview, 300
        program files, 305
        thermal properties, 303–304
        vents, ceilings, 302
        vents, doors, 301
Ceiling jets
    described, 15
    flame extensions under ceilings, 76–77
    temperatures and velocities, 74–76
Centerline temperature, plumes, 65
Centerline velocity, plumes, 65
CFAST (Consolidated Fire and Smoke Transport Model), 276, 292
CFD, *see* Computational fluid dynamics
Chemical heat of combustion, 30, 31
CIB (International Council for Building Research), 284
CO (carbon monoxide)
    estimating production in fires, 246–247
    yield as function of equivalence ratio and temperature, 248
    yields in compartment fires, 251
Combustible wall lining materials, 126
Combustion and Flame, 286
Combustion products
    combustion efficiency, 25, 31–32
    computer submodels for, 274–275
    conservation equation for species
        compartment application, 240–242
        control volume formulation, 238–240
    fuel chemistry
        energy release rates, 26
        factors influencing fire development, 21–22
        mixture fraction, 227, 233–234
        specific yields, 230–238
        stoichiometry and species, 229–230
    heat of combustion
        described, 26, 30–31
        equation for, 232
        table of values, 33
    nature of, 228–229
    process of combustion, 12–14

307

species concentrations estimations, 251–253
terminology, 227–228
yield estimation from experiments
    bench-scale test, 242–244
    compartment fires data, 250–251
    hood experiment data, 244–250
Compartment fires
    carbon monoxide yields in, 251
    conservation equations, 240–242
    pressure rise conservation equations
        closed volume, 192
        leaky compartment, 192–195
    ventilation sources as openings, 23
    yield estimations, experimental, 250–251
Complete heat of combustion, 30
Computational fluid dynamics (CFD)
    combustion submodels, 274–275
    described, 269–270
    internet resources for models, 275
    model for, 3–4
    radiation submodels, 271–274
    turbulence submodels, 270–271
Computer modeling
    computational fluid dynamics
        combustion submodels, 274–275
        described, 269–270
        radiation submodels, 271–274
        turbulence submodels, 270–271
    deterministic, 3–4, 257
    internet resources
        CFD programs, 275
        evacuation programs, 277
        special-purpose programs, 276–277
        two-zone programs, 276
    probabilistic, for building fire safety, 256–257
    two-zone
        building geometry limitations, 268–269
        conservation equations, 258–262
        described, 258
        embedded submodels, 267
        heat transport submodels, 266–267
        mass transport submodels, 263–266
        source term submodels, 262–263
        unresolved phenomena, 267
Conduction heat transfer
    described, 142–143
    submodel for, 267
Cone Calorimeter, 171
Configuration factor, 141, 159–163
Conservation equations
    control volume
        energy, 187–190
        mass, 184
        species, 238–240
        thermodynamical system definition, 183
        thermodynamic properties, 184–187
    described, 182–183
    pressure rise, closed volume
        assumptions, 192
        equation derivation, 190–192

pressure rise, leaky compartment
    assumptions, 195
    equation derivation, 192–195
smoke control, large spaces
    boundary heat losses, 211
    energy, 210–211
    gas temperature, 211
    mass, 210
    mass and energy balances, 210–211
    mechanical ventilation, 215–218
    natural ventilation, 212–215
smoke filling, ceiling leaks
    equations for, 200–202
    limitations, 203–204
smoke filling, floor level leaks
    calculation methods, 199
    comparison with experiments, 200
    described, 196–197
    energy, 197–198
    equation for, 199
    gas temperature estimates, 202–203
    heat release rate, 198
    limitations, 203–204
    mass, 197
    time to fill, 198, 199
smoke filling, large spaces
    assumptions, 204–205
    calculation methods, 207–208
    comparison with experiments, 208–209
    energy, 206–207
    mass, 205–206
species and combustion products
    compartment application, 240–242
    control volume formulation, 238–240
summary, 218–221
terminology, 182
two-zone computer modeling
    assumptions, 259–260
    control volume, 258–259
    energy, 261
    mass, 260
    species, 260–261
    summary, 262
    ventilation-limited condition, 261–262
Consolidated Compartment Fire Model version VENTS
    (CCFM), 276
Consolidated Fire and Smoke Transport Model (CFAST),
    276, 292
Convective heat transfer
    coefficient, estimates of, 148–149
    described, 143
    energy release rates, 64
    fire plume to ceilings, 149–154
    level of heat flux likely, 149
    to surfaces, submodel for, 266–267
CPlot program, 292

D

Decay phase

described, 18
design fire, 44
Density factor, 99
Design fire
 background, 39
 complexity in, 44
 decay phase, 18, 44
 determining, 26
 growth phase, 39–43
 steady phase, 43–44
DETACT model, 257
DETACT-QS (Detector Actuation-Quasi Steady), 276
DETACT-T2 (Detector Actuation-Time squared), 276
Detectors and suppression devices activation models, 257
Deterministic models
 CFD, 3–4, 257
 hand-calculation categories
  combustion energy evolved, 5
  fire-induced environmental factors, 5–6
  heat transfer, 6
 two-zone, 4, 258
Diffuse surfaces and radiation exchange, 160
Diffusion flames, 48, 50
Dufour effect, 143

E

Eddies
 dissipation concept model, 274
 simulation turbulence model, 271
Education in fire safety engineering, 285
Effective absorption coefficient, 166, 167
Effective heat of combustion, 30, 31
Effective heat transfer coefficient, 120–123
Electrical circuit analogy, heat transfer, 172
ELVAC (Elevator Evacuation), 276
Emissivity
 and absorptivity, models for, 273
 cylindrical flame, 170
 definition, 141, 144
 spectral, in radiative heat transfer, 164–165
Energy conservation equations
 control volume, 187–190
 smoke control, large spaces, 210–211
 smoke filling, floor level leaks, 197–198
 smoke filling, large spaces, 206–207
 two-zone computer modeling, 261
Energy evolved, 5
Energy flow rate equation, 56
Energy release rates
 described, 25–26
 design fire
  background, 39
  complexity in, 44
  decay phase, 18, 44
  determining, 26
  growth phase, 39–43
  steady phase, 43–44
 dimensionless parameter for, 51, 198
 equation for, 30

factors controlling
 burning rate, 27–28
 enclosure effects, 28
 flame height as function of, 52
 free burn measurements
  described, 28
  example products, 34–38
  pool fires, 32–34, 35
  techniques and parameters, 29–32
  T-squared fire, 38–39
 fuels, characteristic values for, 26
 smoke-filling conservation equation, 198
 terminology, 25–26
Engineering models
 deterministic
  CFD, 3–4, 257
  hand-calculation categories, 5–6
  two-zone, 4, 258
 probabilistic, 3, 256–257
Enthalpy
 described, 182
 equation for specific, 186
Entrainment
 coefficient, 60
 definition, 260
 in mass transport submodel, 263
 rate equation, 57
Equivalence ratio
 CO yield and temperature, 248
 definition, 227
 fuel mixture fraction and, 233–234
 global
  described, 244–245
  limitations to, 248, 250
  plumes, 245
  upper layer, 245, 248
ERM, 257, 277
EUROCODE method, 136–138
EVACNET, 257, 277
Evacuation models, 257, 277, 288
EXITT, 257
EXODUS, 257, 288
Experiments in fire dynamics
 HAZARD I program, 276
  CEdit, 290–293
  CFAST, 292
  CPlot, 292
  in scale rooms, 298
 large-scale, 296–297
 1/3-scale room, 293–296
Extinction-absorption coefficient, 32, 141–142

F

Fabric reflectivities, 144, 146
FASTlite, 276, 287
Fick's Law, 238
Film gage, 148
FIRDEMND (Handheld Hosestream Suppression Model), 276

Fire and Materials, 286
Fire development/growth
    described, 14–17
    factors influencing, 21–24
        bounding surfaces, 23–24
        fuel, 21–22
        geometry, 22–23
        ignition source, 21
        openings, 23
    stages in
        enclosure temperatures, 17–18
        flow through openings, 19
    terminology, 11–12, 17–18, 19–20
Fire-induced environmental factors, 5–6
Fire laboratories, 283
Fire load, 115, 128
Fire plumes, *see also* Flames
    bounded plumes, 71–72
    ceiling jets
        described, 15
        flame extensions, 76–77
        temperatures and velocities, 74–76
    characteristics of, 52–54
    convective heat transfer to ceilings, 149–154
    described, 15
    enclosure geometry and, 23
    flame characteristics
        buoyancy, 48–49
        diffusion, 48
        flame height, 49–52
        limitations, 52
        turbulence, 49
    global equivalence ratio, 245
    ideal
        assumptions, 54–55
        constants determination, 58–61
        continuity equation for mass, 57
        initial considerations, 55–57
        momentum and buoyancy equations, 57–58
    line source plumes, 72–73
    mass flow rates, 65
    plume equations from experiments
        Heskestad, 63–65, 69
        McCaffrey, 67–69
        Thomas, 69–70
        Zukoski, 62–63
    radius, 54, 65
    temperature differences, 60
    temperature of, 53, 60
    terminology, 48
    turbulent characteristics, 52–54
Fire Protection Engineering Tools (FPETool), 276
Fire safety engineering
    engineering education, 285
    organizations, 284
Fire Technology, 286
FIRST (Fire Simulation Technique), 276
First law of thermodynamics, 183, 187
Flames, *see also* Fire plumes
    candle, 12–14
    characteristics
        buoyancy, 48–49

        diffusion, 48
        flame height, 49–52
        limitations, 52
        turbulence, 49
    emissivity in cylindrical, 170
    extensions under ceilings, 76–77
    height, 49–52
    laminar flamelet model, 274–275
    premixed, 48
    processes in, 13–14
    temperature as function of, 21, 22
    turbulent, 52–54
Flammability Apparatus, 242, 243
Flashover
    CO yields and, 251
    described, 11, 16, 18
    predicting time to, 124–125
Flow coefficient, 81, 89, 91
Forced flow effects, 266
Fourier's Law, 143
FPETool (Fire Protection Engineering Tools), 276
Free burn measurements
    described, 28
    example products, 34–38
    pool fires, 32–34
    techniques and parameters, 29–32
    T-squared fire, 38–39
Free burn rates, 27
Free burn tests, 30
Froude number, 50
Fuel chemistry and combustion products
    energy release rates, 26
    factors influencing fire development, 21–22
    mixture fraction, 227, 233–234
    specific yields
        concentrations related to fuel fraction, 235–238
        equivalence ratio, 233–234
        fuel mixture fraction, 233–234
        for fuels (table), 231
        heat of combustion, 232
        mass optical density, 232–233
        of species, 230–232
    stoichiometry and species, 229–230
Fuel-controlled fire
    described, 11–12, 20, 43
    fuel package and fire growth, 23
Fully developed fire, 12, 16, 18

G

Gardon gage, 148
Gases
    gas law, 89
    properties in radiative heat transfer, 164–165
    specific heat and gas constant, 187
Gasification, heat of, 26, 27
Gas temperatures (ventilated enclosure)
    described, 15
    estimates during smoke filling, 202–203, 211
    post-flashover stage
        definitions, 128–129
        described, 117

energy and mass balance, 129–132
EUROCODE method, 136–138
hand-calculation method, 135–136
Magnusson and Thelandersson method, 132–135
temperature-time curves, 127–128
pre-flashover stage
calculation methods, 123–124
combustible wall lining materials, 126
described, 116–117
effective heat transfer coefficient, 120–123
energy balance, 117–119
experimental/statistical correlation, 120
limitations to methodology, 124
mechanically ventilated compartments, 125–126
predicting time to flashover, 124–125
wall and corner fires, 126
terminology, 115–116
Geometry of the enclosure
bounding surfaces, 23–24
influence on fire growth, 22–23
wall and corner fires
bounded plumes, 71–72
pre-flashover stage, 126
Global equivalence ratio (GER)
described, 244–245
limitations to, 248, 250
plumes, 245
upper layer, 245, 248
Grashof Number, 152
Gray gas/gray body assumption
described, 142, 144
radiative heat transfers, 164
Growth of fire, see Fire development/growth
Growth phase in design fire, 39, 42–43
Growth rate factor, 38

H

Hand-calculation models, 4–5
energy evolved, 5
fire-induced environmental factors, 5–6
heat transfer, 6
Handheld Hosestream Suppression Model (FIRDEMND), 276
HAZARD I program, 276
CEdit
fire specifications, 293
second compartment, 292–293
single component, 290–292
user's guide, 301–305
CFAST, 292
CPlot, 292
internet address, 287
overview, 290
scaled room simulations, 298
shell program, 290
Heat, 189
Heat flux
enclosure geometry and, 23
level in convective heat transfer, 149

losses to boundaries, 211
measurements, 147–148
Heat of combustion
described, 26, 30–31
equation for, 232
table of values, 33
Heat of evaporation, 27
Heat of gasification, 26, 27
Heat release rate, see Energy release rates
Heat transfer, 6
background, 142
coefficient, 120–123
computer submodels for, 266–267
convective
coefficients, 148–149
fire plume to ceilings, 149–154
level of heat flux likely, 149
described, 16
enclosure applications
electric circuit analogy, 172
sensor at ceiling level, 172–176
sensor at floor level, 176–178
measurements
of heat flux, 147–148
of temperature, 145–147
modes of, 142–144
radiative
basic principles of, 159–170
calculation methods, 155–159
CFD submodel for, 267
described, 155
spectral effects, 144
thermal, 142–143
transfer equation in computer models, 271–274
terminology, 141–142
Height of neutral plane and mass flow rates
cold air through vent, 98
height of neutral plane, 98
hot gases out of vent, 97
pressure difference across vent, 96
sign convention, 95–96
velocity in vent, 97
Heskestad plume, 63–65, 69
Hood experiments
CO yield prediction, 246–247
equations for normalized yields, 245–246
global equivalence ratio
described, 244–245
limitations to, 248, 250
plumes, 245
upper layer, 245, 248
Hot layer, 15–16
Hybrid probabilistic models, 256
Hydrodynamic pressure, 82, 84
Hydrostatic pressure, 82, 84

I

IAFSS (International Association for Fire Safety Science), 281, 282, 286
Ideal gas law, 186–187
Ignition, 14–15, 17, 21

Intensity, radiative, 159–160, 163
Internal energy, 182, 185
International Association for Fire Safety Science (IAFSS), 281, 282, 286
International Council for Building Research (CIB), 284
International Organization for Standardization (ISO), 284
Internet resources
    computer modeling
        CFD programs, 275
        evacuation programs, 277
        special-purpose programs, 276–277
        two-zone programs, 276
    education, university level, 285
    evacuation modeling, 288
    fire modeling, 287
    fire safety organizations, 280–281
    international organizations, 284
    laboratories, 283
    magazines, 286
    tools, 282
ISO 834, 127
ISO (International Organization for Standardization), 284

J

JASMINE, 275
Jet flames, 49
Journal of Applied Fire Science, 286
Journal of Fire Science, 286
Journal of Hazardous Materials, 286

K

κ-ε turbulence model, 271
Kirchhoff's Law, 144

L

Laboratories, fire, 283
Laminar flamelet combustion model, 274–275
Large eddy simulation turbulence model, 271
Large-scale room
    experiments, 296–297
    HAZARD I simulation, 298
LAVENT, 276
Line source plumes, 72–73

M

Magnusson and Thelandersson method
    assumptions, 132
    energy release rate, 132–133
    temperature–time curves, 133–134
MAH method, 116
Mass conservation equations
    control volume, 184
    smoke control, large spaces, 210
    smoke filling, floor level leaks, 197
    smoke filling, large spaces, 205–206
    two-zone computer modeling, 260
Mass flow rate, 54, 56

Mass fraction of a species, 228
Mass loss rate, *see* Burning rate
Mass optical density, 232–233
Mass plume flow, 60
Mass transfer
    in a candle flame, 14
    computer submodels for, 263–266
Mattress energy release rates, 36
McCaffrey plume, 67–69
Mean beam length
    corrector, 32
    definition, 142
    equation for, 170
    for gas body shapes, 171
Mean flame height, 49–50, 64
Mechanically ventilated compartments
    pre-flashover stage and, 125–126
    pressure differences due to, 83
    smoke control, large spaces
        lower layer pressurization, 216–218
        upper layer ventilation, 215–216
Mixing of gas layers, *see* Well-mixed case
Modak's method, 155–156, 274
Molecular diffusion, 48
Momentum, 56, 57–58
Monochromatic emissive power, 143

N

Narrow-band radiation models, 274
National Fire Protection Association (NFPA), 280
National Institute of Standards and Technology (NIST), 280
Navier-Stokes equations, 3
Network probabilistic models, 256
Neutral plane, 82, 84, 90
Nominal temperature-time curves
    described, 116
    post-flashover fire, 127–128
Normalized yield of species, 228
North American Emergency Response Guidebook, 282
Nozzle, 91

O

1/3-scale room
    experiments, 293–296
    HAZARD I simulation, 298
One-zone case, *see* Well-mixed case
Opening factor, 116, 129
Openings
    fire development, factors influencing, 23
    flow through, 19
    pressure profiles, 99–100
Orifice, 91
Oxygen consumption calorimetry, 29
Oxygen starvation, 16

P

Parametric fire exposure (EUROCODE method), 136–138
Planck's Law, 143

Plumes, *see* Fire plumes
Point-wise equations, 183
Pool fires
    free burn measurements, 32–34
    Modak's method, 155–156
    radiative heat transfers, 155
Post-flashover fire
    definition, 12
    described, 19
    gas temperatures
        definitions, 128–129
        described, 116, 117
        energy and mass balance, 129–132
        EUROCODE method, 136–138
        hand-calculation method, 135–136
        Magnusson and Thelandersson method,
            132–134
        temperature-time curves, 127–128
    mixed case applied to, 95
Prantl Number, 148
Pre-flashover fire
    definition, 12
    described, 19
    gas temperatures
        calculation methods, 123–124
        combustible wall lining materials, 126
        described, 116–117
        effective heat transfer coefficient, 120–123
        energy balance, 117–119
        experimental/statistical correlation, 120
        limitations to methodology, 124
        mechanically ventilated compartments, 125–126
        predicting time to flashover, 124–125
        wall and corner fires, 126
Premixed flame, 48
Pressure profiles
    described, 19
    example problem
        flow velocity, 87–88
        hydrodynamic pressure, 86–87
        hydrostatic pressure, 86
        temperature and density relationship, 88–89
    mass flow rate through vents, 89–91
    pressure characteristics, 83–85
    pressure differences
        in a building, 82–83
        caused by fire, 82
        due to mechanical ventilation, 83
    stages of development, 19, 92–95
    stratified case flow rates
        into/out of vents, 103–107
        through ceiling vents, 107–111
    terminology, 81–82
    well-mixed case flow rates
        burning rate, 101–102
        and neutral plane height, 95–99
        out of an opening, 99–100
Pressure rise conservation equations
    closed volume
        assumptions, 192
        equation derivation, 190–192

    leaky compartment
        assumptions, 195
        equation derivation, 192–195
Pressure work, 188
Probabilistic models, 3, 256–257

R

Radiative heat transfer
    basic principles
        configuration factors, 159–163
        gray gas assumption, 164
        intensity Transfer Equation, 163
        net radiative outward rate, 167–170
        real gas properties, 164–165
        spectral absorptivity, 163–164
        spectral emissivity, 164–165
        through intervening medium, 166–167
    calculation methods
        cylindrical flame to target, 156–159
        Modak's, 155–156
    CFD submodel for, 267
    described, 155
    pool fires, 155
    spectral effects, 144
    thermal, 142–143
    transfer equation in computer models, 271
        radiative properties, 273–274
        solution methods, 272–274
Radiative intensity, 159–160, 163
Radius of a plume, 54, 65
Rayleigh Number, 148, 152
Reciprocity relation, radiative heat exchange, 161
Reflectivity, 143, 144, 146
Reynolds Number, 91
Risk Analysis, 286
Risk Management Quarterly Newsletter, 286

S

Scaled-room experiments, 298
Seebeck effect, 143
SFPE (Society of Fire Protection Engineers), 281
Shaft work, 188
Shape factor, 141
Shear work, 188
Simulated natural fire exposure, 116, 128
Simulation based probabilistic models, 256–257
Simulation of Fires in Enclosures (SOFIE), 275
SIMULEX, 257, 277, 288
SMARTFIRE, 275
Smoke control conservation equations
    boundary heat losses, 211
    energy, 210–211
    gas temperature, 211
    mass, 210
    mass and energy balances, 210–211
    mechanical ventilation
        lower layer pressurization, 216–218
        upper layer, 215–216

natural ventilation, upper layer
    calculation methods, 213–215
    comparison with experiments, 215
    mass flow rate, 213
    pressure differences, 212–213
Smoke filling conservation equations
  ceiling leaks
    equations for, 200–202
    limitations, 203–204
  floor level leaks
    calculation methods, 199
    comparison with experiments, 200
    described, 196–197
    energy, 197–198
    equation for, 199
    gas temperature estimates, 202–203
    heat release rate, 198
    limitations, 203–204
    mass, 197
    time to fill, 198, 199
  large spaces
    assumptions, 204–205
    calculation methods, 207–208
    comparison with experiments, 208–209
    energy, 206–207
    mass, 205–206
Smoke gas explosion, 17
Society of Fire Protection Engineers (SFPE), 281
SOFIE (Simulation of Fires in Enclosures), 275
Species yield, *see* Combustion products
Specific heat
  described, 182
  equation for, 186
  gas constant and, 187
Specific internal energy, 185
Spectral effects of radiation, 144
  absorptivity, 273
  emissivity, 164–165
  radiative heat transfers, 163–164
Standard temperature-time curve, 127
Static pressure head, 85
Statistical probabilistic models, 256
Steady phase in design fire, 43–44
Stefan-Boltzmann constant, 131, 143
Stoichiometry, 228, 229–230
Stratified case mass flow rates
  described, 82
  into/out of vents
    as function of height, 105–106
    position of neutral plane/smoke layer, 106–107
    sign convention, 103–105
    in through the vent, 106
  through ceiling vent
    height of neutral plane, 109
    limitations to methodology, 110–111
    mass flow rates, 108
    pressure differences, 107–108
    in terms of temperature, 109
    velocities, 108
Symbols, dimensions, and units, 8–9

T

Temperature, gas, *see* Gas temperatures
Temperature of fire
  in enclosures, 17–18
  within flames, 13, 149
  as function of height, 21, 22
  heat transfer measurements of, 145–147
  plume
    difference with height, 60
    distribution in, 53
Temperature-time curves
  gases, 133–134
  nominal
    described, 116
    post-flashover fire, 127–128
  standard, 127
Termocouples, 145–147
Thermal inertia, 24
Thermal penetration time, 121
Thermal radiation, 143–144
Thomas plume, 69–70
Transfer Equation, 163
Transmissivity, 143
Trash bags energy release rates, 36
T-squared fire, 38–39
Turbulence
  CFD models for, 270–271
  described, 49
  diffusion flame, 238
  fire plume characteristics, 52–54
Two-zone computer modeling, 4
  building geometry limitations, 268–269
  conservation equations
    assumptions, 259–260
    control volume, 258–259
    energy, 261
    mass, 260
    species, 260–261
    summary, 262
    ventilation-limited condition, 261–262
  described, 258
  embedded submodels, 267
  heat transport submodels, 266–267
  internet resources, 276
  mass transport submodels, 263–266
  source term submodels, 262–263
  unresolved phenomena, 267

U

United States Fire Administration (USFA), 281
Units, dimensions, and symbols, 7–9
Upholstered furniture energy release rates, 36

V

Velocity of a plume, 53
Vent flows
  described, 16
  example problem

flow velocity, 87–88
hydrodynamic pressure, 86–87
hydrostatic pressure, 86
temperature and density relationship, 88–89
horizontal partitions, submodel for, 264–265
mass flow rate, 89–91
pressure characteristics, 83–85
pressure differences
    in a building, 82–83
    caused by fire, 82
    due to mechanical ventilation, 83
pressure profile stages, 92–95
stratified case flow rates
    into/out of vents, 103–107
    through ceiling vents, 107–111
terminology, 81–82
vertical partitions, submodel for, 264
well-mixed case flow rates
    burning rate, 101–102
    and neutral plane height, 95–99
    out of an opening, 99–100
Ventilation-controlled fire
definition, 12
described, 18, 20, 43
gas temperatures, *see* Gas temperatures
mechanical smoke control
    lower layer pressurization, 216–218
    upper layer, 215–216
natural smoke control
    calculation methods, 213–215
    comparison with experiments, 215

mass flow rate, 213
pressure differences, 212–213
pre-flashover stage, 125–126
Ventilation factor, 23, 116
View factor, 141
Virtual origin, 64

W

Wall and corner fires
bounded plumes, 71–72
pre-flashover stage and, 126
Well-mixed case, 82
burning rate, 101–102
mass flow rate out of opening, 99–100
mass flow rates and height of neutral plane
    cold air through vent, 98
    height of neutral plane, 98
    hot gases out of vent, 97
    pressure difference across vent, 96
    sign convention, 95–96
    velocity in vent, 97
mass transport submodels, 265–266
Wide-band radiation models, 274
Williamsson, Babrauskas and, 135–136
Wood pallets energy release rates, 35
Work, 188–189

Z

Zukoski plume, 62–63

Printed by Publishers' Graphics Kentucky